MULTIPLE ACCESS SYSTEMS FOR NEXT-GENERATION COMMUNICATIONS

Theory and Practice of Multiple Access Systems

ADVANCED SERIES IN ELECTRICAL AND COMPUTER ENGINEERING

The purpose of this series is to publish work of high quality by authors who are experts in their respective areas of electrical and computer engineering. Each volume contains the state-of-the-art coverage of a particular area, with emphasis throughout on practical applications. Sufficient introductory materials will ensure that a graduate and a professional engineer with some basic knowledge can benefit from it.

Published:

Vol. 24: *Multiple Access Systems for Next-Generation Communications: Theory and Practice of Multiple Access Systems*
by Kyung Sup Kwak

Vol. 23: *Graph Theory and Mobile Communications*
by Masakazu Sengoku

Vol. 22: *Silicon Systems for Wireless LAN*
edited by Zoran Stamenković, Gildas Leger and Alberto Bosio

Vol. 21: *Application of Omics, AI and Blockchain in Bioinformatics Research*
edited by Jeffrey J. P. Tsai and Ka-Lok Ng

Vol. 20: *Computational Methods with Applications in Bioinformatics Analysis*
edited by Jeffrey J. P. Tsai and Ka-Lok Ng

Vol. 18: *Broadband Matching: Theory and Implementations (Third Edition)*
by Wai-Kai Chen

Vol. 17: *Practical Signal Processing and Its Applications: With Solved Homework Problems*
by Sharad R. Laxpati and Vladimir Goncharoff

Vol. 16: *Design Techniques for Integrated CMOS Class-D Audio Amplifiers*
by Adrian I. Colli-Menchi, Miguel A. Rojas-Gonzalez and Edgar Sanchez-Sinencio

Vol. 15: *Active Network Analysis: Feedback Amplifier Theory (Second Edition)*
by Wai-Kai Chen

Vol. 14: *Linear Parameter-Varying System Identification: New Developments and Trends*
by Paulo Lopes dos Santos, Teresa Paula Azevedo Perdicoúlis, Carlo Novara, Jose A. Ramos and Daniel E. Rivera

For the complete list of titles in this series, please visit
http://www.worldscientific.com/series/asece

Advanced Series in Electrical and Computer Engineering – Vol. 24

MULTIPLE ACCESS SYSTEMS FOR NEXT-GENERATION COMMUNICATIONS

Theory and Practice of Multiple Access Systems

Kyung Sup Kwak
Inha University, South Korea

World Scientific

NEW JERSEY • LONDON • SINGAPORE • BEIJING • SHANGHAI • HONG KONG • TAIPEI • CHENNAI • TOKYO

Published by

World Scientific Publishing Co. Pte. Ltd.
5 Toh Tuck Link, Singapore 596224
USA office: 27 Warren Street, Suite 401-402, Hackensack, NJ 07601
UK office: 57 Shelton Street, Covent Garden, London WC2H 9HE

Library of Congress Cataloging-in-Publication Data
Names: Kwak, Kyung Sup, author.
Title: Multiple access systems for next-generation communications : theory and practice of multiple access systems / Kyung Sup Kwak, Inha University, South Korea.
Description: New Jersey : World Scientific, 2025. | Series: Advanced series in electrical and computer engineering ; vol. 24 | Includes bibliographical references and index. | Contents: Stochastic model in multi-access systems -- Statistical channel model -- Fixed orthogonal multiple access systems -- Random multiple access systems -- Non-orthogonal multiple access systems -- Emerging technologies with NOMA system -- Fundamentals of RSMA system -- Intelligent reflecting surface-assisted rate-splitting CR-NOMA system.
Identifiers: LCCN 2024039487 | ISBN 9789819801060 (hardcover) | ISBN 9789819801077 (ebook for institutions) | ISBN 9789819801084 (ebook individuals)
Subjects: LCSH: Multiple access protocols (Computer network protocols) | Wireless communication systems.
Classification: LCC TK5103.2 .K944 2025 | DDC 004.6/2--dc23/eng/20250110
LC record available at https://lccn.loc.gov/2024039487

British Library Cataloguing-in-Publication Data
A catalogue record for this book is available from the British Library.

For any available supplementary material, please visit
https://www.worldscientific.com/worldscibooks/10.1142/14048#t=suppl

Desk Editors: Selas Hamilton/Julio Hong

Typeset by Stallion Press
Email: enquiries@stallionpress.com

Preface

The advent of 6G mobile communications marks a transformative milestone, promising unprecedented connectivity and serving as the foundation of the Internet of Everything. With objectives such as ultra-reliable connectivity, unparalleled throughput, and extensive coverage, 6G demands advanced interference management and efficient utilization of wireless resources, necessitating a reevaluation of physical layer (PHY) and multiple access (MA) strategies.

While earlier generations (1G to 5G) relied on orthogonal multiple access (OMA) techniques such as TDMA, FDMA, and CDMA, the introduction of multi-input multi-output (MIMO) and space-division multiple access (SDMA) in 4G and 5G significantly enhanced capacity. For 6G, non-orthogonal multiple access (NOMA) and Rate-Splitting Multiple Access (RSMA) have emerged as critical innovations. Power-domain NOMA utilizes superposition coding (SC) and successive interference cancellation (SIC) to manage interference, while RSMA offers greater flexibility by splitting messages into common and private components, bridging traditional methods like SDMA and NOMA.

This book provides a comprehensive foundation in conventional multiple access systems and explores emerging technologies such as NOMA and RSMA. It offers valuable insights into advancements in multiple access techniques for wireless mobile communications and networks, catering to postgraduate students and researchers in the field.

This work represents the culmination of my teaching and research activities at Inha University. I am deeply grateful to my former graduate students, postdoctoral researchers, and collaborators for their invaluable contributions. Special thanks go to Professors S. M. Riazul Islam,

Hongwu Liu, Qinghai Yang, Tao Han, Zhiquan Bai, M. S. Chowdhury, and K. Ashrafuzzaman, as well as Professor B. Clerckx, whose pioneering work on RSMA significantly enriched this book. I also extend my gratitude to the institutions and publishers for permitting the use of content from my prior publications.

About the Author

Kyung Sup Kwak, Ph.D., is a distinguished professor at Inha University, Korea, and a globally recognized expert in wireless communications and information technology. After earning his doctorate from the University of California, San Diego, he began his career at Hughes Network Systems and the IBM Network Analysis Center in the United States before transitioning to academia.

At Inha University, Dr. Kwak has held key leadership roles, including dean of the Graduate School of Information Technology and Telecommunications and director of the UWB Wireless Communications Research Center, a national IT research hub. In 2008, he was appointed Inha Fellow Professor (IFP), followed by his appointment as Inha Hanlim Fellow Professor in 2015. He has also served as president of the Korean Institute of Communication & Information Sciences (KICS) in 2006 and the Korean Society of Intelligent Transport Systems (KITS) in 2009.

Widely honored for his contributions to research and industry, Dr. Kwak has received numerous prestigious awards, including the Haedong Scientific Achievement Award and national commendations from Korea's Minister of Information & Communication, Prime Minister, and President.

Dr. Kwak has authored or co-authored over 400 peer-reviewed journal papers and 13 books and has contributed extensively to IEEE conferences. His research interests include multiple access systems, UWB communication systems, wireless body area networks, and nano/molecular communication networks.

Contents

Chapter 1

Introduction

1.1 Purpose of the Book

This book offers a comprehensive exploration of both conventional and emerging multiple access systems, along with their practical applications. It introduces advanced technologies such as non-orthogonal multiple access (NOMA) and rate-splitting multiple access (RSMA), which are at the forefront of achieving reliable and low-latency communication for massive Internet of Things (mIoT) applications. These cutting-edge methods are theoretically proven and poised to become the integral components of 5G networks, with ongoing discussions regarding their adoption as the new standard for 6G multiple access.

Designed for postgraduates and researchers in the field, the book provides a theoretical foundation of conventional multiple access systems and their real-world applications. By focusing on emerging technologies like NOMA and RSMA, it offers valuable insights into the latest advancements in wireless communication research and development.

1.2 What and Why Are Multiple Access Systems?

The advent of 6G mobile communications marks a significant paradigm shift, promising unprecedented connectivity standards and positioning itself as the cornerstone of the Internet of Everything. 6G aims to deliver unparalleled throughput, ultra-reliable connectivity, heterogeneous quality of service (QoS), extensive coverage, and the convergence of various technologies. As such, the development of 6G requires a more efficient utilization of wireless resources and the implementation of advanced interference management techniques, leading to a reevaluation and enhancement of physical layer (PHY) and multiple access (MA) strategies for wireless communication systems.

Over the decades, the evolution of cellular networks from 1G to 5G has witnessed the emergence of novel MA techniques aimed at mitigating multi-user interference. Orthogonal multiple access (OMA) techniques have been predominant, employing orthogonal domains such as time-division multiple access (TDMA), frequency-division multiple access (FDMA), and code-division multiple access (CDMA) to allocate resources to users. The advent of 4G and 5G networks saw the integration of MIMO communication and space-division multiple access (SDMA) to address increasing capacity demands, leveraging spatial resources and multi-antenna processing to serve multiple users simultaneously.

To further enhance spectral efficiency, MA has evolved toward NOMA, where users share the same time–frequency resources through power or code domains. Power-domain NOMA utilizes superposition coding at the transmitter and successive interference cancellation (SIC) at the receiver to manage multi-user interference effectively. Additionally, RSMA emerges as a novel approach in 6G networks, splitting user messages into common and private parts to better manage interference levels. RSMA offers flexibility in adapting to various interference scenarios, bridging the gap between traditional approaches like SDMA and NOMA.

The radio channel facilitates broadcast communication, but for mobile communications, efficient access control is necessary. Designing these systems involves balancing capacity, complexity, and QoS, influenced by the method of MA employed. Various multi-user communication systems exist, categorized by their approach to sharing the communication channel. The choice of method depends on system design priorities and characteristics.

Multiplexing concepts and techniques are explored, highlighting the unique challenges of wireless environments compared to wired networks. Various media access control mechanisms are introduced to address these challenges, each with its own advantages and disadvantages, tailored to wireless environments. The performance of digital communication hinges on the signal-to-noise ratio (SNR) of the received signal, influenced by complex wireless channel conditions. Mathematically modeling wireless channels is essential for system design and validation. Channels are categorized into large-scale and small-scale models, crucial for understanding and designing mobile communication systems. Fading in communication systems can lead to increased errors, but it also enables improved utilization of ALOHA networks through capture effects. Exploring the fading model, fading statistics probabilities are derived, influencing the channel capacity of ALOHA systems. Capture channels allow receivers to successfully receive packets even with simultaneous transmissions, necessitating control mechanisms for stability. Stabilizing slotted ALOHA systems over capture channels through transmission control algorithms impacts throughput. The IEEE 802.15.6 standard for wireless body area networks (WBANs) adopts a slotted ALOHA-based protocol, offering an analytical model to estimate the saturation throughput in heterogeneous networks. Analyzing the slotted carrier sense multiple access with collision avoidance (CSMA/CA) algorithm in the wireless personal area networks (WPANs), a model is presented for contention access periods in beacon-enabled medium access control (MAC) protocols. The model's simplicity and accuracy offer potential for broader application and extension.

As 5G and beyond evolve, mobile cellular networks encounter challenges in spectrum utilization and accommodating diverse applications. Delving into NOMA as a solution, the discussion highlights its ability to achieve superior spectral efficiency and manage numerous connections, along with its advantages and challenges, shaping future wireless communication systems. As 6G networks emerge, RSMA offers a promising MA scheme, addressing challenges in resource utilization and interference management. RSMA splits user messages, facilitating partial decoding of interference and seamless transitions between different MA approaches, presenting a flexible solution for evolving wireless networks.

1.3 Outline of the Book

Chapter 2 explores the stochastic model in multi-access systems, covering probability and random variables, random processes, Poisson and Markov processes, and queueing systems, including Little's theorem and M/G/1 queueing systems.

Chapter 3 examines statistical channel models, covering the radio wave environment, SNR, large-scale propagation models (path loss, shadowing, outage probability), and small-scale propagation models (frequency-selective and non-selective fading channels, signal correlation in NLoS and

LoS channels). It also discusses the probability distribution of the received SNR, including Rayleigh, Rician, and Nakagami-m channels.

Chapter 4 discusses fixed orthogonal multiple access systems, covering resource allocation, multiplexing, and various access techniques, including duplexing, frequency-division and time-division multiplexing (TDM), fixed-assignment TDM/TDMA, and code-division multiplexing. It also examines the performance of FDMA and TDMA in terms of bit rate and message delays.

Chapter 5 delves into the various aspects of random multiple access systems, covering fundamentals, stabilization techniques, capture models, and performance analyses in practical network scenarios, such as WBANs and WPANs.

Chapter 6 delves into NOMA systems. It explores the fundamentals of NOMA, including PD-NOMA and CD-NOMA, along with key technologies like superposition coding and SIC. The chapter discusses resource allocations, covering user pairing, clustering, and power allocation, before addressing performance issues, such as outage probability, achievable rate, and spectral efficiency. Emerging technologies with NOMA are examined. Challenges in NOMA are identified, including dynamic user pairing and impact of interference. Implementation issues are discussed, such as decoding complexity and signaling overhead, leading to the conclusion and discussion.

Chapter 7 examines NOMA systems in enabling applications, discussing coordinated transmission for cooperative NOMA systems and heterogeneous networks for NOMA systems. It covers system models, outage probability, performance analysis, and summaries for both topics.

Chapter 8 delves into the fundamentals of RSMA. It explores the concepts of OMA/SDMA/NOMA, including orthogonal, space division, and non-orthogonal multiple access techniques. The chapter then delves into the concept of RSMA, detailing the two-user interference channel and rate-splitting for it. The architecture of RSMA systems is examined, covering scenarios such as two-user downlink RSMA, two-user uplink RSMA, and K-user downlink and uplink RSMA. Performance issues such as system complexity and performance metrics are discussed, followed by an exploration of RSMA properties. Emerging technologies and challenges in RSMA implementation are analyzed, including enabling applications and future research directions. The chapter concludes with insights into standardization and implementation.

Chapter 9 discusses intelligent reflecting surface-enabled rate-splitting CR-NOMA systems. It begins with an introduction, followed by a system model. The chapter examines outage performance, including channel gain distribution analysis and outage probability analysis. Performance results are presented, leading to the conclusion.

Chapter 2

Stochastic Model in Multi-access Systems

2.1 Probability and Random Variables [1–3]

2.1.1 *Probability*

A basic notion in probability theory is random experiment; an experiment whose outcome cannot be determined in advance. The set of all possible outcomes of an experiment is called the sample space of that experiment, and we denote it by S. An event is a subset of a sample space and is said to occur if the outcome of the experiment is an element of that subset. We shall suppose that for each event E of the sample space S, a number $P(E)$ is defined and satisfies the following three axioms:

(i) $0 \leq P(E) \leq 1$;
(ii) $P(S) = 1$;
(iii) for any sequence of events $E_1, E_2, \ldots$ that are mutually exclusive, i.e., events for which $E_i \cap E_j = \emptyset$ when $i \neq j$ (where $\emptyset$ is the null set),

$$P\left(\bigcup_{i=1}^{\infty} E_i\right) = \sum_{i=1}^{\infty} P(E_i).$$

We refer to $P(E)$ as the probability of the event E. Some simple consequences of axioms (i), (ii), and (iii) are as follows:

(1) If $E \subset F$, then $P(E) \leq P(F)$;
(2) $P(E^c) = 1 - P(E)$, where E^c is the complement of E;
(3) $P(\bigcup_{i=1}^{n} E_i) = \sum_{i=1}^{n} P(E_i)$ when the E_i are mutually exclusive;
(4) $P(\bigcup_{i=1}^{n} E_i) \leq \sum_{i=1}^{n} P(E_i)$.

The inequality (4) is known as Boole's inequality.

2.1.2 *Random variables*

Consider a random experiment having a sample space S. A random variable X is a function that assigns a real value to each outcome in S. For any set of real numbers A, the probability that X will assume a value that is contained in the set A is equal to the probability that the outcome of the experiment is contained in $X^{-1}(A)$. That is, $P\{X \in A\} = P(X^{-1}(A))$, where $X^{-1}(A)$ is the event consisting of all points $s \in S$ such that $X(s) = A$.

The cumulative distribution function (CDF) F of the random variable X is defined for any real number x by

$$F(x) = P\{X \leq x\} = P\{X \in (-\infty, x]\}$$

We shall denote $1 - F(x)$ by $\overline{F}(x)$, and so,

$$\overline{F}(x) = P\{X > x\}$$

A random variable is called continuous if there exists a function $f(x)$, called the probability density function (PDF), such that

$$P\{X\ is\ in\ B\} = \int_B f(x)dx$$

for every set B. Since $F(x) = \int_{-\infty}^{x} f(x)dx$, it follows that

$$f(x) = \frac{d}{dx}F(x)$$

The joint distribution function F of two random variables X and Y is defined by

$$F(x, y) = P\{X \leq x, Y \leq y\}$$

The distribution functions of X and Y, respectively,

$$F_X(x) = P\{X \leq x\} \quad \text{and} \quad F_Y(y) = P\{Y \leq y\}$$

can be obtained from $F(x, y)$ by making use of the continuity property of the probability operator. Specifically, let y_n, $n \geq 1$, denote an increasing sequence converging to ∞. Then as the events $\{X \leq x, Y \leq y_n\}$, $n \geq 1$, are increasing and

$$\lim_{n\to\infty} \{X \leq x, Y \leq y_n\} = \bigcup_{n=1}^{\infty} \{X \leq x, Y \leq y_n\} = \{X \leq x\}$$

it follows from the continuity property that

$$\lim_{n\to\infty} P\{X \leq x, Y \leq y_n\} = P\{X \leq x\}$$

or, equivalently,

$$F_X(x) = \lim_{y\to\infty} F(x, y)$$

Similarly,

$$F_Y(y) = \lim_{x\to\infty} F(x, y)$$

The random variables X and Y are said to be independent if

$$F(x, y) = F_X(x)F_Y(y)$$

for all x and y.

The random variables X and Y are said to be jointly continuous if there exists a function $f(x, y)$, called the joint PDF, such that

$$P\{X\ is\ in\ A, Y\ is\ in\ B\} = \int_A \int_B f(x, y)dxdy$$

for all sets A and B.

The joint distribution of any collection $X_1, X_2, \ldots, X_n$ of random variables is defined by

$$F(x_1, x_2, \ldots, x_n) = P\{X_1 \leq x_1, \ldots, X_n \leq x_n\}$$

Furthermore, n random variables are said to be independent if

$$F(x_1, x_2, \ldots, x_n) = F_{X_1}(x_1)F_{X_2}(x_2)\ldots F_{X_n}(x_n)$$

where

$$F_{X_i}(x_i) = \lim_{\substack{x_j\to\infty \\ j\neq i}} F(x_1, x_2, \ldots, x_n)$$

Now, a random variable X is said to be discrete if its set of possible values is countable. The CDF $F_X(x)$ of a discrete random variable X can be obtained by $F_X(x) = P\{X \leq x\} = \sum_{x_k \leq x} P_X(x_k)$. Suppose that the jumps in $F_X(x)$ of a discrete random variable X occur at the points

$x_1, x_2, \ldots,$ where the sequence may be either finite or countably infinite, and we assume $x_i < x_j$ if $i < j$:

$$F_X(x_i) - F_X(x_{i-1}) = P(X \leq x_i) - P(X \leq x_{i-1}) = P(X = x_i)$$

Let $p_X(x_i) = P(X = x_i)$. The function is called the probability mass function (PMF) of the discrete random variable X.

The PMF has the following properties:

(i) $0 \leq p_X(x_k) \leq 1$, for $k = 1, 2, \ldots,$
(ii) $p_X(x) = 0$ if $x \neq x_k$ $(k = 1, 2, \ldots),$
(iii) $\sum_k p_X(x_k) = 1$.

The PDF/PMF and CDF of a number of well-known random variables are given in the following:

- **Bernoulli distribution:** A random variable X is called a Bernoulli random variable with parameter p with $0 \leq p \leq 1$ if its PMF is given by

$$p_X(k) = P(X = k) = p^k(1 - p)^{1-k} \quad \text{for } k = 0, 1$$

The CDF $F_X(x)$ of the Bernoulli random variable X is given by

$$F_X(x) = \begin{cases} 0 & x < 0 \\ 1 - p & 0 \leq x < 1 \\ 1 & x \geq 1 \end{cases}$$

- **Binomial distribution:** A random variable X is called a Binomial random variable with parameter (n, p) with $0 \leq p \leq 1$ if its PMF is given by

$$p_X(k) = P(X = k) = \binom{n}{k} p^k(1 - p)^{n-k} \quad \text{for } k = 0, 1, \ldots$$

where

$$\binom{n}{k} = \frac{n!}{k!(n - k)!}$$

Note that the Bernoulli random variable is just a binomial random variable with parameters $(1, p)$.

- **Poisson distribution:** A random variable X is called a Poisson random variable with parameter $\lambda > 0$ if its PMF is given by

$$p_X(k) = P(X = k) = \frac{\lambda^k}{k!}e^{-\lambda} \quad \text{for } k = 0, 1, \ldots$$

The CDF $F_X(x)$ of the Poisson random variable X is given by

$$F_X(x) = P\{X \leq x\} = \sum_{k \leq x} p_X(x = k) = e^{-\lambda} \sum_{k=0}^{x} \frac{\lambda^k}{k!} \quad \text{for } n \leq x < n+1$$

- **Uniform distribution:** A random variable X is called a uniform random variable over (a, b) if its PDF is given by

$$f_X(x) = \begin{cases} \dfrac{1}{b-a} & a < x < b \\ 0 & \text{otherwise} \end{cases}$$

The corresponding CDF of X is

$$F_X(x) = \begin{cases} 0 & x < a \\ \dfrac{x-a}{b-a} & a \leq x < b \\ 1 & x \geq b \end{cases}$$

- **Exponential distribution:** A random variable X is called an exponential random variable with parameter $\lambda > 0$ if its PDF is given by

$$f_X(x) = \begin{cases} \lambda e^{-\lambda x} & x \geq 0 \\ 0 & x < 0 \end{cases}$$

The corresponding CDF of X is

$$F_X(x) = \begin{cases} 1 - e^{-\lambda x} & x \geq 0 \\ 0 & x < 0 \end{cases}$$

- **Gaussian (normal) distribution:** A random variable X is called a Gaussian random variable with parameter (μ, σ^2) if its PDF is given by

$$f_X(x) = \frac{1}{\sqrt{2\pi\sigma^2}} e^{-\frac{(x-\mu)^2}{2\sigma^2}}$$

The corresponding CDF of X is

$$F_X(x) = \frac{1}{\sqrt{2\pi\sigma^2}} \int_{-\infty}^{x} e^{-\frac{(x-\mu)^2}{2\sigma^2}} dx = \frac{1}{\sqrt{2\pi}} \int_{-\infty}^{\frac{x-\mu}{\sigma}} e^{-\frac{y^2}{2}} dy$$
$$= 1 - Q\left(\frac{x-\mu}{\sigma}\right)$$

where $Q(x) = \frac{1}{2\pi} \int_x^{\infty} e^{-\frac{y^2}{2}} dy$ is called the Q-function.

2.1.3 *Expectation*

The expectation or mean of the random variable X, denoted by $E[X]$, is defined by

$$E[X] = \int_{-\infty}^{\infty} x dF(x) = \begin{cases} \int_{-\infty}^{\infty} x f(x) dx & \text{if } X \text{ is continuous} \\ \sum_x x P(X = x) & \text{if } X \text{ is discrete} \end{cases}$$

Let $h(X)$ be any function of X. Then $h(X)$ is itself a random variable and the expectation of $h(X)$ is

$$E[h(X)] = \int_{-\infty}^{\infty} x dF_h(x)$$

where $F_h(x)$ is the distribution function of $h(X)$. However, it can be shown as

$$E[h(X)] = \int_{-\infty}^{\infty} h(x) f(x) dx$$

The variance of the random variable X is defined by

$$\text{Var}[X] = E[(X - E[X])^2] = E[X^2] - (E[X])^2$$

Two jointly distributed random variables X and Y are said to be uncorrelated if their covariance defined by

$$\text{Cov}(X, Y) = E\{[X - E[X]][Y - E[Y]]\} = E[XY] - E[X]E[Y] = 0$$

It follows that independent random variables are uncorrelated. However, the converse need not be true.

An important property of expectations is that the expectation of a sum of random variables is equal to the sum of the expectations:

$$E\left[\sum_{i=1}^{n} X_i\right] = \sum_{i=1}^{n} E[X_i]$$

The corresponding property for variances is that

$$\mathrm{Var}\left[\sum_{i=1}^{n} X_i\right] = \sum_{i=1}^{n} \mathrm{Var}[X_i] + 2\sum_{i<j}\sum \mathrm{Cov}(X_i, Y_j)$$

2.1.4 *Conditional expectation*

If X and Y are discrete random variables, the conditional PMF of X, given $Y = y$, is defined, for all y such that $P\{Y = y\} > 0$, by

$$P\{X = x|Y = y\} = \frac{P\{X = x, Y = y\}}{P\{Y = y\}}$$

The conditional distribution function of X, given $Y = y$, is defined by

$$F(x|y) = P\{X \le x|Y = y\}$$

and the conditional expectation of X, given $Y = y$, is defined by

$$E(X|Y = y) = \int x dF(x|y) = \sum_{x} x P\{X = x|Y = y\}$$

If X and Y have a joint PDF $f(x, y)$, the conditional PDF of X, given $Y = y$, is defined for all y such that

$$f_Y(y) > 0 \ \text{ by } \ f_Y(x|y) = \frac{f(x, y)}{f_Y(y)}$$

and the conditional probability distribution function of X, given $Y = y$, is defined by

$$F(x|y) = P\{X \le x|Y = y\} = \int_{-\infty}^{x} f(x|y)dx$$

The conditional expectation of X, given $Y = y$, in this case, is defined by

$$E(X|y) = \int_{-\infty}^{x} xf(x|y)dx$$

Thus, all definitions are exactly as in the unconditional case except that all probabilities are now conditional on the event that $Y = y$.

Let us denote by $E(X|Y)$ that the function of the random variable Y whose value at $Y = y$ is $E(X|Y = y)$. An extremely useful property of the conditional expectation is that for all random variables X and Y,

$$E[X] = E[E(X|Y)] = \int E[X|Y = y]dF_Y(y)$$

when the expectations exist.

If Y is a discrete random variable, then

$$E[X] = E[E(X|Y)] = \sum_y E[X|Y = y]P[Y = y]$$

While if Y is continuous with density $f(y)$, then

$$E[X] = E[E(X|Y)] = \int E[X|Y = y]f(y)dy$$

We now consider the case where X and Y are both discrete random variables:

$$\begin{aligned}
E[X] = E[E(X|Y)] &= \sum_y E[X|Y = y]P[Y = y] \\
&= \sum_y \sum_x xP[X = x|Y = y]P[Y = y] \\
&= \sum_y \sum_x xP[X = x, Y = y] \\
&= \sum_y x \sum_x P[X = x, Y = y] \\
&= \sum_x xP(X = x) = E[X]
\end{aligned}$$

Thus, we see that $E[X]$ is a weighted average of the conditional expected value of X given that $Y = y$, with each of the terms $E[X|Y = y]$ being weighted by the probability of the event on which it is conditioned.

As defined the conditional expectation of X given the value of Y, we can also define the conditional variance of X given that $Y = y$, which is

$$\text{Var}[X|Y] = E[[X - E(X|Y)]^2|Y]$$

which can be written as

$$\text{Var}[X|Y] = E[X^2|Y] - (E[X|Y])^2$$

Finally, it is shown that

$$\text{Var}[X] = E[\text{Var}[X|Y]] + \text{Var}(E[X|Y])$$

2.1.5 *Moment-generating and characteristic functions*

The moment-generating function of X is defined by

$$\psi_X(t) = E[e^{tX}] = \int e^{tX} dF(x)$$

All the moments of X can be successively obtained by differentiating $\psi_X(t)$ and then evaluating at $t = 0$. That is,

$$\psi'_X(t) = E[Xe^{tX}], \psi''_X(t) = E[X^2e^{tX}], \ldots, \psi^n_X(t) = E[X^n e^{tX}]$$

Evaluating at $t = 0$ yields

$$\psi^n_X(0) = E[X^n]$$

It should be noted that we have assumed that it is justifiable to interchange the differentiation and integration operations. This is usually the case. When a moment-generating function exists, it uniquely determines the distribution. This is quite important because it enables us to characterize the distribution.

As the moment-generating function of a random variable X need not exist, it is theoretically convenient to define the characteristic function of X by

$$\phi_X(t) = E[e^{itX}] = \int e^{itX} dF(x) \quad \text{where } i = \sqrt{-1}$$

It can be shown that $\phi_X(t)$ always exists and, like the moment-generating function, uniquely determines the distribution of X.

We may also define the joint moment-generating function of the random variables $X_1, X_2, \ldots, X_n$ by

$$\psi_X(t_1, t_2, \ldots, t_n) = E[e^{\sum_{i=1}^n t_i X}]$$

or the joint characteristic function by

$$\phi_X(t_1, t_2, \ldots, t_n) = E[e^{i \sum_{i=1}^{n} t_i X}]$$

It may be proven that the joint moment-generating function (when it exists) or the joint characteristic function uniquely determines the joint distribution.

When dealing with random variables that only assume non-negative values, it is sometimes more convenient to use Laplace transforms rather than characteristic functions. The Laplace transform of the distribution F is defined by

$$\mathcal{L}_X(s) = \int_0^\infty e^{-sx} dF(x)$$

This integral exists for complex variables $s = \sigma + i\omega$, where $\sigma, \omega > 0$. As in the case of characteristic functions, the Laplace transform uniquely determines the distribution. We may also define Laplace transforms for arbitrary functions in the following manner: The Laplace transform of the function g, denoted by $\mathcal{L}_g(s)$, is defined by

$$\mathcal{L}_{g(X)}(s) = \int_0^\infty e^{-sx} dF(g(x))$$

$$\mathcal{L}_{g(X)}(s) = \int_0^\infty e^{-sx} g(x) dx$$

provided that the integral exists. It can be shown that $\mathcal{L}_g(s)$ determines $g(x)$ up to an additive constant.

2.1.6 *Some probability inequalities*

Markov's Inequality: If X is a nonnegative random variable, then for any $a > 0$,

$$P[X \geq a] \leq E[X]/a$$

Proof. Let $I[X \geq a]$ be 1 if $x \geq a$ and 0 otherwise. Then, it is easy to see since $X \geq 0$. Then $aI[X \geq a] \leq X$. Taking expectations yields the result. □

Chebyshev Inequality: If X is a random variable with a finite mean μ and variance σ^2, then for any value $k > 0$,

$$P[|X - \mu| \geq k] \leq \sigma^2/k^2$$

Proof. From Markov's Inequality, for $Y = (X - \mu)^2$ and $a = k^2$, we have $P[Y \geq a] \leq E[Y]/a$ means $P[(X - \mu)^2 \geq k^2] \leq E[(X - \mu)^2]/k^2$. Now, since $(X - \mu)^2 \geq k^2$ if and only if $|X - \mu| \geq k$, thus $P[|X - \mu| \geq k] \leq E[(X - \mu)^2]/k^2 = \sigma^2/k^2$. □

Chernoff Bounds: Let X be a random variable with moment-generating function $M(t) = E[e^{tX}]$. Then for $a > 0$,

$$P[X \geq a] \leq e^{-ta}M(t) \quad \text{for all } t > 0$$

$$P[X \leq a] \leq e^{-ta}M(t) \quad \text{for all } t < 0$$

Proof. For $t > 0$,

$$P[X \geq a] = P[e^{tX} \geq e^{ta}] \leq E[e^{tX}]e^{-ta} = e^{-ta}M(t)$$

where the inequality follows from Markov's Inequality. The proof for $t < 0$ is similar.

Since the Chernoff bounds hold for all t in either the positive or negative quadrant, we obtain the best bound on $P[X \geq a]$ by using that t minimizes $e^{-ta}M(t)$; □

Jensen's Inequality: If f is a convex function, then

$$E[f(X)] \geq f(E[X])$$

provided the expectation exists.

Proof. We will give a proof under the supposition that f has a Taylor series expansion. Expanding about $\mu = E[X]$ and using the Taylor series with a remainder formula yields for some ξ in the interval between x and μ,

$$\begin{aligned} f(x) &= f(\mu) + f'(\mu)(x - \mu) + \frac{f''(\mu)(x - \xi)^2}{2} \\ &\geq f(\mu) + f'(\mu)(x - \mu) \end{aligned}$$

with $f''(\xi) \geq 0$ by convexity.

Hence,

$$f(X) \geq f(\mu) + f'(\mu)(X - \mu)$$

Taking expectations gives that

$$E[f(X)] \geq f(\mu) + f'(\mu)E(X - \mu) = f(\mu) = f(E[X])$$

□

2.1.7 *Limit theorems*

Some of the most important results in probability theory are in the form of limit theorems. The two most important are as follows:

Strong law of large numbers: If $X_1, X_2, \ldots$ are independent and identically distributed with mean μ, then

$$P\{\lim_{n\to\infty}(X_1 + \cdots + X_n)/n = \mu\} = 1$$

Central limit theorem: If $X_1, X_2, \ldots$ are independent and identically distributed with mean μ and variance σ^2, then

$$\lim_{n\to\infty} P\left\{\frac{X_1 + \cdots + X_n - n\mu}{\sigma\sqrt{n}} \le a\right\} = \int_{-\infty}^{a} \frac{1}{\sqrt{2\pi}} e^{-x^2/2} dx$$

Thus, if we let $S = \sum_{i=1}^{n} X_i$ where $X_1, X_2, \ldots$ are independent and identically distributed, then the strong law of large numbers states that, with probability 1, S_n/n will converge to $E[X_i]$, whereas the central limit theorem states that S_n will have an asymptotic normal (Gaussian) distribution as $n \to \infty$.

2.2 Random Processes [4–6]

2.2.1 *Definitions*

A random (stochastic) process $X = \{X(t), t \in T\}$ is a collection of random variables. That is, for each t in the index set T, $X(t)$ is a random variable. We often interpret t as time and call $X(t)$ the state of the process at time t. If the index set T is a countable set, we call X a discrete-time random process, and if T is a continuum, we call it a continuous-time process.

Any realization of X is called a sample path. A continuous-time random process $\{X(t), t \in T\}$ is said to have independent increments if for all $t_0 < t_1 < t_2 < \cdots < t_n$, the random variables

$$X(t_1) - X(t_0), X(t_2) - X(t_1), \ldots, X(t_n) - X(t_{n-1})$$

are independent. It is said to possess stationary increments if $X(t + s) - X(t)$ has the same distribution for all t. That is, it possesses independent increments if the changes in the processes' value over non-overlapping time intervals are independent, and it possesses stationary increments if the distribution of the change in value between any two points depends only on the distance between those points.

2.2.1.1 *nth-order densities*

A random process is defined to be completely characterized if the probability densities for $X(t_1), X(t_2), \ldots, X(t_n)$ are known for all times $t_1, t_2, \ldots, t_n$ and all n. nth-order density functions of random variables $X(t) \equiv X_t$ are denoted by, for all t,

$$f_{X_t}(x_1, x_2, \ldots, x_n) \quad \text{or} \quad f(x_1, x_2, \ldots, x_n; t_1, t_2, \ldots, t_n)$$

Then, the first-order density functions and the second-order density functions are expressed, respectively, by

$$f_{X_t}(x) \quad \text{or} \quad f(x; t) \quad \text{and} \quad f_{X_t}(x_1, x_2) \quad \text{or} \quad f(x_1, x_2; t_1, t_2)$$

2.2.1.2 *Mean and variance of random process*

The mean of a random process, $X(t)$, is defined as a function of time t by

$$\mu_X(t) = E[X(t)] = \int_{-\infty}^{\infty} x f_{X_t}(x)$$

When the mean of X_t does not depend on t, the $\mu_X(t)$ is constant and denoted by μ_X.

The variance of the random process $X(t)$ is similarly defined by

$$\sigma_X^2(t) = E[(X(t) - \mu_X(t))^2] = E[X_t^2] - \mu_X^2(t)$$

2.2.1.3 *Autocorrelation and autocovariance function*

The autocorrelation function of a real random process $X(t)$ is, in general, a two-dimensional function of the variables at t_1 and t_2 is defined by

$$R_{XX}(t_1, t_2) = E[X(t_1)X(t_2)]$$

For a complex random process $X(t)$, the autocorrelation function is defined by

$$R_{XX}(t_1, t_2) = E[X(t_1)X^*(t_2)]$$

where * stands for complex conjugate.

Note that since $X(t_1)X(t_2) = X(t_2)X(t_1)$, $R_{XX}(t_1, t_2) = R_{XX}(t_2, t_1)$

Now, if $X(t)$ is wide-sense stationary, then the autocorrelation function depends only on the time difference $\tau = t_2 - t_1$, $R_{XX}(t_1, t_2) = R_{XX}(t_1, t_1 + \tau) = R_{XX}(\tau)$ with $t_2 = t_1 + \tau$, and is an even function, $R_{XX}(-\tau) = R_{XX}(\tau)$. For the complex process, $R_{XX}(t_1, t_2) = R^*_{XX}(t_2, t_1)$ and $R_{XX}(-\tau) = R^*_{XX}(\tau)$.

The autocovariance function of a real random process $X(t)$ is defined by

$$\begin{aligned} C_{XX}(t_1, t_2) &= E[(X(t_1) - \mu_X(t_1))(X(t_2) - \mu_X(t_2))] \\ &= R_{XX}(t_1, t_2) - \mu_X(t_1)\mu_X(t_2) \end{aligned}$$

For a complex random process, the autocovariance function is defined by

$$\begin{aligned} C_{XX}(t_1, t_2) &= E[(X(t_1) - \mu_X(t_1))(X(t_2) - \mu_X(t_2))^*] \\ &= R_{XX}(t_1, t_2) - \mu_X(t_1)\mu^*_X(t_2) \end{aligned}$$

2.2.1.4 *Stationarity of random processes*

A random process $X(t)$ is called strong sense (strict sense) stationary if the sets of random variables $X(t_1), X(t_2), \ldots, X(t_n)$ and $X(t_1 + \delta)$, $X(t_2 + \delta), \ldots, X(t_n + \delta)$ have the same probability functions of all $t_1, t_2, \ldots, t_n, \delta$, and n. That is, for all $t_1, t_2, \ldots, t_n, \delta$, and n, we have

$$\begin{aligned} & f(x_1, x_2, \ldots, x_n; t_1, t_2, \ldots, t_n) \\ & = f(x_1, x_2, \ldots, x_n; t_1 + \delta, t_2 + \delta, \ldots, t_n + \delta) \end{aligned}$$

A random process $X(t)$ is called Nth-order stationary if for all $n \leq N$ for a fixed number N, the following is satisfied, a weaker form of stationarity:

$$\begin{aligned} & f(x_1, x_2, \ldots, x_n; t_1, t_2, \ldots, t_n) \\ & = f(x_1, x_2, \ldots, x_n; t_1 + \delta, t_2 + \delta, \ldots, t_n + \delta) \end{aligned}$$

A random process $X(t)$ is called wide sense (weaker) stationary if

(1) $\mu_X(t) = \mu$, the mean is a constant, thus independent of the time;
(2) $R_{XX}(t_1, t_2) = R_{XX}(t_2 - t_1) = R_{XX}(\tau)$, the autocorrelation function depends only on time difference $\tau = t_2 - t_1$.

Note that we can say that if (1) is satisfied, $X(t)$ is stationary in mean, and if (2) is satisfied, is stationary in autocorrelation.

2.2.1.5 *Power spectral density of random process*

For a stationary process in autocorrelation, $X(t)$, the power spectral density is defined as the Fourier transform of the autocorrelation function:

$$S_{XX}(f) = \int_{-\infty}^{\infty} R_{XX}(\tau)e^{-j2\pi f\tau}d\tau$$

A wide-sense stationary random process has the following properties:

(1) $\boldsymbol{R_{XX}(-\tau) = R_{XX}(\tau)}$: Even function of τ;
(2) $\boldsymbol{|R_{XX}(\tau)| \leq R_{XX}(0)}$: Bounded function of τ;
(3) $\boldsymbol{R_{XX}(\tau) \leftrightarrow S_{XX}(\tau)}$: Fourier transform pairs;
(4) $\boldsymbol{S_{XX}(f) \geq 0}$: Power spectral density is real and non-negative;
(5) $\boldsymbol{R_{XX}(0) = \frac{1}{2\pi}\int_{-\infty}^{\infty} S_{XX}(f)df}$: Average power in $X(t)$;
(6) $\boldsymbol{R_{XX}(\tau) = C_{XX}(\tau) + \mu_X^2}$: Relation between autocovariance and autocorrelation.

2.2.1.6 *Ergodicity of random process*

A random process $X(t)$, possibly non-stationary, is mean ergodic (ergodic in the mean) if and only if, in mean square sense,

$$\lim_{T\to\infty} \frac{1}{2T} \int_{-\infty}^{\infty} X(t)dt = M_X$$

where M_X is in general a random variable, not the mean of $X(t)$.

A random process $X(t)$, possibly non-stationary, is autocorrelation ergodic (ergodic in the autocorrelation) if and only if

$$\lim_{T\to\infty} \frac{1}{2T} \int_{-\infty}^{\infty} X(t+\tau)X(t)dt = R_X(\tau)$$

where $R_X(\tau)$ is in general a random variable, not the autocorrelation function of $X(t)$.

Note that ergodicity enables the determination of statistical parameters of the random process, such as means and correlations, through the time average of a single sample path, which is crucial when dealing with random access communication systems.

Mean ergodicity: Let $X(t)$ be a continuous-time wide-sense stationary random process with finite mean μ_X and covariance $C_X(\tau)$. $X(t)$ is mean ergodic if

$$\lim_{T\to\infty} \frac{1}{2T} \int_{-2T}^{2T} \left(1 - \frac{|\tau|}{2T}\right) C_X(\tau) d\tau = 0$$

Let X_n be a discrete-time wide-sense stationary random process with finite mean μ_X and covariance $C_X(n)$. X_n is mean ergodic if

$$\lim_{n\to\infty} C_X(n) = 0$$

2.2.2 *Poisson process*

Definition (Counting process). A random process $\{N(t),\ t \geq 0\}$ is said to be a counting process if $N(t)$ satisfies the following:

(i) $N(t) \geq 0$;
(ii) $N(t)$ is integer-valued;
(iii) $N(s) \leq N(t)$ if $s \leq t$;
(iv) $N(t) - N(s)$ represents the number of events that has occurred in the interval $(s, t]$, for $s \leq t$.

Note that a counting process is said to possess independent increment if the numbers of events in two disjoint intervals are independent.

Definition (Poisson process). The counting process $\{N(t),\ t \geq 0\}$ is said to be a Poisson process with the parameter $\lambda \geq 0$ if

(i) $N(0) = 0$;
(ii) $N(t)$ has stationary and independent increments;
(iii) $P[N(h) = 1] = \lambda h - o(h)$;
(iv) $P[N(h) \geq 2] = o(h)$.

Here, a function $f(\cdot)$ is said to be $o(h)$ if $\lim_{h\to 0} f(h)/h = 0$.

Definition. The counting process $\{N(t), t \geq 0\}$ is said to be a Poisson process with the rate $\lambda \geq 0$ if

(i) $N(0) = 0$;
(ii) $N(t)$ has stationary and independent increments;
(iii) the distribution of the number of events in any interval is Poisson distributed with mean λt; i.e., for all $t, s \geq 0$,

$$P[N(t+s) - N(s) = n] = e^{-\lambda t} \frac{(\lambda t)^n}{(n)!}$$

Note that $E[N(t] = \lambda t$ and $\text{var}[N(t)] = \lambda t$.
Note that the above two definitions are equivalent.

Lemma 1. *Let X_n denote a binomial random variable with parameter (n, p). If $np \to \lambda$ as $n \to \infty$, then $P[X_n = i] \to e^{-\lambda}(\lambda^i/i!)$.*

Proof. Since X_n is a binomial random variable with (n, p), let $\lambda = np$, then

$$P[X_n = i] = \frac{n!}{(n-i)i!} p^i (1-p)^{n-i}$$
$$= \frac{n!}{(n-i)i!} \left(\frac{\lambda}{n}\right)^i \left(1 - \frac{\lambda}{n}\right)^{n-i}$$
$$= \frac{n(n-i)\dots(n-i+1)}{n!} \frac{\lambda^i}{i!} \frac{(1-\lambda/n)^n}{(1-\lambda/n)^i}$$

But for large n, we have

$$\left(1 - \frac{\lambda}{n}\right)^n \simeq e^{-\lambda}, \quad \frac{n(n-i)\dots(n-i+1)}{n!} \simeq 1 \quad \text{and} \quad \left(1 - \frac{\lambda}{n}\right)^i \simeq 1$$

Hence,

$$P[X_n = i] \simeq e^{-\lambda} \frac{\lambda^i}{i!}$$

We know that the Poisson distribution is a limiting version of the binomial distribution so that $N(t)$ have a Poisson process with mean λt. □

Lemma 2. *For a Poisson process with rate $\lambda \geq 0$, then*

$$P[N(h) = 0] = e^{-\lambda t} \quad \text{for } h \geq 0$$

Lemma 3. *For a Poisson process, let T_1 be the time of the first event occurred. For $n > 1$, let T_n denote the elapsed time between $(n-1)$st and nth event. The sequence is called the sequence of inter-arrival times. Then, the inter-arrival times $\{T_i, i = 1, 2, \dots\}$ are independent exponential random variables each with mean $1/\lambda$.*

Lemma 4. *Now, let $S_n = \sum_{i=1}^{n} T_i$. Then S_n is the arrival time of the nth event, which is called the waiting time until the nth event. S_n is a gamma distribution with parameters (n, λ). The PDF of S_n is given by*

$$f_{S_n}(x) = \lambda e^{-\lambda x} \frac{(\lambda x)^{n-1}}{(n-1)!} \quad \text{for } x \geq 0$$

2.2.3 *Markov process*

Consider a random process $\{X(t), t \geq I\}$ taking values in a set S. Here, I is called the index set and S is called the state space.

The $\{X(t), t \geq I\}$ is a Markov process if for any choice of n, $t_k \in I$ and $i.j, i_k \in S$,

$$P[X(t_{n+1}) \leq j|X(t_n) = i, X(t_{n-1}) = i_{n-1}, \ldots, X(t_1) = i_1, X(t_0) = i_0] \\ = P[X(t_{n+1}) \leq j|X(t_n) = i]$$

This means that the random process is a Markov process if the future is independent of the past when the present is known.

If now the state space S is discrete, the process is a Markov chain (MC) if for any choice of n, $t_k \in I$ and $i.j, i_k \in S$,

$$P[X(t_{n+1}) = j|X(t_n) = i, X(t_{n-1}) = i_{n-1}, \ldots, X(t_1) = i_1, X(t_0) = i_0] \\ = P[X(t_{n+1}) = j|X(t_n) = i]$$

Note that the process or chain is homogeneous if the conditional probabilities $P[X(t_{n+1}) \leq j|X(t_n) = i]$ (or $P[X(t_{n+1}) = j|X(t_n) = i]$) depend on the time difference $t_{n+1} - t_n$ only. Otherwise, it is non-homogeneous.

Note that if the state set contains a finite or an infinite number of elements, the chain is classified to be finite or infinite, respectively.

2.2.3.1 *Discrete-time Markov chains*

Now, consider the sampled values $X_n = X(t_n)$ of $\{X(t), t \geq I\}$, $n = 0, 1, 2, \ldots,$. Consider a sequence of discrete random variables $X_0, X_1, X_2, \ldots, X_n, \ldots$ and assume that the set of possible values of these random variables is $S = \{0, 1, 2, \ldots, M\}$. Then X_n can be interpreted as the state of a system at time n. We say that when $X_n = i$, then the system is in state i at time n.

Definition. If

$$P[X_{n+1} = j|X_n = i, X_{n-1} = i_{n-1}, \ldots, X_1 = i_1, X_0 = i_0] \\ = P[X_{n+1} = j|X_n = i] = p_{ij}(n)$$

then $\{X_i,\ i = 1, 2, \ldots\}$ forms a discrete-time MC.

MC has the property that the conditional distribution of any future state, X_{n+1}, given the past $X_{n-1}, \ldots, X_1, X_0$ and the present state, X_n, is independent of the past state and depends only on the present state.

The values $\{p_{ij}(n), 0 \le i \le M, 0 \le j \le N\}$ are called the one-step transition probabilities and satisfy

$$p_{ij}(n) \ge 0, \quad \sum_{j=0}^{N} p_{ij}(n) = 1 \quad \text{for } i = 1, 2, \ldots, M$$

If $p_{ij}(n)$ is independent of time n, it is homogeneous; otherwise, it is non-homogeneous.

Note that with $p_0(i_0) = P[X_0 = i_0]$, the joint distribution has the expression

$$\begin{aligned} &P[X_{n+1} = i_{n+1}, X_n = i_n, X_{n-1} = i_{n-1}, \ldots, X_1 = i_1, X_0 = i_0] \\ &\quad = P[X_0 = i_0] \prod_{k=0}^{n} P[X_{k+1} = i_{k+1} | X_k = i_k] \end{aligned}$$

Now, define a matrix $\mathbf{P}(n)$, the ijth element of which is given by

$$p_{ij}(n) \equiv P[X_{n+1} = j | X_n = i]$$

Definition. The matrix $\mathbf{P}(n) = \{p_{ij}(n)\}$ is called the one-step transition matrix as follows:

$$\mathbf{P}(n) = \begin{bmatrix} p_{00}(n) & p_{01}(n) & \ldots & p_{0N}(n) \\ p_{10}(n) & p_{11}(n) & \ldots & p_{1N}(n) \\ p_{M0}(n) & p_{M1}(n) & \ldots & p_{MN}(n) \end{bmatrix}$$

where the dimensionality of the matrix depends on the size of the state space: finite or infinite. The graphical way to display the contents of the transition matrix is a state transition diagram. In Fig. 2.1, the state is denoted as a circle, and the transition probability is labeled on the connecting line from state i to state j.

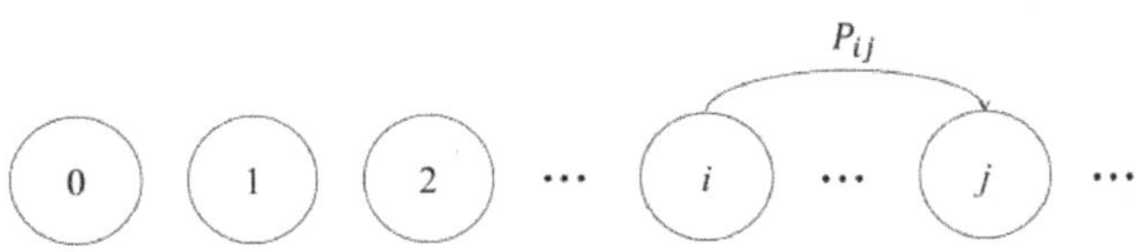

Fig. 2.1. State transition diagram.

Lemma 1. *For a homogeneous MC, we have*

$$P[X_{n+1} = i_{n+1}, X_n = i_n, X_{n-1} = i_{n-1}, \ldots, X_1 = i_1, X_0 = i_0] = p_0(i_0) \prod_{k=0}^{n} p_{i_k i_{k+1}}$$

2.2.3.2 *Probability mass function*

2.2.3.2.1 Marginal probability: $P[X_n = j]$

Lemma 1. *For any $n \geq 1$, we have the marginal PMF as*

$$\begin{aligned} P[X_n = j] &= \sum_{i_{n-1}} \sum_{i_{n-2}} \cdots \sum_{i_0} P[X_n = j, X_{n-1} = i_{n-1}, \ldots, X_1 = i_1, X_0 = i_0] \\ &= \sum_{i_{n-1}} \sum_{i_{n-2}} \cdots \sum_{i_0} p_{i_{n-1}j}(n-1) p_{i_{n-2}i_{n-1}}(n-2) \ldots p_{i_0 i_1}(0) \end{aligned}$$

Lemma 2. *For any $n \geq 1$, we have*

$$\begin{aligned} P[X_n = j] &= \sum_i P[X_n = j | X_{n-1} = i] P[X_{n-1} = i] \\ &= \sum_i p_{ij}(n-1) P[X_{n-1} = i] \end{aligned}$$

Lemma 3. *Let $\overline{p}(n)$ be a low vector, $\overline{p}(n) = (\ldots, P[X_n = j], \ldots)$ in the matrix $\mathbf{P}(n)$. If the MC is homogeneous, then for any $n \geq 1$, we have*

$$\bar{p}(n) = \bar{p}(0)\mathbf{P}^n$$

because of $\mathbf{P}(n-1) = \mathbf{P}$ due to being homogeneous. Here, $\mathbf{P}^n$ is the nth power of matrix $\mathbf{P}$.

2.2.3.2.2 Conditional probability: $P[X_n = j | X_0 = i]$

Lemma 1. *For any $n \geq 1$, we have*

$$\begin{aligned} P[X_n = j | X_0 = i] &= \sum_{i_{n-1}} \sum_{i_{n-2}} \cdots \sum_{i_1} p_{i_{n-1}j}(n-1) p_{i_{n-2}i_{n-1}}(n-2) \ldots p_{i_0 i_1}(0) \end{aligned}$$

Lemma 2. *For any $n \geq 1$, we have*

$$P[X_n = j|X_0 = i] = \sum_k p_{kj}(n-1)P[X_{n-1} = k|X_0 = i]$$

Lemma 3. Chapman–Kolmogorov equation: *Now, define n-step transition probabilities as follows:*

$$p_{ij}^{(n)} = P[X_{n+m} = j|X_m = i]$$

Then, we have

$$p_{ij}^{(n)} = \sum_k p_{ik}^{(l)} p_{kj}^{(n-l)} \quad \text{for all } 0 < l < n$$

which is called the Chapman–Kolmogorov equation.

Proof.

$$p_{ij}^{(n)} = P[X_n = j|X_0 = i] = \sum_k P[X_n = l, X_l = k|X_0 = i]$$

$$= \sum_k P[X_n = j|X_l = k, X_0 = i]P[X_l = k|X_0 = i]$$

$$= \sum_k p_{ik}^{(l)} p_{kj}^{(n-l)}$$

Note that

$$p_{ij}^{(n+m)} = \sum_k p_{ik}^{(n)} p_{kj}^{(m)}$$

and in matrix form

$$\mathbf{P}^{(n+m)} = \mathbf{P}^{(n)}\mathbf{P}^{(m)}$$

□

2.2.4 *Limiting distribution*

Definitions. Now, consider a homogeneous MC with the transition probabilities

$$p_{ij}^{(n)} = P[X_{n+k} = j|X_k = i] \quad \text{for all } k.$$

- State j is said to be **accessible** from state i if $p_{ij}^{(n)} > 0$ for at least one $n > 0$.
- State j **communicates** with state i if both states are **accessible** from each other, writing $i \leftrightarrow j$.

- Two states that communicate are said to be in the **same class**. All states in a communication class have the same **class**.
- The MC is **irreducible** if there is only one communicating **class**, i.e., all states communicate with each other.

Now, consider $p_{ii}^{(n)}$, the probability that the MC will return to state i after n steps.

- State i is said to have **period** d if $p_{ii}^{(n)} = 0$ whenever n is not divisible by d and n is the greatest integer with this property (if $p_{ii}^{(n)} = 0$ for all $n > 0$, then the period of the state i is defined infinity). If a state has period 1, the state is said to be **aperiodic**. The periodicity is shown to be a class property.
- State j is **recurrent** if the MC returns to it infinitely often (or with probability 1). Otherwise, the state is transient.

Recurrence: Let's consider $p_{jj}^{(n)} = P[X_{n+k} = j|X_k = j]$ for all k.

Then, state j is recurrent if

$$\sum_{n=1}^{\infty} p_{jj}^{(n)} = \infty$$

State j is transient if

$$\sum_{n=1}^{\infty} p_{jj}^{(n)} < \infty$$

All states in a communicating class are either recurrent or transient:

- **Ergodicity:** Now, consider $\lim_{n\to\infty} \frac{1}{n}\sum_{i=1}^{n} X_i$ for an irreducible, aperiodic and positive recurrent MC. If $P\{\lim_{n\to\infty} \frac{1}{n}\sum_{i=1}^{n} X_i = M\} = 1$ for a constant value M, then the MC is ergodic.
- **Stationarity:** If $p_j = \sum_i p_{ij} p_i$ exists for all $j = 0, 1, 2, \ldots$, the probability distribution $\{p_j, j = 0, 1, 2, \ldots\}$ is said to be stationary for the MC.

  Note that if the initial probability distribution of X_0, $p_j = P[X_0 = j]$, for $j = 0, 1, 2, \ldots$ is a stationary distribution, then $P[X_n = j] = p_j$ is the stationary distribution.

2.2.4.1 *Limiting probabilities*

Lemma 1. *If the discrete-time MC is irreducible and aperiodic, then the limiting probability* $\lim_{n\to\infty} P[X_n = j] \equiv \pi_j$ *exists for all* $j \in S$. *Moreover, for all* $i \in S$, $\lim_{n\to\infty} P[X_n = j | X_0 = i] \equiv \pi_j$. *This result does not preclude the probability that* $\pi_j = 0$.

Lemma 2. *If the discrete-time MC is irreducible, aperiodic, and positive recurrent, then the system has a unique solution with* $\pi_j > 0$, *for all* $j \in S$, *such that*

$$\pi_j = \lim_{i\in S} p_{ij}\, \pi_i \quad \text{with} \quad \sum_{i\in S} \pi_i = 1$$

Lemma 3. *Now, if the discrete-time MC is irreducible, aperiodic, and positive recurrent with period d,*

$$\text{then } \lim_{n\to\infty} \tfrac{1}{d} P[X_n = j | X_0 = i] = \pi_j$$

where $\{\pi_j\}$ *is the solution to the system* $\pi_j = \sum_{i\in S} p_{ij}\pi_i$ *with* $\sum_{i\in S} \pi_i = 1$.

The limiting distribution (probabilities) $\{\pi_k\}$ *of a stationary MC can be obtained as the unique solution of*

$$\pi_k = \sum_i p_{ik}\pi_i \quad k \geq 0 \quad \text{and} \quad \sum_k \pi_k = 1$$

or in matrix form

$$\overline{\pi} = \overline{\pi}\mathbf{P} \quad \text{with} \quad \sum_k \pi_k = 1$$

2.2.5 *Drift*

Consider a discrete-time MC $\{X_n, n = 0, 1, \ldots\}$, with $S = \{0, 1, \ldots\}$. Define the drift function at state i as

$$D_i = E[X_{n+1} - X_n | X_n = i]$$

The drift is a measure of the average change of the MC.

It is noted that

$$D_i = E[X_{n+1} - X_n | X_n = i] = E[X_{n+1} | X_n = i] - E[X_n | X_n = i] = \sum_{j\in S} j p_{ij} - i$$

and so $\sum_{j\in S} j p_{ij} = D_i + i$.

Lemma. Positive recurrence: Consider an irreducible, aperiodic discrete-time MC $\{X_n, n = 0, 1, \ldots\}$, with $S = \{0, 1, \ldots\}$.

The MC is **positive recurrent** if there exists a positive integer m, such that

$$\sum_{j=0}^{\infty} p_{ij} q_j \geq q_i - 1$$

for all $i \geq m$, has a non-negative solution $\{q_j\}$, with

$$s_i \equiv \sum_{j=0}^{\infty} p_{ij} q_j < \infty, \quad \text{for all } i < m - 1$$

Proof. It suffices to prove that $\pi_j > 0$ for one state j. Since the MC is irreducible and aperiodic, the limit

$\pi_j \equiv \lim_{n\to\infty} E[X_n = j | X_0 = i]$ exists for all $j \in S$. Now, fix a $j \in S$. Define the collection of sequences $\{q_i^{(n)}, n = 0, 1, \ldots\}$ for $i \leq j$ as follows: Set $q_i^{(1)} = q_i$, and compute

$$q_i^{(n+1)} = \sum_{j=0}^{\infty} p_{ij}^{(n)} q_j$$

We have

$$\begin{aligned} q_i^{(n+1)} &= \sum_{j=0}^{\infty} \sum_{k=0}^{\infty} p_{ij}^{(n-1)} p_{ik} q_k \\ &\leq \sum_{j=0}^{m-1} p_{ij}^{(n-1)} s_j + \sum_{j=m}^{\infty} p_{ij}^{(n-1)} (q_j - 1) \\ &\leq \sum_{j=0}^{m-1} (1 + s_j) p_{ij}^{(n-1)} + q_i^{(n)} - 1 \end{aligned}$$

Since $q_i^{(1)}$ is finite, we conclude that the entire sequence $\{q_i^{(n)}\}$ is finite, for all i and n. From the above, we get

$$q_i^{(n+1)} \leq \sum_{j=0}^{m-1} (1 + s_j) \sum_{l}^{n-1} p_{ij}^{(l)} + q_i^{(1)} - (n - 1)$$

So,

$$\frac{q_i^{(n+1)}}{n} \leq \sum_{j=0}^{m-1} (1+s_j) \frac{1}{n} \sum_{l}^{n-1} p_{ij}^{(l)} + \frac{q_i^{(1)}}{n} - \frac{(n-1)}{n}$$

Let $n \to \infty$ and rearranging terms, we get

$$\lim_{n\to\infty} \frac{q_i^{(n+1)}}{n} \leq \sum_{j=0}^{m-1} (1+s_j)\pi_j \geq 1$$

which shows that at least one (and thus all) π_j must be strictly positive. □

2.2.6 *Stability criteria*

Stability Lemma [8]. *Consider an irreducible, aperiodic discrete-time MC* $\{X_n, n = 0, 1, \ldots\}$, *with* $S = \{0, 1, \ldots\}$ *and drift* $|D_i| < \infty$.

The MC is **positive recurrent** if there exist an integer $m \geq 0$ and a positive $d > 0$ such that

$$D_i \leq -d, \quad \text{for all } i \geq m$$

Then the MC has a stationary distribution and is called stable. This result provides a sufficient condition for positive recurrent, based on the transition matrix information only, without the need to find any further probabilities.

Proof. Consider the sequence $\{q_i, i = 0, 1, \ldots\}$, define as $q_i = i/d$. For all $i \geq m$, we have from the definition of drift D_i

$$\sum_{j=0}^{\infty} p_{ij} q_j = \sum_{j=0}^{\infty} p_{ij} \frac{j}{d} = \frac{D_i + i}{d} \leq q_i - 1$$

So $D_i \leq (q_i - 1)d - i \leq -i$ since $(q_i - 1)d \leq 0$ □

Instability Lemma [9]. *Consider an irreducible, aperiodic discrete-time MC* $\{X_n, n = 0, 1, \ldots\}$, *with* $S = \{0, 1, \ldots\}$ *and drift* $|D_i| < \infty$.

The MC is **null recurrent** if there exist integers $m \geq 0$ and $k > 0$ such that

$D_i > 0$, for all $i \geq m$, and $p_{ij} = 0$, for all $i, j \in S$, with $0 \leq j \leq i - k$.

Then the MC does not have a stationary distribution, i.e., $p_j = 0$ for all j. This result implies that the transitions to states "around 0" are not possible.

For a discrete-time MC, we are interested in whether the chain has a stationary distribution $\{p_j\}$ or not. The sign of D_i indicates whether the state tends to increase ($D_i > 0$) or decrease ($D_i < 0$), starting at i. The chain will be stable if the drift is negative for all large enough states.

2.3 Queueing Systems [6–10]

We consider queueing systems where customers arrive at random times to obtain service. In the context of a multiple access communication system, customers represent packets assigned to a communication channel (link) for transmission. The service time corresponds to the packet transmission time and is equal to L/C seconds, where L is the packet length in bits and C is the link transmission capacity in bits/seconds. Note that packet lengths can be taken to include frame headers and trailers. In real wired networks, customers represent active calls in the telephone and circuit switched networks and again the service time corresponds to the duration of the call.

The probability distribution of the time between two successive arrivals, i.e., the inter-arrival times, and the probability distribution of the customers' service time are given. We are typically interested in estimating quantities such as the average number of customers in the system (the number of customers either waiting in the queue or undergoing service) and the average delay time per customer (the time a customer spends waiting in the queue plus the service time). These quantities will be estimated in terms of information, such as the customer arrival rate and service rate. In many cases, the customer arrival and service rates are not sufficient to determine the delay characteristics of the system. We may need more statistical information about customer inter-arrival and service times.

First of all, we introduce a number of important concepts and results.

2.3.1 *Little's theorem*

Define the following:

$N(t)$ = number of customers in the systems at time t,
$A(t)$ = number of customers arrived in the interval $[0, t]$,

T_i = time spent (waiting time) in the system by the ith arriving customer,
$\lambda(t)$ = arrival rate at time t.

Now, we consider the time average of arrival rate λ_t, customer delay T_t, and customer size N_t in the system up to time t, as follows:

$$\lambda_t = \frac{A(t)}{t}, \quad T_t = \frac{\sum_{i=0}^{A(t)} T_i}{A(t)} \quad \text{and} \quad N_t = \frac{1}{t}\int_0^t N(\tau)d\tau \tag{2.1}$$

Little's formula: If the steady-state time averages of λ_t, T_t, and N_t exist, then we have

$$N = \lambda T \tag{2.2}$$

Little's result holds for almost every queueing system if the system is in a steady state.

Proof [7]. Just consider a queueing system which has the first-in first-out (FIFO) Poisson arrivals with a single server and a general service distribution. Note that the probability that there are n customers in the system when a customer departs is equal to the probability that n customers arrive during the time spent in the queueing system by the customer. That is to say, the probability of n customers in the queueing system at a departure instant is equal to the probability of there being n customers in the queueing system at any time. For a Poisson process, the probability of n arrivals in any interval of t, with rate λ, is

$$p_n(t) = \frac{(\lambda t)^n}{n!}e^{-\lambda t} \tag{2.3}$$

Assume that the CDF of the waiting time (T_t), $F_T(t)$, exists, then

$$p_n(t) = \int_0^\infty p[n \text{ arrivals } in\ [0, T]|T = t]dF_T(t)$$

Using the mean function, $E[N(t)] = \sum_{n=0}^{\infty} np_n(t)$, we have

$$\begin{aligned} E[N(t)] = \sum_{n=0}^{\infty} np_n(t) &= \int_0^\infty \sum_{n=0}^{\infty} np[n \text{ arrivals } in\ [0, T]|T = t]dF_T(t) \\ &= \int_0^\infty dF_T(t) \sum_{n=0}^{\infty} n\frac{(\lambda t)^n}{n!}e^{-\lambda t} \end{aligned}$$

$$
\begin{aligned}
&= \int_0^\infty (\lambda t) e^{-\lambda t} dF_T(t) \sum_{n=1}^{\infty} \frac{(\lambda t)^{n-1}}{(n-1)!} \\
&= \int_0^\infty (\lambda t) e^{-\lambda t} dF_T(t) e^{\lambda t} \\
&= \int_0^\infty (\lambda t) dF_T(t) = \lambda \int_0^\infty t dF_T(t) = \lambda E[T_t] \square
\end{aligned}
$$

2.3.2 *M/M/1 queueing system*

The M/M/1 queueing system consists of a single queueing station with a single server (in a wireless communication context, a single transmission channel). Customers (in a wireless communication context, packets/messages) arrive according to a Poisson process with rate λ and the probability distribution of the service time is exponential with mean $1/\mu$ seconds. Note that thc first letter indicates the nature of the arrival process (M stands for memoryless, which here means a Poisson process, i.e., exponentially distributed inter-arrival times, G stands for a general distribution of inter-arrival times, and D stands for deterministic inter-arrival times), and the second letter indicates the nature of the probability distribution of the service times. We assume that inter-arrival times and service times are statistically independent of each other. Finally, the last number indicates the number of servers.

We first consider the assumptions on the arrival and service statistics of the M/M/1 system.

2.3.2.1 *Arrival statistics: Poisson process*

An arrival process $\{A(t),\ t \geq 0\}$ is a Poisson process with rate λ if we have the following:

(i) $A(t)$ is a counting process that represents the total number of arrivals occurred from in $[0, t]$

$A(0) = 0$, and for $s < t$, $\{A(t) - A(s)\}$ implies the number of arrivals in the interval $(s, t]$.

(ii) The numbers of arrivals that occur in disjoint time intervals are independent.

(iii) The number of arrivals in any interval of length τ is Poisson distributed with parameter $\lambda\tau$. That is, for all $t, \tau > 0$

$$P[A(t+\tau) - A(t) = n] = \frac{(\lambda\tau)^n}{n!}e^{-\lambda\tau}, \quad n = 1, 2, \ldots$$

Here, $E[A(\tau)] = \lambda\tau$, which is the mean number of arrivals in the interval of τ and so λ means the mean arrival rate per unit time.

The Poisson process has the following properties:

(i) Inter-arrival times are independent and exponentially distributed with parameter λ. That is, if t_n denotes the time of nth arrival, the intervals $\tau_n = t_{n+1} - t_n$ have the probability distribution

$$F_{\tau_n}(s) = P[\tau_n \leq s] = 1 - e^{-\lambda s}, \quad s \geq 0$$

and are mutually independent. The corresponding PDF is $f(\tau) = \lambda e^{-\lambda\tau}$

(ii) For every $t > 0$ and small $\delta > 0$,

$$P[A(t+\delta) - A(t) = 0] = 1 - \lambda\delta + o(\delta)$$

$$P[A(t+\delta) - A(t) = 1] = \lambda\delta + o(\delta)$$

$$P[A(t+\delta) - A(t) = 2] = o(\delta)$$

where we denote by $o(\delta)$ a function of δ such that

$$\lim_{\delta\to 0}\frac{(\delta)}{\delta} = 0$$

Here, the Taylor series is used: $e^{-\lambda\delta} = 1 - \lambda\delta - \frac{(\lambda\delta)^2}{2} - \cdots$
This means that the probability that more than two arrivals in a very short period is almost zero.

2.3.2.2 *Service statistics*

Our assumption regarding the service process is that the customer service times have an exponential distribution with parameter μ, i.e., if s_n is the service time of customer, the time of the nth customer has the CDF,

$$F_{s_n} = P[s_n \leq s] = 1 - e^{-\mu s} \tag{2.4}$$

The PDF of s_n is $f(s) = \mu e^{-\mu s}$. Furthermore, the service times s_n are mutually independent and also independent of all inter-arrival times.

The parameter μ is called the service rate and represents the rate in customers served per unit time at which the server operates when busy.

An important fact regarding the exponential distribution is its memoryless property, which can be expressed as

$$P[\tau_n > \gamma + t | \tau_n > t] = P[\tau_n > \gamma] \quad \text{for } \gamma, t \geq 0$$

$$P[s_n > \gamma + t | s_n > t] = P[s_n > \gamma] \quad \text{for } \gamma, t \geq 0$$

for the inter-arrival and service times τ_n and s_n, respectively. This means that the additional time needed to complete a customer's service in progress is independent of when the service started. Similarly, the time up to the next arrival is independent of when the previous arrival occurred. Verification of the memoryless property follows:

$$P[\tau_n > \gamma + t | \tau_n > t] = \frac{P[\tau_n > \gamma + t]}{P[\tau_n > t]} = \frac{e^{-\lambda(\gamma + t)}}{e^{-\lambda t}} = e^{-\lambda\gamma} = P[\tau_n > t]$$

2.3.2.3 *Markov chain formulation*

The memoryless property together with the independence assumptions on inter-arrival and service times imply that once we know the number $N(t)$ of customers in the system at time t, the time at which customers will arrive or complete service in the future are independent of the arrival times of the customers presently in the system and of how much service the customer currently in service (if any) has already received. This means that the future number of customers depends on past numbers only through the present number, i.e., $\{N(t), t \geq 0\}$ is a continuous-time MC.

Now, let for a small positive number, $\delta > 0$, $N_k = N(k\delta)$ is the number of customers in the system at $t = k\delta$. Then $\{N_k, \ k = 0, 1, 2, \ldots\}$ is a discrete-time MC with steady-state occupancy probabilities corresponding to the continuous MC, $\{N(t), t \geq 0\}$. Let p_{ij} be the transition probabilities, then

$$p_{ij} = P[N_{k+1} = j | N_k = i]$$

Note that p_{ij} depends on δ. By using the property (ii) of the Poisson process, we have the following:

$$p_{00} = 1 - \lambda\delta + o(\delta) \tag{2.5}$$

$$p_{ii} = 1 - \lambda\delta - \mu\delta + o(\delta) \quad \text{for } i \geq 1 \tag{2.6}$$

$$p_{ii+1} = \lambda\delta + o(\delta) \quad \text{for } i \geq 0 \tag{2.7}$$

$$p_{ii-1} = \mu\delta + o(\delta) \quad \text{for } i \geq 1 \tag{2.8}$$

$$p_{ij} = o(\delta) \quad \text{for} \quad i \quad \text{and} \quad j \neq i, i+1, i-1$$

In order to justify these equations, note that, when at a state $i \geq 1$, the probability of 0 arrivals and 0 departures in a δ interval $I_k = (k\delta, \ (k+1)\delta)$, is $e^{-\lambda\delta}e^{-\mu\delta}$, which is because the number of arrivals and the number of departures are Poisson distributed and independent of each other. Expanding this in power series in δ,

$$P\{0 \text{ customer arrive and } 0 \text{ depart in } I_k\}$$
$$= e^{-\lambda\delta}e^{-\mu\delta} = 1 - \lambda\delta - \mu\delta + o(\delta) \tag{2.9}$$

Similarly, we have that

$$P\{0 \text{ customer arrive and } 1 \text{ depart in } I_k\}$$
$$= e^{-\lambda\delta}(\mu\delta e^{-\mu\delta}) = \mu\delta + o(\delta)$$
$$P\{1 \text{ customer arrive and } 0 \text{ depart in } I_k\}$$
$$= \lambda\delta e^{-\lambda\delta}e^{-\mu\delta} = \lambda\delta + o(\delta)$$

These probabilities add up to one plus $o(\delta)$. Thus, the probability of more than one arrival or departure is negligible for δ small. It follows that for $i \geq 1$, p_{ii}, which is the probability of an equal number of arrivals and departures in I_k, is within $o(\delta)$ of the value in equation (2.9), this verifies equation (2.6). Equations (2.5), (2.7), and (2.8) are verified in the same way.

The state transition diagram for the MC p_{ij} is shown in Fig. 2.2, where we have omitted the terms $o(\delta)$. The state n corresponds to n customers in the system.

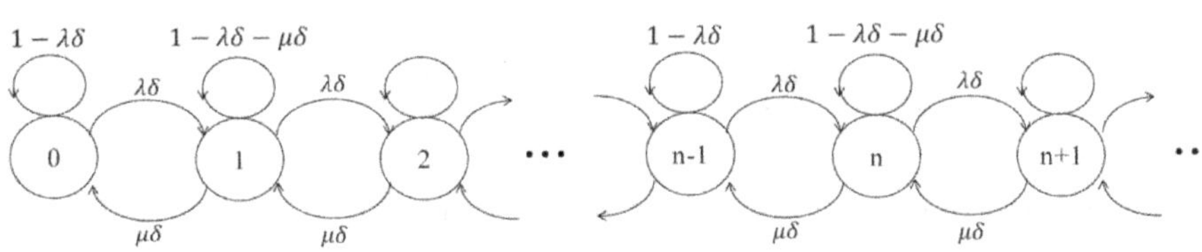

Fig. 2.2. Discrete-time Markov chain for M/M/1 system.

2.3.2.4 *Stationary distribution*

Consider now the steady-state probabilities

$$p_n = \lim_{k\to\infty} P[N_k = n] = \lim_{t\to\infty} P[N(t) = n] \tag{2.10}$$

Note that during any time interval, the total number of transitions from state n to $n+1$ must differ from the total number of transitions from $n+1$ to n by at most one. Equivalently, in steady state, the probability that the system is in state n and makes a transition to $n+1$ at the next transition interval is the same as the probability that the system is in state $n+1$ and makes a transition to n,

$$p_n\lambda\delta + o(\delta) = p_{n+1}\mu\delta + o(\delta)$$

By taking the limit in this equation as $\delta \to 0$, we have

$$p_n\lambda = p_{n+1}\mu$$

which is called the global balance equation between two states $\{0, 1, 2, \ldots, n\}$ and $\{n+1,\ n+2, \ldots\}$.

This equation can be written as

$$p_{n+1} = \rho p_n \quad \text{with} \quad \rho = \lambda/\mu \quad \text{for } n = 0, 1, 2, \ldots$$

It follows that

$$p_{n+1} = \rho^{n+1} p_0 \tag{2.11}$$

If $\rho < 1$ (the service rate exceeds the arrival rate), the probabilities p_n are all positive and add up to unity, so

$$1 = \sum_{n=0}^{\infty} p_n = \sum_{n=0}^{\infty} \rho^n p_0 = \frac{p_0}{1-\rho}$$

Then the final probability distribution is obtained as

$$p_n = (1-\rho)\rho^n \quad \text{for } n = 0, 1, 2, \ldots$$

The mean number of customers in the system in steady state can be calculated as follows:

$$N_S = \lim_{t\to\infty} E[N(t)] = \sum_{n=0}^{\infty} np_n = \sum_{n=0}^{\infty} n\rho^n(1-\rho)$$
$$= \rho(1-\rho)\frac{d}{d\rho}\sum_{n=0}^{\infty}\rho^n = \frac{\rho}{1-\rho} = \frac{\lambda}{\mu-\lambda}$$

Using Little's formula, the total average delay per customer is

$$W_S = \frac{N}{\lambda} = \frac{\rho}{\lambda(1-\rho)} = \frac{1}{\mu-\lambda} \tag{2.12}$$

The mean waiting time in the queue is

$$W_Q = \frac{1}{\mu-\lambda} - \frac{1}{\mu} = \frac{\rho}{\mu-\lambda} \tag{2.13}$$

The mean number of customers in the queue is

$$N_Q = \lambda W_Q = \frac{\rho^2}{1-\rho} \tag{2.14}$$

2.3.3 *Occupancy distribution upon arrival*

Consider a probabilistic characterization of a queueing system as seen by an arriving customer.

The steady-state occupancy probability upon arrival is

$$a_n = \lim_{t\to\infty} P[N(t) = n \mid \text{an arrival occurred just after time } t]$$

For M/M/1 system, it is shown that

$$a_n = p_n \quad \text{for } n = 0, 1, 2, \ldots$$

Let $A(t, t+\delta)$ be the event that an arrival occurs in the interval $(t, t+\delta)$. Let

$$p_n(t) = P[N(t) = n]$$
$$a_n = P[N(t) = n \mid \text{an arrival occurred just after time } t]$$

Using Bayes' rule,

$$\begin{aligned}a_n &= \lim_{\delta\to 0} P[N(t) = n|A(t, t+\delta)] \\ &= \lim_{\delta\to 0} \frac{P[N(t) = n,\ A(t, t+\delta)]}{P[A(t, t+\delta)]} \\ &= \lim_{\delta\to 0} \frac{P[A(t, t+\delta)|N(t) = n]P[N(t) = n]}{P[A(t, t+\delta)]}\end{aligned}$$

Since the event $A(t, t+\delta)$ is independent of the number in the system $N(t)$ at time t,

$$P[A(t, t+\delta)|N(t) = n] = P[A(t, t+\delta)]$$

Finally,

$$a_n(t) = P[N(t) = n] = p_n(t) \tag{2.15}$$

Taking the limit as $t \to \infty$, we obtain

$$a_n = p_n \tag{2.16}$$

In fact, this holds very general conditions for queueing systems with Poisson arrivals regardless of the distribution of the service times. This fact holds for the case that inter-arrival times and service times are independent.

2.3.4 *Occupancy distribution upon departure*

Let us consider the distribution of the number of customers in the system just after a departure has occurred, i.e., the probabilities

$$d_n(t) = P[N(t) = n \mid \text{a departure occurred just before time } t]$$

The corresponding steady-state values are denoted as

$$d_n = \lim_{t\to\infty} d_n(t) \quad \text{for } n = 0, 1, 2 \ldots$$

It turns out that

$$d_n = a_n \quad \text{for } n = 0, 1, 2 \ldots \tag{2.17}$$

under very general assumptions; the only essential requirement is that the system reaches a steady state with all n having positive steady-state probabilities and that $N(t)$ changes in unit increments. (These assumptions certainly hold for a stable M/M/l system ($\rho < 1$), but they also hold for most stable single-queue systems of interest.)

For any sample path of the system and for every n, the number in the system will be n infinitely often (with a probability one). This means that for each time the number in the system increases from n to $n+1$ due to an arrival, there will be a corresponding future decrease from $n+1$ to n due to a departure. Therefore, in the long run, the frequency of transitions from n to $n+1$ out of transitions from any k to $k+1$ equals the frequency of transitions from $n+1$ to n out of transitions from any $k+1$ to k, which implies $d_n = a_n$. Therefore, in steady state, the system appears statistically identical to an arriving and a departing customer. When arrivals are Poisson, we saw earlier that $a_n = p_n$, so, in this case, both an arriving and a departing customer in steady state see a system that is statistically identical to the one seen by an observer looking at the system at an arbitrary time.

2.3.5 *Pollaczek–Khinchin (P–K) formula* [7]

Assume the arrival process is Poisson with parameter λ. Consider a random process $\{X(t),\ t \geq 0\}$ and its imbedded process $\{X_k = X(t_k),\ k = 0, 1, 2 \ldots\}$.

Define the following:

X_i = number of customers in the system at the ith departure point,

A_{i+1} = number of customers who arrived during the service time of the $(i+1)$st customer,

S_i = service time of the ith customer.

We assume that the service times are independently and identically distributed (i.i.d) with an arbitrary distribution, $B(\tau)$.

If the PDF of S exists,

$$B(\tau) = P[S \leq \tau] = \int^{\tau} f_S(t)dt$$

If the queueing system has a steady-state solution, then we can obtain the following relationship for all $i > 0$,

$$X_{i+1} = \begin{cases} X_i - 1 + A_{i+1} & \text{for } X_i > 1 \\ A_{i+1} & \text{for } X_i = 0 \end{cases} \tag{2.18}$$

Using the unit step function, $U(X_i) = 1$ if $X_i > 0$ and $U(X_i) = 0$ if $X_i = 0$, we can express

$$X_{i+1} = X_i - U(X_i) + A_{i+1} \tag{2.19}$$

The mean steady-state system size (the number of customers in the system), called the P–K formula, is as follows:

$$N_S = \rho + \frac{\rho^2 + \lambda^2\sigma_s^2}{2(1-\rho)} \tag{2.20}$$

where $\sigma_s^2 = \text{Var}[S] = E[(S - E[S])^2]$ and $E[S] = \frac{1}{\mu}$

Proof. Taking the expectation of both sides of equation (2.19) yields

$$E[X_{i+1}] = E[X_i] - E[U(X_i)] + E[A_{i+1}]$$

Since it is assumed that a steady-state solution exists, we have

$$E[X_{i+1}] = E[X_i] \quad \text{and so}$$
$$E[U(X_i)] = E[A_{i+1}]$$

But $E[A_i] = \int_0^\infty E[A_i|S = \tau]dB(\tau) = \int_0^\infty \lambda\tau dB(\tau) = \lambda E[S]$

Now, squaring and taking the expectation of both sides of equation (2.19)

$$\begin{aligned} E[X_{i+1}^2] = E[X_i^2] + E[U^2(X_i)] + E[A_{i+1}^2] - 2E[X_i U(X_i)] \\ - 2E[A_{i+1}U(X_i)] + 2E[A_{i+1}X_i] \end{aligned}$$

Since

$$E[X_{i+1}^2] = E[X_i^2], \quad U^2(X_i) = U(X_i) \quad \text{and } X_i U(X_i) = X_i,$$

$$\begin{aligned} \text{then, } 0 = \; & E[U(X_i)] + E[A_{i+1}^2] - 2E[X_i] \\ & - 2E[A_{i+1}]E[U(X_i)] + 2E[A_{i+1}X_i], \end{aligned}$$

i.e., $0 = \rho + E[A_{i+1}^2] - 2E[X_i] - 2\,\rho^2 + 2\rho E[X_i]$,

i.e., $0 = \rho + E[A_{i+1}^2] - 2E[X_i](1-\rho) - 2\,\rho^2$

So

$$E[X_i] = \frac{\rho - 2\rho^2 + E[A_{i+1}^2]}{2(1-\rho)} \tag{2.21}$$

Note that

$$\begin{aligned} \text{Var}[A_i] &= E[\text{Var}[A_i|S]] + \text{Var}[E[A_i|S]] \\ &= E[\lambda S] + \text{Var}[\lambda S] = \lambda E[S] + \lambda \text{Var}[S] = \rho + \lambda\sigma_s^2 \end{aligned}$$

and

$$E[A_i^2] = \rho + \lambda^2\sigma_s^2 + \rho^2 \tag{2.22}$$

Finally, we have the P–K formula.

Another way to prove the P–K formula using the mean residual service time is as follows [10]:

Define the following:

W_i = waiting time in the queue of the ith customer;
R_i = residual service time seen by the ith customer.

Note that if customer j is already being served when i arrives, R_i implies the remaining time until customer j's service time is complete. If no customer is in service, then

R_i = zero;
X_i = service time of the ith customer;
N_i = number of customers found waiting in the queue by the ith customer upon arrival.

Then we have

$$W_i = R_i + \sum_{j=i-N_i}^{i-1} X_j \tag{2.23}$$

□

P–K formula: If the long-term time averages are the same as the ensemble averages in probability one and the steady-state averages exist, then

$$E[W_i] = E[R_i] + E[X_i]E[N_i]$$

That is, $W_Q = R + \frac{1}{\mu}N_Q$ with $R = \frac{1}{2}\lambda E[X_i^2]$ and $E[X_i] = \frac{1}{\mu}$
Finally,

$$W_Q = \frac{\lambda E[X_i^2]}{2(1-\rho)} \tag{2.24}$$

Proof. Now, using Little's formula, we have
$N_Q = \lambda W_Q$ and $W_Q = R + \rho W_Q$, and then

$$W_Q = \frac{R}{1-\rho}$$

where $\rho = \lambda/\mu$ is the utilization factor.

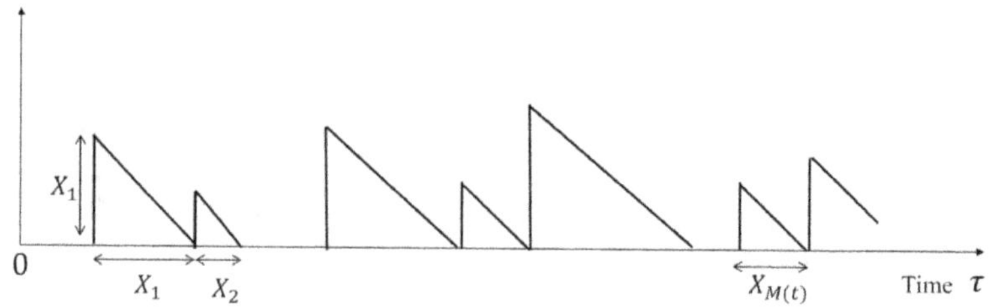

Fig. 2.3. Distribution of the residual service time.

Now, let us prove $R = \frac{1}{2}\lambda E[X_i^2]$ by a graphical argument. Define the residual service time $r(\tau)$, the remaining time for completion of a customer in service at τ. Now, the service duration for a customer i has X_i and decays linearly to be completed at time t, i.e., $r(t) = 0$ (see Fig. 2.3). Then the average time of $r(\tau)$ in the interval [0, t] is

$$\frac{1}{t}\int_0^t r(\tau)d\tau = \frac{1}{t}\sum_{i=1}^{M(t)} \frac{1}{2}X_i^2 \tag{2.25}$$

where $M(t)$ is the number of service completions within $[0, t]$. The above equation can be written as

$$\frac{1}{t}\int_0^t r(\tau)d\tau = \frac{1}{2}\frac{M(t)}{t}\frac{\sum_{i=1}^{M(t)} X_i^2}{M(t)}$$

Assuming that the limits exist, then one can obtain the following:

$$\lim_{t\to\infty}\frac{1}{t}\int_0^t r(\tau)d\tau = \frac{1}{2}\lim_{t\to\infty}\frac{M(t)}{t}\lim_{t\to\infty}\frac{\sum_{i=1}^{M(t)} X_i^2}{M(t)}$$

Assuming that the time averages converge to the ensemble averages, then finally

$$R = \frac{1}{2}\lambda E[X^2] \tag{2.26}$$

□

2.3.6 *M/G/1 queueing system*

We consider a single-server queueing system with a Poisson arrival process with λ. However, the customer service time has a general distribution, not necessarily exponential as in the M/M/1 system. We assume that customers' service times are identically distributed, mutually independent, and independent of the inter-arrival times. Define the following:

X_i = service time for ith customer,
$E[X_i]$ = mean service time of ith customer $= 1/\mu$,
$E[X_i^2]$ = second moment of service time,
$\sigma_X^2 = \text{Var}[X_i]$ = variance of service time.

Then using the P–K formula, the mean customer waiting time in the queue is

$$W_Q = \frac{\lambda E[X_i^2]}{2(1-\rho)} \tag{2.27}$$

and the mean total customer waiting time in the queue and in service is

$$W_S = \frac{1}{\mu} + \frac{\lambda E[X_i^2]}{2(1-\rho)} \tag{2.28}$$

Now, using Little's formula, the mean number of customers N_Q in the queue and N_S in the system are

$$N_Q = \lambda W_Q = \frac{\lambda^2 E[X_i^2]}{2(1-\rho)} \tag{2.29}$$

$$N_S = \lambda W_S = \frac{\lambda}{\mu} + \frac{\lambda^2 E[X_i^2]}{2(1-\rho)} \tag{2.30}$$

We can easily apply P–K formula to other queueing systems. For the M/M/1 system, service times are exponential with $E[X^2] = 2/\mu^2$, and thus, the mean customer waiting time in the queue is

$$W_Q = \frac{\rho}{\mu(1-\rho)} \tag{2.31}$$

For the M/D/1 system, service time is deterministic (not random) and has $E[X^2] = 1/\mu^2$, then the waiting time in the queue is

$$W_Q = \frac{\rho}{2\mu(1-\rho)} \tag{2.32}$$

From the P–K formula and Little's formula, the mean number of customers in the queue is

$$N_Q = N_S - \rho \tag{2.33}$$

The total mean waiting time (in queue and in service) is

$$W_S = W_Q + 1/\mu \tag{2.34}$$

Since the M/D/1 case yields the minimum possible value of $E[X^2]$ for given μ, it follows that the values of W_Q, W_S, N_Q, and N_S for an M/D/1 queue are lower bounds to the corresponding quantities for an M/G/l queue of the same λ and μ. It is interesting to note that W_Q and N_Q for the M/D/1 queue are exactly one half their values for the M/M/1 queue of the same λ and μ. The values of W_S and N_S for M/D/1, on the other hand, range from the same as M/M/l for small ρ to one half of M/M/1 as ρ approaches 1. The reason is that the expected service time is the same in the two cases, and for ρ small, most of the waiting occurs in service, whereas for ρ large, most of the waiting occurs in the queue.

2.4 Summary

This chapter provides a comprehensive overview of stochastic modeling in multi-access systems, covering probability and random variables, random processes including Poisson and Markov processes, and queueing systems such as M/M/1 and M/G/1, offering insights into the various aspects of system behavior and performance analysis.

References

[1] K. C. Chung, *A Course on Probability Theory*, Academic Press, New York, 1974.

[2] S. Ross, *A First Course in Probability*, Macmillan, New York, 1994.

[3] Y. Viniotis, *Probability and Random Processes for Electrical Engineers*, McGraw Hill, 1998.

[4] S. Ross, *Stochastic Processes*, John Wiley & Sons, Inc. 1996.

[5] S. S. Karlin and H. M. Taylor, *A First Course in Stochastic Processes*, Academic Press, New York, 1976.

[6] R. W. Wolff, *Stochastic Modeling and The Theory of Queues*, Prentice-Hall, Inc. 1989.

[7] D. Gross and C. M. Harris, *Fundamentals of Queueing Theory*, John Wiley & Sons, Inc. 1998.

[8] A. G. Pakes, Some conditions for ergodicity and recurrence of Markov chain, *Oper. Res.*, 17, 1059–1061, 1969.
[9] M. Kaplan, A Sufficient condition for non-ergodicity of a Markov Chain, *IEEE Trans. Infor. Theory*, 25, 470–471, Jul 1979.
[10] D. Bertsekas and R. Gallager, *Data Networks*, Prentice Hall International, Inc. 1992.

Chapter 3

Statistical Channel Model

The performance of digital communication in a physical layer is determined by the signal-to-noise ratio power (SNR) of the received signal [1–3]. In wireless mobile communication, the transmitted signal is received in a complex form due to phenomena, such as scattering, diffraction, and reflection. The wireless mobile communication channel, which represents the relationship between the transmitted and received signals, takes on a complex form due to transmission through multiple paths [4, 5]. The purpose of this chapter is to mathematically model such complex channels. Through channel modeling, it is much more cost-effective to validate system designs through computer simulations using mathematically generated channel models similar to real channel environments before completing hardware designs rather than designing transmission and reception systems based on channel characteristics and then measuring performance in actual systems. In modeling wireless mobile channels, they are broadly categorized into two distinct forms: large-scale channel models and small-scale channel models. Large-scale channel models refer to models concerning average received power, while small-scale channel models concern models for dynamically changing received power [13–15]. We understand how to utilize these channel models in the design of mobile communication systems.

3.1 Introduction

3.1.1 *Radio wave environment*

The purpose of communication is to transmit information signals so that they are received by the recipient without errors. However, during communication, there is always thermal noise modeled as white Gaussian noise, and the wireless channel representing the space between the transmitter and receiver distorts the received signal in a very different form from the transmitted signal, making it difficult to detect what signal was transmitted. Therefore, if we can mathematically model the wireless channel and understand its characteristics, we can know the form of the distorted signal at the receiving end, making signal detection easier.

The wireless channel, as shown in Fig. 3.1, generates numerous paths due to scattering, diffraction, reflection, and so on, and each path has a different delay time accordingly. Due to these different delay times, the signal is propagated with different powers, and the received signal thus takes on a very complex form.

Considering the wireless channel environment with complex wireless mobile communication, depicted in Fig. 3.1, as a system similar to Fig. 3.2(a), and representing it as a mathematical function $f[\cdot]$, the received signal $r(t)$ for the transmitted signal $s(t)$ can be expressed as follows:

$$r(t) = f[s(t)] \tag{3.1}$$

For instance, if the wireless channel is modeled as a Linear Time-Invariant (LTI) system, the function $f[\cdot]$ can be considered as the convolution of the input signal s(t) with the impulse response function $h(t)$. Specifically, if we denote the response function to the impulse function as $h(t) = f[\delta(t)]$, then in a linear time-invariant system, the received signal

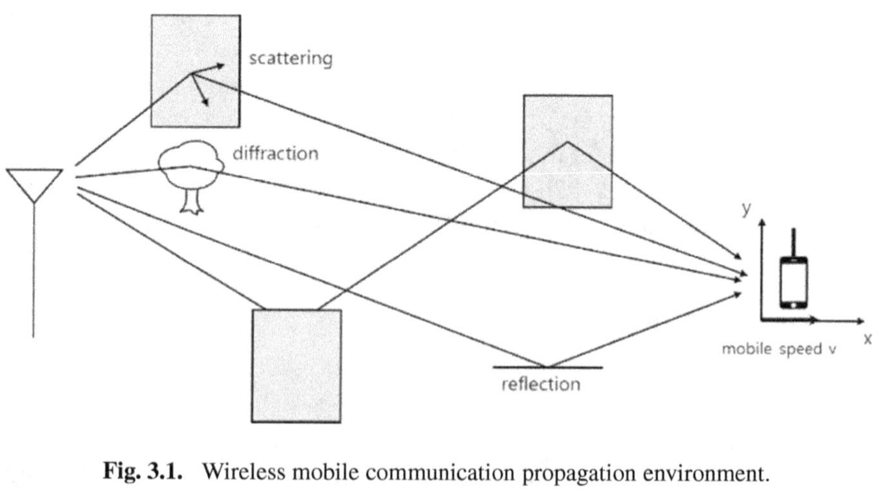

Fig. 3.1. Wireless mobile communication propagation environment.

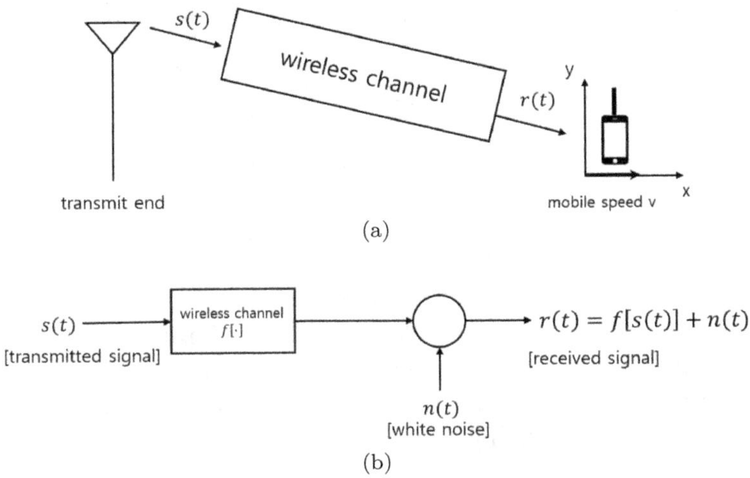

Fig. 3.2. Wireless mobile channel system model.

can be expressed as follows:

$$r(t) = h(t) * s(t) = \int_{-\infty}^{t} h(\tau)s(t-\tau)d\tau \tag{3.2}$$

where $*$ denotes the convolution operation.

Considering thermal noise, which is always present and modeled as white Gaussian noise due to heat generation in electronic devices, the received signal passed through the wireless channel can be represented as shown in Fig. 3.2(b).

To model the wireless mobile channel, let's define the following with respect to power. If we denote the received signal as $r(t)$, its envelope can be represented as $|r(t)|$ and the instantaneous power as $P(t) = |r(t)|^2$. Then, the average power can be calculated as follows:

$$P_{\text{avg}} = \lim_{T\to\infty} \frac{1}{2T} \int_{-T}^{T} P(t)dt \tag{3.3}$$

3.1.2 *Signal-to-noise ratio*

The performance of digital communication, measured by Bit Error Rate (BER) or Symbol Error Rate (SER), is influenced by the Signal-to-Noise power Ratio (SNR) of the received signal. For instance, the BER, $P_b(\gamma)$

with γ = SNR, for Binary Phase Shift Keying (BPSK) modulation can be expressed as follows:

$$\mathrm{BER} = P_b(\gamma) = Q(\sqrt{2\mathrm{SNR}}) = Q(\sqrt{2\gamma}) \tag{3.4}$$

where $Q(x)$ represents the Gaussian Q-function defined as

$$Q(x) = \frac{1}{2\pi}\int_x^{\infty} \mathrm{e}^{-t^2/2}dt \tag{3.5}$$

Therefore, in digital communication, the performance is determined by the SNR of the received signal regardless of the waveform of the transmitted signal. Similarly, in various modulation techniques, SNR of the received signal remains the most significant factor affecting BER or SER performance. Therefore, examining the SNR of the received signal is crucial in system design.

In wireless communication, SNR is distinguished into Carrier-to-Noise power Ratio (CNR). CNR represents the ratio of the power intensity before signal dcmodulation to the power intensity of noise. Therefore, it is dimensionless and often represented in decibels (dB). If we denote it as Γ, it is defined as follows:

$$\mathrm{CNR} = \Gamma = \frac{P_r}{N} \tag{3.6}$$

where P_r is the power of the received signal and N is the power of the noise signal. Noise signal typically represents thermal noise, which always exists due to heat generation in electronic devices. To understand the noise power value in equation (3.6), let's first examine the noise power spectral density and noise figure.

The power spectral density of noise is proportional to temperature and can be expressed as

$$N_0 = kT_0 \tag{3.7}$$

where k is the Boltzmann constant, which is 1.38×10^{-23} W/K, and T_0 represents temperature in Kelvin. For example, if the indoor temperature is 300 K (17° C), the noise power spectral density would be $kT_0 = -174$ dBm/Hz.

The modem structure of the transceiver comprises radio frequency (RF) parts and analog and digital modems. The performance of the RF section is characterized by the gain (G), representing the amplification or loss of the RF circuit blocks, and the noise figure (F), representing the ratio of SNR

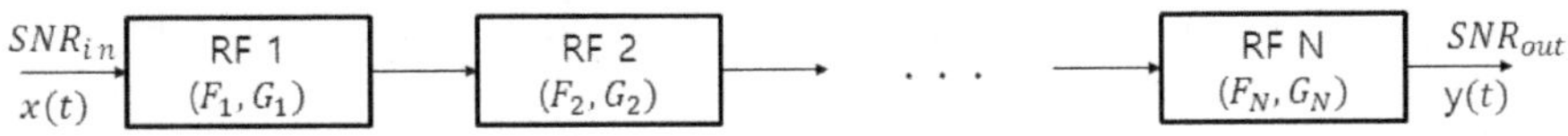

Fig. 3.3. Serial RF block connection structure.

between the input and output signals of the RF circuit blocks, is given as follows [1]:

$$F = \frac{(\mathrm{SNR})_{\mathrm{in}}}{(\mathrm{SNR})_{\mathrm{out}}} \tag{3.8}$$

Here, the noise figure of the RF modem can be expressed as the ratio of the input SNR of the first circuit block to the output SNR of the last circuit block. The calculation of the noise figure for multiple RF circuit blocks connected in series can be achieved using the Friis formula [1]. Here, $\mathrm{SNR_{in}}$ is the input SNR to the RF section, and $\mathrm{SNR_{out}}$ is the output SNR of the RF section. If the RF blocks are connected in series as depicted in Fig. 3.3, and each RF block is represented by its noise figure F and gain value G, then the total noise figure value can be expressed as follows [1]:

$$F = \frac{(\mathrm{SNR})_{\mathrm{in}}}{(\mathrm{SNR})_{\mathrm{out}}} = F_1 + \frac{F_2 - 1}{G_1} + \frac{F_3 - 1}{G_1 G_2} + \cdots + \frac{F_N - 1}{G_1 G_2 \cdots G_{N-1}} \tag{3.9}$$

On the receiving side, filters are used to receive signals in the frequency band of the information signal. The bandwidth of the filter $\mathrm{B_f}$ is inversely related to the symbol duration T_s, however, when using the commonly used Root Raised Cosine (RRC) filter, it becomes $B_f > 1/T_s$, and B_f may differ from the signal's bandwidth $B_s = 1/T_s$. In such cases, the noise power value N can be expressed as follows [2]:

$$N = kT_0 B_f F \tag{3.10}$$

The CNR value of the received signal is defined as the link budget, which can be expressed as follows [2]:

$$\Gamma = \frac{P_r}{N} = \frac{P_r}{kT_0 B_f F} \tag{3.11}$$

At this point, the energy-to-noise ratio (E_s/N_0) per decoded symbol is expressed as follows [2]:

$$\frac{E_s}{N_0} = \Gamma \times \frac{B_f}{R_s} = \frac{P_r}{kT_0 F R_s} \tag{3.12}$$

where R_s represents the symbol transmission rate in the baud rate. Furthermore, in equation (3.12), E_s/N_0 directly influences the performance of digital communication and is equal to γ in equation (3.4).

Channel modeling refers to understanding how the received signal power, represented by P_r in equation (3.12) governing the SNR value, changes in wireless and mobile scenarios, and then mathematically modeling this change. Through such channel modeling, it becomes possible to predict the received signal power when the transmitted signal remains constant, thereby allowing for the prediction of communication distance. Additionally, appropriate signal processing techniques can be employed to enhance performance metrics, such as BER/SER.

Traditionally, wireless channels are categorized into large-scale channel models and small-scale channel models. These models are independent of each other. Large-scale channel models characterize the average power of received signals, while small-scale channel models characterize the instantaneous power of received signals. Section 3.2 discusses large-scale channel modeling, focusing on modeling the average power value of received signals, while Section 3.3 addresses small-scale channel modeling, focusing on modeling the instantaneous power value of received signals.

3.2 Large-Scale Propagation Model

Propagation causes the power of the received signal to attenuate with distance. The power of the received signal can be distinguished between the average power value expressed in equation (3.3) and the instantaneous power value measured at each moment. The large-scale propagation model refers to the model for the average received power defined in equation (3.3), which is categorized into path loss and shadowing.

Path loss represents the loss of average received power with distance. For instance, if the transmitted power is P_t, understanding how the average power of the received signal measured at points 100 m and 200 m away differs exactly expresses path loss.

If only path loss is considered, points at the same distance from the transmitter would have the same average received power. However, in reality, the average received power measured at points at the same distance differs and exhibits the characteristics of a random variable. The phenomenon where the average received power measured at locations with the same distance but different environments is modeled as a random variable is called shadowing. For example, if the transmitted power is P_t, measuring the average power at various locations 100 m away would result in different powers due to the differing environments of each location, such as scattering, diffraction, and reflection, depicted in Fig. 3.1. These differently measured average power values are random variables, and the probabilistic model for them is referred to as shadowing.

3.2.1 *Path loss*

In wireless communication, the average received power at a distance d away from the transmitter when the transmitting power is P_t can be expressed as follows [3]:

$$P_r(d) = \kappa \frac{P_t}{d^\beta} \tag{3.13}$$

where

$$\kappa = \left(\frac{\lambda}{4\pi}\right)^2 \frac{G_T G_R}{L_{R_x}} \tag{3.14}$$

where λ represents the wavelength of the transmitted signal and β is the path loss exponent. For free space as in Fig. 3.4, β has a value of 2, where there is nothing but the transmitter and receiver. For real environments,

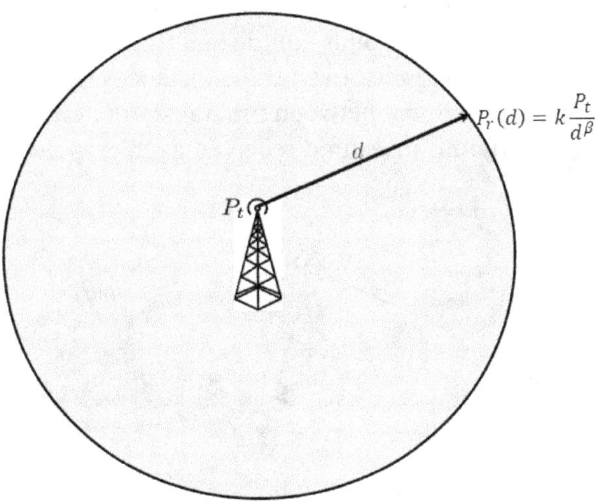

Fig. 3.4. Path loss in free space.

β has a value of 3~8, which includes obstacles such as buildings and trees between the transmitter and receiver.

In equation (3.14), G_T and G_R represents the antenna gains at the transmitter and receiver, respectively. L_{R_x} represents implementation losses in the receiver hardware, which refer to losses incurred due to imperfect hardware. Expressing the average received power value of equation (3.13) in milli-decibel [dBm] units, it is as follows.

$$\begin{aligned} P_r(d)_{\mathrm{dBm}} &= 10\log_{10}[P_r(d)\times 10^3] \\ &= P_{t(\mathrm{dBm})} + 10\log_{10}(\kappa) - 10\beta\log_{10}(d) \end{aligned} \tag{3.15}$$

Modeling path loss involves measuring the path loss exponent β values in equations (3.13) and (3.15). To achieve this, the following approach is taken. Let's denote the average received power of the signal at a certain distance d_0 as $P_r(d_0)_{\mathrm{dBm}}$. Then, the received power at distance d can be calculated using equation (3.15) as follows:

$$P_r(d)_{\mathrm{dBm}} = P_r(d_0)_{\mathrm{dBm}} - 10\beta\log_{10}\left(\frac{d}{d_0}\right) \tag{3.16}$$

3.2.2 *Shadowing*

Consider two points at different distances, as shown in Fig. 3.5. If we model the average received power only with path loss expressed by equation (3.13), the received average power at points equidistant from the transmitter should be the same. However, in reality, even if the distances from the transmitter are the same, the environment between the transmitter and receiver is different, leading to different measured received average powers. Therefore,

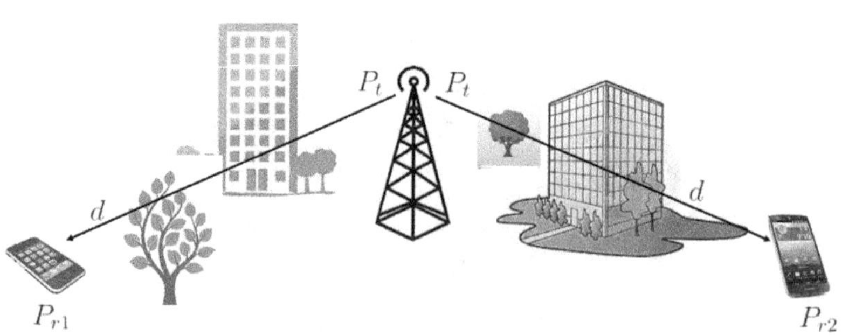

Fig. 3.5. Shadowing environment.

modeling the received average power in wireless mobile communication channels based solely on path loss is not sufficient. This phenomenon, where the received average power varies due to differences in the propagation environment despite the same distance, is known as shadowing, and modeling of this phenomenon is necessary.

To explain this, let's express the average received power value in equation (3.13) as follows [4, 5]:

$$P_r(d)_{\text{dBm}} = P_{t(\text{dBm})} + 10\log_{10}(\kappa) - 10\beta\log_{10}(d) + \chi_s(\text{dB}) \qquad (3.17)$$

In equation (3.17), $\chi_s(\text{dB})$ represents the shadowing factor, which is a random variable following a Gaussian normal distribution. The shadowing factor is known to follow a Gaussian normal distribution based on actual measurements. $\chi_s(\text{dB})$ is expressed as a Gaussian random variable with a mean of 0 and a variance of σ_s^2:

$$\chi_s(\text{dB}) \sim N(0, \sigma_s^2) \qquad (3.18)$$

Let's represent the received power in equation (3.17) as follows:

$$P_r(d)_{\text{dBm}} = \mu_s(d)_{\text{dB}} + \chi_s(\text{dB}) \qquad (3.19)$$

In equation (3.19), $\mu_s(d)_{\text{dB}}$ represents the received average power measured only considering path loss at a distance d from the transmitter. According to equation (3.17), it can be expressed as follows:

$$\mu_{s(\text{dBm})}(d) = P_{t(\text{dBm})} + 10\log_{10}(\kappa) - 10\beta\log_{10}(d) \qquad (3.20)$$

Figure 3.6 illustrates that the average power values of received signals due to shadowing vary by a Gaussian random variable $\chi_s(\text{dB})$ when expressed in decibels. In Fig. 3.6, the average of all measured average power values at points equidistant from the transmitter is $\mu_s(d)_{\text{dB}} = E[P_r(d)_{\text{dBm}}]$, which is modeled solely by path loss. This value is referred to as the area mean power. $P_r(d)_{\text{dBm}}$ represents the local mean power at a specific location. Therefore, the average received power $P_r(d)_{\text{dBm}}$ follows a Gaussian-normal distribution with a mean of $\mu_s(d)_{\text{dB}}$ and a variance of σ_s^2, given by the probability density function:

$$f_{P_r(\text{dB})}(x) = \frac{1}{\sqrt{2\pi\sigma_s^2}} \exp\left[-\frac{(x - \mu_s(d)_{\text{dB}})^2}{2\sigma_s^2}\right] \qquad (3.21)$$

As seen in equation (3.21), the average power value [Watts] follows a log-normal distribution in decibel units. This is called a log-normal distribution.

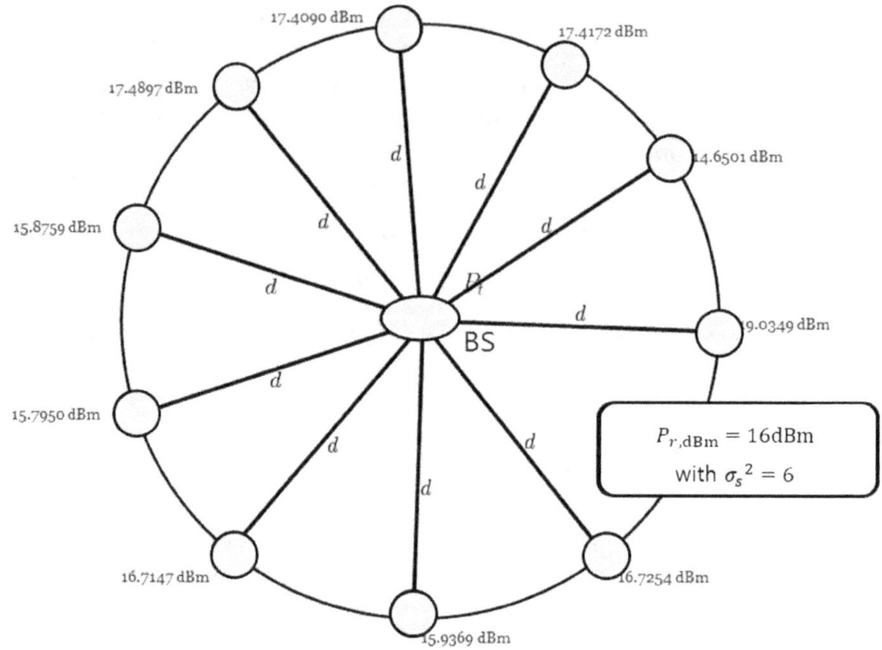

Fig. 3.6. Shadowing phenomenon.

In cellular mobile systems, the area mean power is measured not at all distances but at a certain reference distance. For example, in Fig. 3.6, measuring the mean values at various points at a distance of d_0 will result in each value being the local mean power at that specific location. The average of these power values will give the area mean power at distance d_0. The area mean power measured at the reference distance is denoted as $\mu_s(d_0)_{\text{dB}}$. Equation (3.19) can be rewritten as follows:

$$P_r(d)_{\text{dBm}} = \mu_{s(\text{dB})}(d_0) - 10\beta \log_{10}\left(\frac{d}{d_0}\right) + \chi_s(\text{dB}) \tag{3.22}$$

Therefore, based on the area mean power measured at the reference distance, the area mean power at all other locations can be calculated.

3.2.3 *Outage probability*

In consideration of the large-scale fading model, the outage probability is defined as the probability that the Carrier-to-Noise Ratio (CNR) of the received signal falls below a certain threshold. Outage probability serves as a crucial metric in determining the cell size in cellular systems. For instance, in a system where communication is satisfactory when the CNR value exceeds a certain threshold of dB, determining the appropriate radius of a cellular cell with a path loss exponent and a standard deviation value of the shadowing factor becomes a critical issue for service providers. Therefore, understanding the methods for calculating outage probability is highly important.

Let's denote the received CNR value, $\Gamma(d)$, at a user located at a distance from the base station as d. If the threshold is Γ_{th}, then the outage probability is defined as follows [2]:

$$P_{\text{out}}(d) \equiv \Pr[\Gamma(d) \leq \Gamma_{\text{th}}] = \Pr[\Gamma_{(\text{dB})}(d) \leq \Gamma_{\text{th}(\text{dB})}] \tag{3.23}$$

where $\Gamma_{(\text{dB})}(d)$ and $\Gamma_{\text{th}(\text{dB})}$ represent the CNR value and threshold in decibel (dB), respectively.

We can express CNR using equation (3.6) as follows:

$$\Gamma_{(\text{dB})}(d) = \left.\frac{P_r}{N}\right|_{(\text{dB})} = 10\log_{10}\left(\frac{P_r \times 10^3}{N \times 10^3}\right) = P_{r(\text{dBm})} - N_{(\text{dBm})} \tag{3.24}$$

Moreover, through equation (3.19), we understand that $\Gamma(d)$ follows a log-normal (LN) distribution due to the shadowing phenomenon, which can be expressed as follows:

$$\Gamma(d) \sim \text{LN}(\mu_{\Gamma(\text{dB})}, \sigma^2_{\Gamma(\text{dB})}) \tag{3.25}$$

The mean and variance in equation (3.25) can be determined using equations (3.19) and (3.24) as follows:

$$\mu_{\Gamma(\text{dB})} = \mu_{s(\text{dBm})}(d) - N_{(\text{dBm})} \tag{3.26}$$

$$\sigma^2_{\Gamma(\text{dB})} = \sigma_s^2 \tag{3.27}$$

In equation (3.26), $N_{(\text{dBm})}$ represents the noise power in milliwatt (mW) in dBm, and $\mu_{s(\text{dBm})}(d)$ is given as in equation (3.20). σ_s^2 in equation (3.27) denotes the variance of the shadowing factor introduced in equation (3.18). Therefore, for a specific location at a distance d from the transmitter in a shadowing environment, the outage probability, which is lower than the threshold value $\Gamma_{th(dB)}$, can be calculated as follows:

$$\begin{aligned} P_{out}(d) &= \Pr[\Gamma_{(dB)}(d) \leq \Gamma_{th(dB)}] \\ &= \int_0^{\Gamma_{th(dB)}} \frac{1}{\sqrt{2\pi\sigma_s^2}} \exp\left[-\frac{(x-\mu_{\Gamma(dB)}(d))^2}{2\sigma_s^2}\right] dx \\ &= Q\left(\frac{\Gamma_{th(dB)} - \mu_{\Gamma(dB)}(d)}{\sigma_s}\right) \end{aligned} \tag{3.28}$$

When the cell radius is R, the outage probability at the cell boundary can be expressed as follows:

$$P_{out}(R) = Q\left(\frac{M_s}{\sigma_s}\right) \tag{3.29}$$

In equation (3.29), M_s is referred to as the shadowing margin [6], which is given as follows:

$$M_s = \Gamma_{th(dB)} - \mu_{\Gamma(dB)}(R) \tag{3.30}$$

Figure 3.7 depicts the outage probability values based on different shadowing margin values, plotted against various standard deviation values of the shadowing factor.

Equation (3.28) represents the outage probability at a specific location located at a distance d from the transmitter. The outage probability may vary at different locations within the cell. The average outage probability value at all locations within the cell radius R can be referred to as the local mean outage probability [7], which can be calculated as follows:

$$\overline{P}_{out} = \frac{1}{\pi R^2} \int_0^R P_{out}(r) 2\pi r dr = Q(X) - \exp\left[XY + \frac{Y^2}{2}\right] Q(X+Y) \tag{3.31}$$

where X and Y are given as follows:

$$X = \frac{M_s(R)}{\sigma_s}, \quad Y = \frac{2\sigma_s}{\beta(10/\ln 10)} \tag{3.32}$$

3.3 Small-Scale Propagation Model

The small-scale propagation model refers to the random process of modeling the instantaneous power values over time due to delay spread and

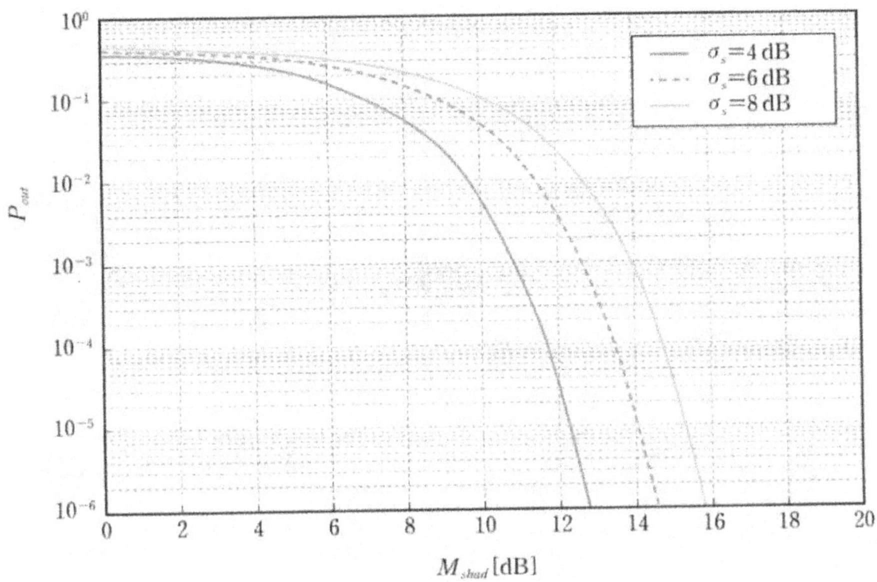

Fig. 3.7. Outage probability at the cell boundary based on the shadowing margin.

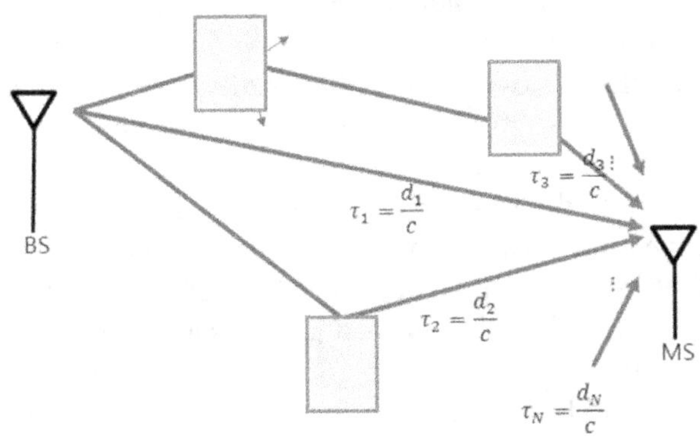

Fig. 3.8. Path delay due to multipath environment.

Doppler frequency shift caused by the multipath nature of the wireless mobile communication channel.

Figure 3.8 serves as an example of a multipath environment where signals transmitted from a base station propagate through numerous paths via scattering, diffraction, and reflection, resulting in the received signal being

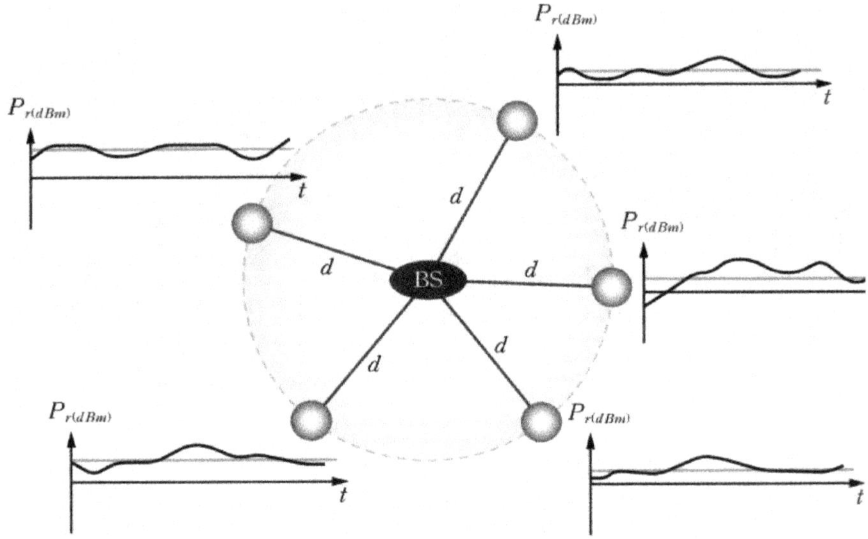

Fig. 3.9. Average power and instantaneous power.

a combination of these paths. The average power and instantaneous power of the combined signal are depicted in Fig. 3.9. In Fig. 3.9, the instantaneous power values (shown in wave) measured at various locations at a distance of d exhibit fluctuations over time. Additionally, the average power values (shown in line) differ slightly at different locations due to shadowing, as explained in the previous section.

Signals transmitted through each path reach the receiver with different delay times depending on the path distance, causing multipath delay spread. In Fig. 3.8, if we denote the distance of each propagation path as d_i [m] and the speed of light as c [m/s], then each path has a path delay of $\tau_i = d_i/c$. Additionally, as the user moves, Doppler frequency shift occurs, causing the received signal to have a complex form. These path delays and Doppler shifts are modeled as a random process, causing the instantaneous power to vary over time.

Furthermore, as depicted in Fig. 3.10, when the mobile user (MS) moves, the angle formed between the incident wave and the direction of MS movement through each multipath is denoted as θ_n. The Doppler frequency shift $f_{D(n)}$ is given as follows:

$$f_{D(n)} = f_m \cos(\theta_n) \tag{3.33}$$

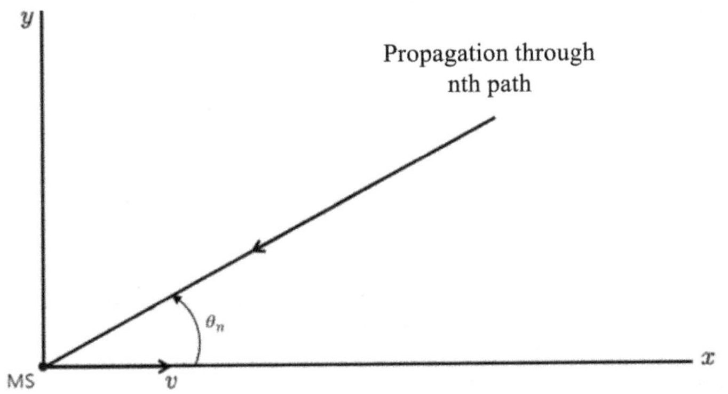

Fig. 3.10. Doppler effect.

where f_m is the maximum Doppler frequency given by

$$f_m = \frac{\upsilon}{\lambda_c} = \frac{\upsilon}{(c/f_c)} \tag{3.34}$$

In equation (3.34), υ [m/s] represents the MS's velocity, while λ_c and f_c denote the wavelength and frequency of the carrier, respectively.

Depending on the severity of the path delay as described above, the small-scale propagation model distinguishes between frequency-selective fading channels and flat fading channels. Furthermore, based on the magnitude of Doppler frequency shift, it categorizes channels into fast-fading and slow-fading channels.

3.3.1 *Frequency-selective fading channel*

To understand the impact of path delay in Fig. 3.8 on the performance of communication systems in a multipath environment, Fig. 3.11 depicts the impulse response of a transmitted impulse signal in a wireless mobile channel, representing a common phenomenon in wireless communication channels. Now, let's transmit a rectangular pulse of duration T into the channel shown in Fig. 3.12(a), the received pulse's duration is extended by $T + \tau_N$, as shown in Fig. 3.12(b). In other words, the length of the received signal's pulse is prolonged by the path delay time compared to the transmitted signal's pulse length.

Now, let's consider the scenario where consecutive pulses are transmitted. Let's simplify the multipath channel as in the case described in

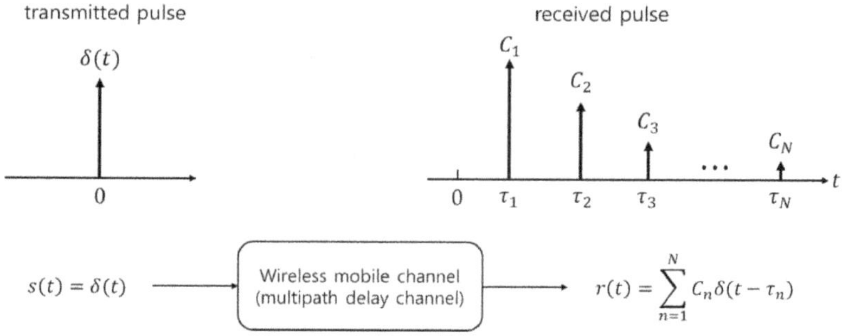

Fig. 3.11. Impulse response function in wireless mobile channel.

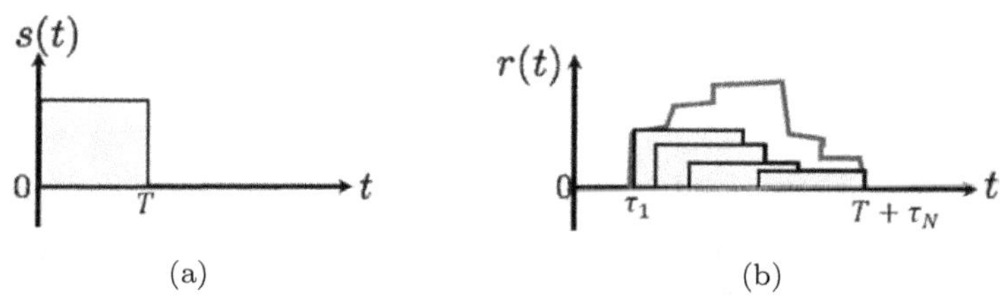

Fig. 3.12. Received rectangular pulse signals due to multipath delay. (a) Transmission of rectangular pulse and (b) received signals due to multipath delay.

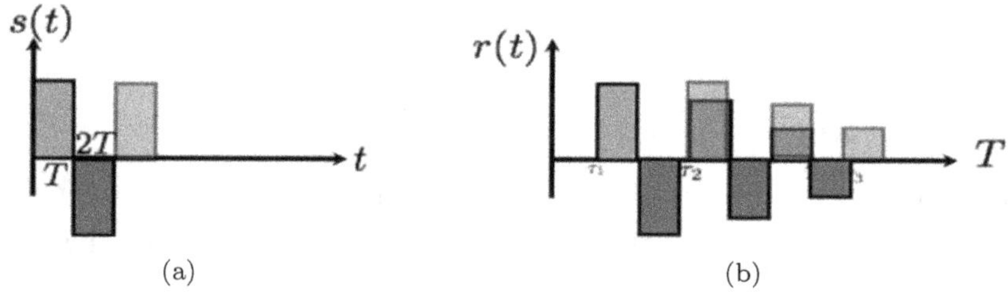

Fig. 3.13. Example of received signals due to three multipath. (a) Transmission of three consecutive rectangular pulse signals and (b) received signals due to three multipath routes.

Fig. 3.11. When three different shapes of pulses are transmitted, as depicted in Fig. 3.13(a), we observe interference occurring due to the overlapping of the first, second, and third pulses. Consequently, the interference between consecutive pulses due to multipath delay leads to Inter-Symbol Interference (ISI). It is evident from Fig. 3.13(b) that ISI becomes more severe as the pulse width decreases, i.e., as T decreases. Thus, multipath delay causes significant performance degradation in wireless mobile communication by inducing ISI.

Now, let's reconsider the issue of ISI caused by path delay in the frequency domain. Assuming that the wireless mobile communication channel is linear and time-invariant (LTI), the channel impulse response $h(t)$ with channel gain c_n can be expressed as follows [3]:

$$h(t) = \sum_{n=1}^{N} c_n \, \delta(t - \tau_n) \tag{3.35}$$

Then, the channel frequency response, obtained by Fourier transforming channel impulse response function, can be written as follows:

$$H(f) = \int_{-\infty}^{\infty} h(t) e^{-j2\pi f t} dt = \sum_{n=1}^{N} c_n e^{-j2\pi f \tau_n} \tag{3.36}$$

For a simpler explanation, let's consider the case where $c_n = e^{-0.1(n-1)}$, $\tau_n = 10^{-5} n$ in Fig. 3.11. In this case, the amplitude spectrum $|H(f)|$ from equation (3.36) is represented as shown in Fig. 3.14.

Now, let's consider two different pulses. If the first pulse has a length of $T = 10^{-3}$ s and the second pulse has a length of $T = 10^{-5}$ s, then their respective frequency responses are depicted as sinc functions in Fig. 3.15. One important point to note is that for low-speed data transmission, longer pulse lengths are preferred, while for high-speed data transmission, shorter

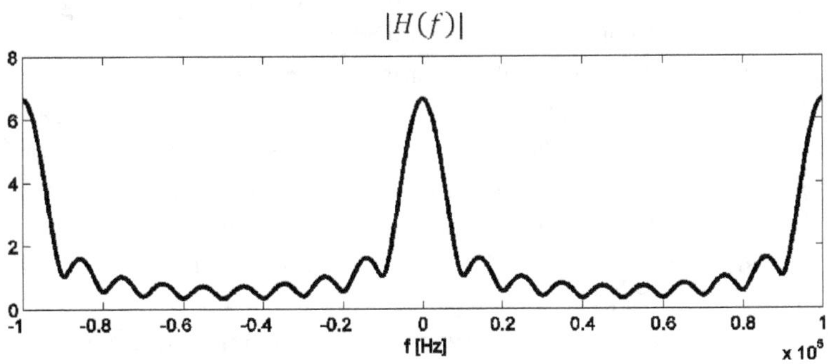

Fig. 3.14. Amplitude spectrum of LTI channel.

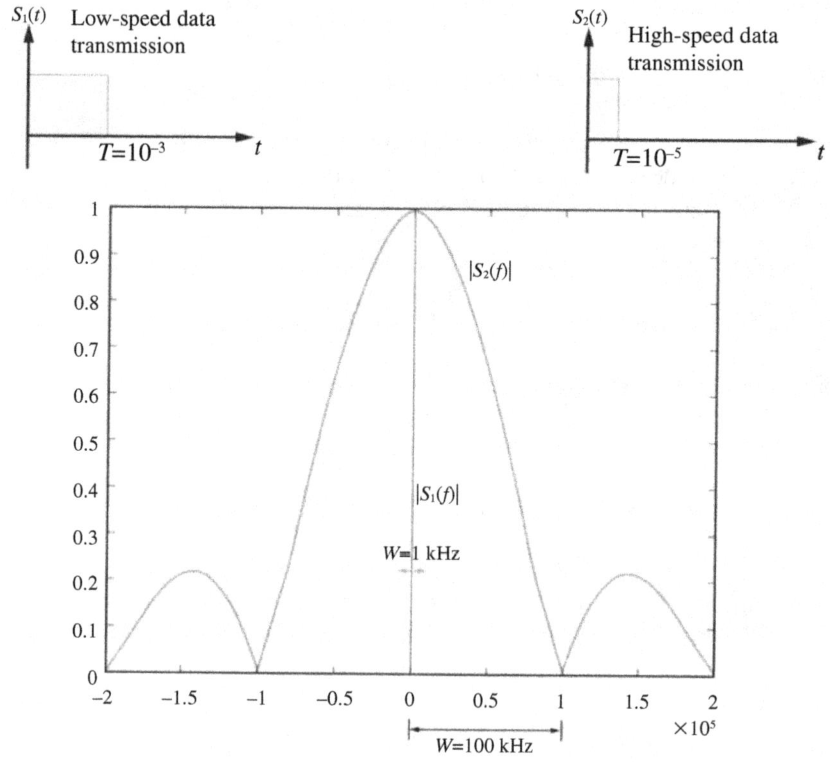

Fig. 3.15. Envelope spectrum of signals with different pulse.

pulse lengths are preferred. In other words, when transmitting digital information using a single pulse, the pulse length can be considered as the symbol length, while the bandwidth of the pulse can be defined as the frequency at which the envelope spectrum becomes first zero in its Fourier transform. Therefore, for the long pulse used for low-speed data transmission, such as $s_1(t)$ in Fig. 3.15, with a duration of $T = 10^{-3}$ s, the bandwidth is approximately 1 kHz, while for the short pulse used for high-speed data transmission, such as $s_2(t)$, with a duration of $T = 10^{-5}$ s, the bandwidth is approximately 100 kHz.

Now, let's examine the received signal when the input signal from Fig. 3.15 passes through the multipath channel depicted in Fig. 3.14. As seen in Fig. 3.16, assuming the channel is an LTI system, the received signal $r(t)$ can be expressed as the convolution of the input signal and

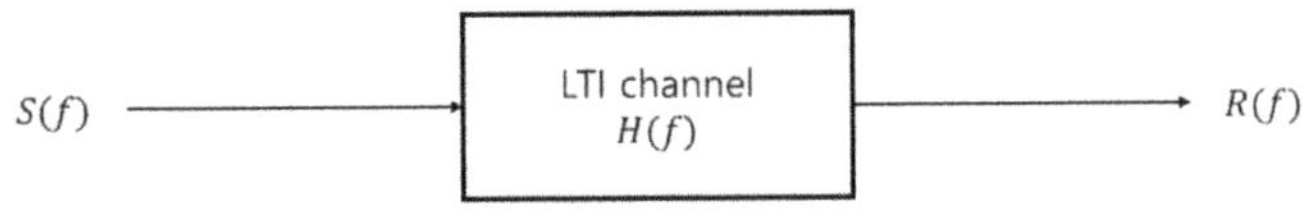

$$r(t) = s(t) * h(t) \leftrightarrow R(f) = S(f)H(f)$$

Fig. 3.16. Relationship between transmitted and received signals in an LTI channel.

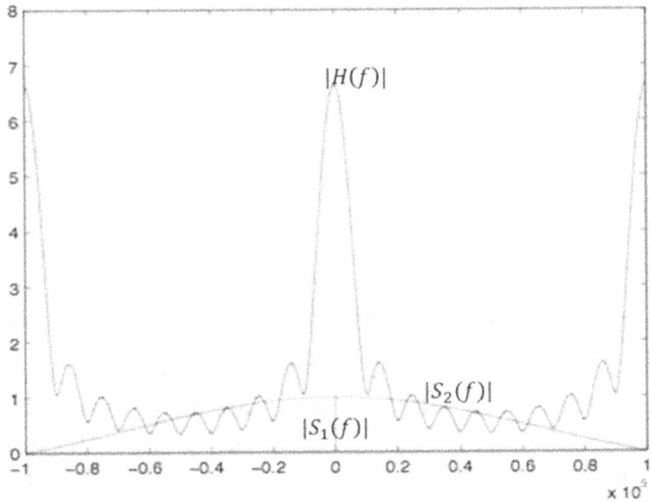

Fig. 3.17. Envelope spectrum of two input signals and channel transfer function.

the impulse response of the channel. In the frequency domain, $R(f)$, the Fourier transform of $r(t)$ can be represented as the multiplication of the frequency domain representations of the input signal and the channel frequency response.

Figure 3.17 illustrates the spectra of two different input signals, $s_1(t)$ and $s_2(t)$, along with the channel transfer function $H(f)$. By multiplying the two transmitted signals and the channel transfer function, we can obtain the Fourier transform function of the received signal. The resulting spectrum of peaks is shown in Fig. 3.18. In Fig. 3.18, the $|R_1(f)|$ represents the spectrum of peaks for the received signal with respect to $s_1(t)$, while the $|R_2(f)|$ represents the spectrum of peaks for the received signal with respect to $s_2(t)$.

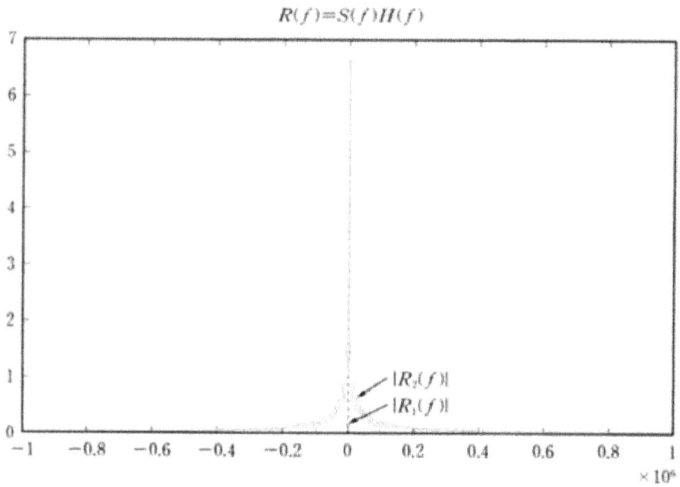

Fig. 3.18. Envelope spectrum of the output function.

From Figs. 3.17 and 3.18, we can infer the following important points:

(1) For low-speed data transmission, i.e., when $s_1(t)$ is the input function, the Fourier transform values are multiplied by a constant value of the channel at specific parts. Consequently, although the magnitude of the received signal's spectrum may differ from that of the input signal's spectrum, their shapes remain the same.
(2) For high-speed data transmission, i.e., when $s_2(t)$ is the input function, the spectrum of the received signal is the result of multiplying the spectrum of the input signal by different values at every frequency component. Therefore, the spectrum of the received signal differs entirely from that of the transmitted signal.

From the above two points, it can be inferred that in the case of high-speed data transmission, as observed in Fig. 3.15, ISI occurs because each frequency component of the input signal reacts differently with the channel, resulting in a completely different shape of the transmitted signal's spectrum, as shown in Fig. 3.18. Here, we define the channel coherence bandwidth as the interval where the channel characteristics remain unchanged. We denote it as B_{coh}. The channel coherence bandwidth is relative to the bandwidth of the transmitted signal and is closely related to the delay spread. For this purpose, if we define the maximum channel delay spread τ_{DS} as the difference between the delay times of the first-arriving multipath signal and the last-arriving multipath signal, then we have

$$\tau_{\mathrm{DS}} = \tau_N - \tau_1 \tag{3.37}$$

The channel coherence bandwidth is approximately inversely proportional to τ_{DS}, as can be observed from Fig. 3.17. Therefore, the channel coherence bandwidth is also defined as follows [9]:

$$B_{\mathrm{coh}} \approx \frac{1}{\tau_{\mathrm{DS}}} \tag{3.38}$$

However, besides equation (3.38), the channel coherence bandwidth is sometimes defined using the Root Mean Square (RMS) of the delay spread of multipath signals. If we denote the average power of the received signal p_n at delay time τ_n, then the RMS delay spread, $\tau_{\mathrm{DS(rms)}}$, is defined as follows [8]:

$$\tau_{\mathrm{DS(rms)}} = \frac{\sum_{n=1}^{N} p_n \tau_n}{\sum_{n=1}^{N} p_n} \tag{3.39}$$

By equation (3.39), the channel coherence bandwidth can also be defined as follows:

$$B_{\mathrm{coh}} = \frac{1}{\tau_{\mathrm{DS(rms)}}} \tag{3.40}$$

Equation (3.38) is primarily used in environments with strong line-of-sight signals, while equation (3.40) is favored in environments with significant scattering. For example, in cellular communications, the channel coherence bandwidth is defined based on equation (3.38), while in indoor environments with strong line-of-sight signals, such as 60 GHz-based wireless LAN systems, equation (3.40) is typically used.

As seen above, when the bandwidth of the transmitted signal is wider than the channel coherence bandwidth, inter-symbol interference (ISI) occurs as each frequency component of the signal reacts differently due to the channel's response. This phenomenon is referred to as a frequency-selective fading channel, illustrated in Fig. 3.19. In Fig. 3.19, if we denote the spectrum of the transmitted signal through Fourier transform as $S(f)$, the spectrum of the received signal, $R(f)$, can be expressed as the multiplication of the channel transfer function $H(f)$. Consequently, each frequency component of the transmitted signal undergoes a different channel response, leading to a different effect of the received signal across frequencies.

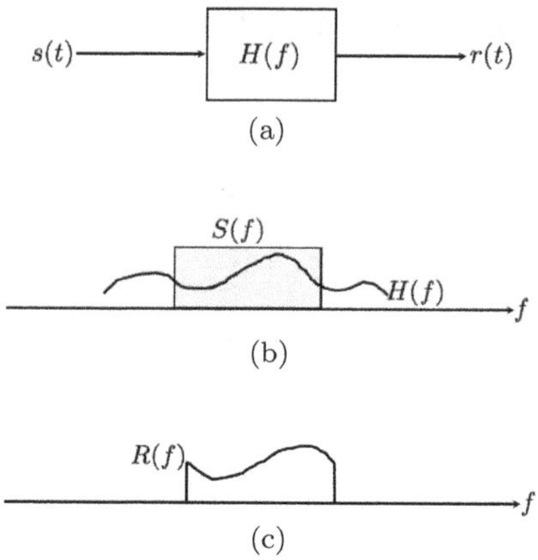

Fig. 3.19. Transmitted and received signals in a frequency-selective fading channel. (a) Input and output signals in a channel modeled as an LTI system (b) relationship between the transmitted signal and the channel response in the frequency domain and (c) Fourier transform function of the received signal.

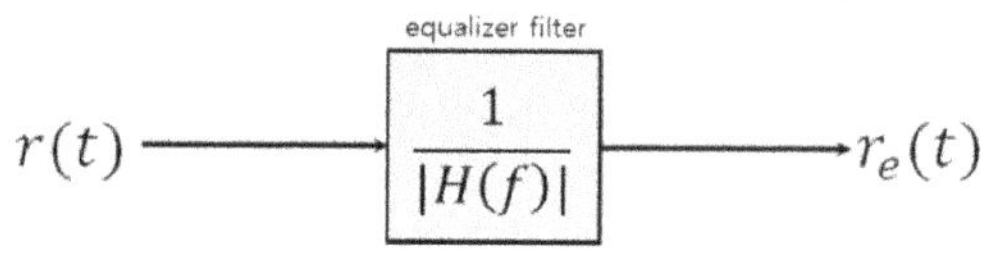

Fig. 3.20. Example of an equalizer filter used to mitigate ISI effects.

In other words, if we denote the frequency bandwidth of the transmitted signal as B_t, the following scenario can be termed as a frequency-selective fading channel:

$$B_t > B_{\text{cor}} \tag{3.41}$$

In a frequency-selective fading channel environment, ISI occurs, and there are two main methods to address this issue.

The first method is applicable when transmitting with a single carrier, where an equalizer is used at the receiver to eliminate the ISI effects. Figure 3.20 illustrates an example of using an equalizer filter to mitigate ISI effects. To design such a filter, knowledge of the channel response is required, which can be obtained through a channel estimator. It is important to note that channel prediction is crucial for the design of wireless communication system receivers, and various methods exist for this purpose.

The second method involves using multiple carriers, thereby achieving the effect of transforming a frequency-selective channel into a frequency-non-selective one.

3.3.2 *Frequency-non-selective fading channel*

When the mobile station moves along the x-axis with a velocity v, and the angle of incidence of the nth signal propagated through multipath is θ_n, the Doppler shift is $f_{D(n)} = f_m \cos(\theta_n)$ as given in equations (3.33) and (3.34), as shown in Fig. 3.10. If we denote the transmitted signal modulated by the carrier wave as $s(t)$, it can be represented using the equivalent low-pass signal $\tilde{s}(t)$ as follows [10]:

$$s(t) = \text{Re}[\tilde{s}(t)e^{j2\pi f_c t}] \tag{3.42}$$

where f_c is the carrier frequency. Now we pass this signal, $s(t)$, into the wireless channel as in Fig. 3.11, then the received signal can be expressed as follows:

$$r(t) = \sum_{n=1}^{N} c_n s(t - \tau_n)|_{f_c \to f_c + f_{D(n)}} \tag{3.43}$$

In equation (3.43), the carrier frequency is shifted to $f_c + f_{D(n)}$ due to Doppler shift. By utilizing equations (3.42) and (3.43), the received signal $r(t)$ can be expressed as follows:

$$r(t) = \text{Re}\left[\sum_{n=1}^{N} c_n \tilde{s}(t - \tau_n)e^{j2\pi(f_c + f_{D(n)})(t - \tau_n)}\right] \tag{3.44}$$

Equation (3.44) can be expressed using an equivalent low-pass signal, which is as follows:

$$r(t) = \text{Re}[\tilde{r}(t)e^{j2\pi f_c t}] \tag{3.45}$$

with

$$\tilde{r}(t) = \sum_{n=1}^{N} c_n \tilde{s}(t - \tau_n)e^{-j\varphi_n(t)}, \tag{3.46}$$

where $\varphi_n(t)$ can be represented as follows using equation (3.44):

$$\varphi_n(t) = 2\pi[(f_c + f_{D(n)})\tau_n - f_{D(n)}t] \tag{3.47}$$

The received signal through the multipath is represented by equation (3.46) and can be illustrated as a block diagram, as shown in Fig. 3.21.

The impact of the wireless mobile channel on the received signal as described in equation (3.46) can be observed:

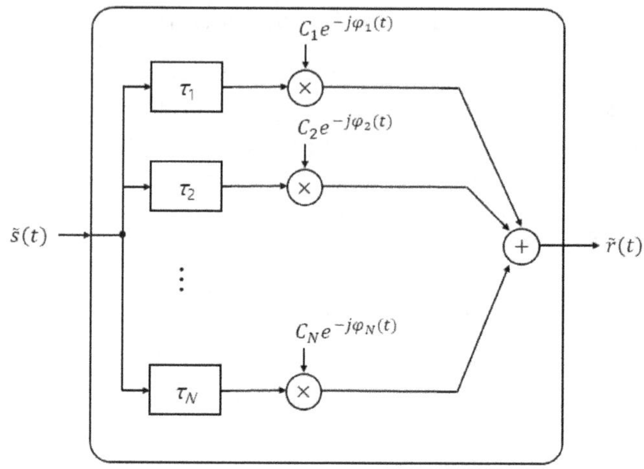

Fig. 3.21. Wireless mobile communication channel block diagram.

(1) The received signal is a sum of multiple signals due to multipath propagation.
(2) Each multipath signal has different fading amplitudes denoted by c_n.
(3) Each path is affected by different phases denoted by φ_n.
(4) Each has different delay times denoted by τ_n.

Here, we assume that c_n and φ_n are random variables and independent. Additionally, we assume that c_n are independent and identically distributed (IID) random variables. Also, the channel phases φ_n are IID random variables, uniformly distributed over $(-\pi, \pi)$.

Now, let us denote the symbol length of the transmitted signal as T_s, then if $\tau_{\mathrm{DS}} > T_s$, it is frequency-selective fading channels, while $\tau_{\mathrm{DS}} \ll T_s$, it is called a frequency-non-selective fading channel or flat fading channel and the effect of ISI can be neglected. In other words, for the flat fading channels, we can assume that $\tau_N \approx \tau_1$. If we denote $\tau_n \equiv \tau_0$, equation (3.46) can be simplified as follows:

$$\tilde{r}(t) = \tilde{s}(t - \tau_0) \sum_{n=1}^{N} c_n e^{-j\varphi_n(t)} = \tilde{s}(t - \tau_0) h(t) \tag{3.48}$$

Furthermore, in such cases, the channel block diagram depicted in Fig. 3.21 can be simplified as shown in Fig. 3.22 for flat fading channels.

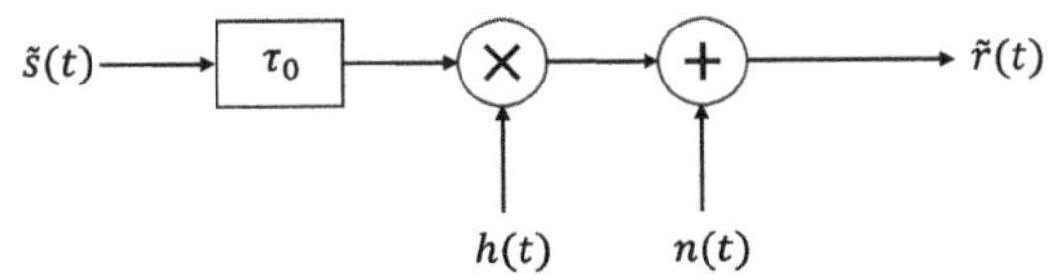

Fig. 3.22. Channel block diagram in flat fading channel.

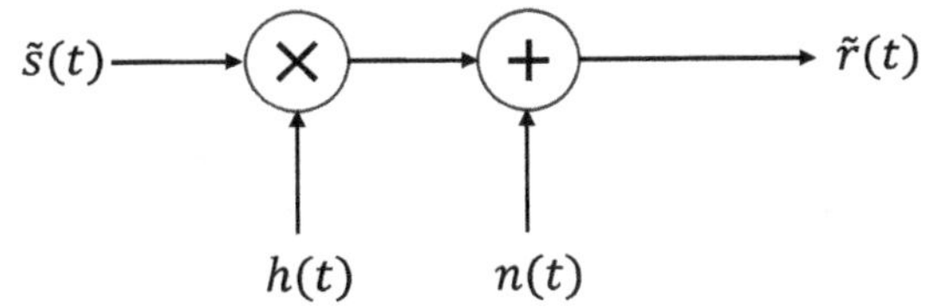

Fig. 3.23. Block diagram for a flat fading channel assuming perfect time synchronization.

In a typical communication system, time delays always occur. Synchronization algorithms for time delay have significantly advanced, allowing for accurate prediction of time delays. Based on this capability, Fig. 3.22. can be simplified into a more concise block diagram, as shown in Fig. 3.23, assuming perfect time synchronization in a flat fading channel.

In the model depicted in Fig. 3.23, we aim to derive the impact of the channel on the received signal in terms of autocorrelation functions and the changing received signal-to-noise ratio (SNR) due to the channel. The received signal is the result of the multiplication of the transmitted signal and the channel, followed by the addition of additive white Gaussian noise (AWGN). The channel, denoted as $h(t)$, varies depending on the channel environment. Specifically, we can distinguish between scenarios where there are obstacles between the transmitter and receiver, causing the propagation of waves characterized by scattering, diffraction, and reflection, as depicted in Fig. 3.1, and scenarios where there are no obstacles, resulting in propagation of waves without the influence of scattering, diffraction, or reflection.

Therefore, to establish a more specific channel model for flat fading channels, let's consider the propagation environment in terms of non-line-of-sight (NLoS) and line-of-sight (LoS) channels, as depicted in Fig. 3.24. The NLoS channel in Fig. 3.24(a) refers to an environment where obstacles between the transmitter and receiver prevent direct propagation of the signal. In contrast, the LoS channel, illustrated in Fig. 3.24(b), represents scenarios where propagation signals can directly reach the receiver

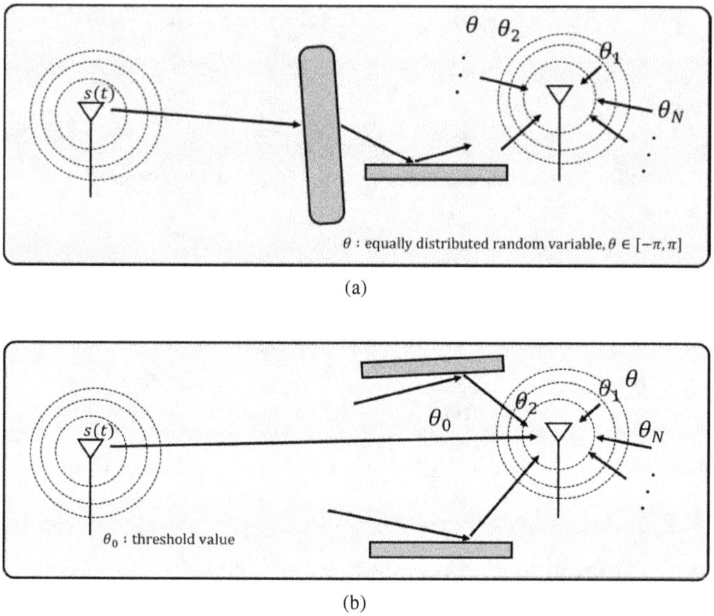

Fig. 3.24. (a) Non-line-of-sight (NLos) and (b) line-of-sight (Los) channel environments.

without obstacles causing scattering, diffraction, or reflection, among other things.

3.3.3 *Signal correlation in NLoS channel*

A non-line-of-sight channel refers to a scenario where the distance between the transmitter and receiver is long, and there is no direct propagation due to obstacles, as illustrated in Fig. 3.24(a).

The NLoS channel environment can be observed in communication between ships, cellular communication in urban areas, and similar scenarios.

To examine the influence of the channel in a flat fading channel, considering the case of the transmitted signal $\tilde{s}(t) = 1$, as shown in Fig. 3.23, The received signal, using Equation (3.45) and (3.48), can be expressed as follows:

$$r(t) = h_I(t)\cos(2\pi f_c t) - h_Q(t)\sin(2\pi f_c t) \tag{3.49}$$

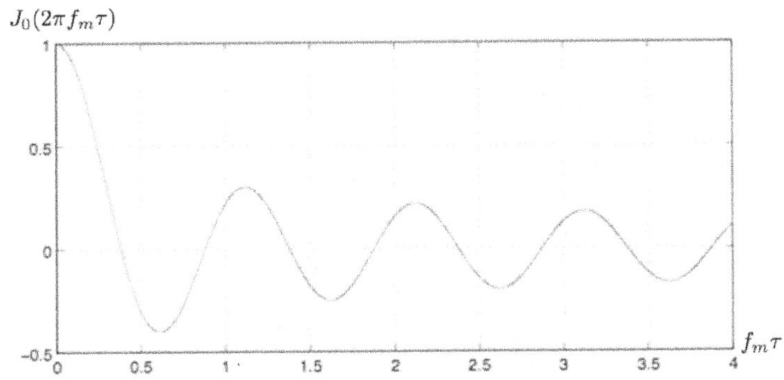

Fig. 3.25. Autocorrelation coefficient in non-line-of-sight channel.

where the in-phase and quadrant component components are

$$h_I(t) = \sum_{n=1}^{N} c_n \cos \varphi_n(t)$$
$$h_Q(t) = \sum_{n=1}^{N} c_n \sin \varphi_n(t) \tag{3.50}$$

When N is large, according to the Central Limit Theorem, $h_I(t)$ and $h_Q(t)$ can be considered Gaussian random processes. Assuming that the received signal $r(t)$ in equation (3.49) is wide-sense stationary, its autocorrelation function $R_{rr}(\tau)$ is as follows [6]:

$$\begin{aligned} R_{rr}(\tau) &= E[r(t)r(t+\tau)] \\ &= E[h_I(t)h_I(t+\tau)]\cos 2\pi f_c t - E[h_I(t)h_Q(t+\tau)]\sin 2\pi f_c t \\ &= R_{h_I h_I}(\tau)\cos 2\pi f_c t - R_{h_I h_Q}(\tau)\sin 2\pi f_c t \end{aligned} \tag{3.51}$$

Note that $R_{h_I h_I}(\tau) = R_{h_Q h_Q}(\tau)$ and $R_{h_I h_Q}(\tau) = -R_{h_Q h_I}(\tau)$.

In equation (3.47), $\varphi_n(t)$ and $\varphi_m(t)$ are assumed to be statistically independent for $n \neq m$, which is reasonable when considering different Doppler shifts and delays. Furthermore, since $f_c\tau_n \gg 1$ in equation (3.47), we can assume that phase $\varphi_n(t)$ follows a uniform distribution over $[-\pi, \pi]$. Based on these characteristics, $R_{h_I h_I}(\tau)$ in equation (3.51) can be expressed using equations (3.33), (3.47), and (3.50) as follows:

$$R_{h_I h_I}(\tau) = E_{\tau,\theta}[h_I(t)h_I(t+\tau)] = \frac{\Omega}{2}E_\theta[\cos(2\pi f_{D(n)}\tau)]$$
$$= \frac{\Omega}{2}E_\theta[\cos(2\pi f_m \tau \cos\theta)] \quad (3.52)$$

where

$$\Omega = E[h_I^2(t)] + E[h_Q^2(t)] = \sum_{n=1}^{N} c_n^2 \quad (3.53)$$

where Ω represents the average channel gain, which indicates the effect of the channel on the received power. In other words, if the transmit power is P_t, then the received average power P_r is as follows:

$$P_r = \Omega P_t \quad (3.54)$$

Ω can be expressed in terms of equation (3.13) and (3.17) as follows:

$$\Omega = k\chi_s d^{-\beta} \quad (3.55)$$

where k is a constant value known from equation (3.14), χ_s is the shadowing factor, d is the distance between the transmitter and receiver, and β is the path loss exponent.

When the distance between the transmitter and receiver is significantly larger than the antenna height, the Angle of Arrival (AoA), θ, in equation (3.52) can be assumed to have a probability density function $f_\theta(\theta) = \frac{1}{2\pi}$, $\theta \in [-\pi, \pi]$ as isotropic scattering. Assuming a two-dimensional model, equation (3.52) can then be rewritten as follows [11]:

$$R_{h_I h_I}(\tau) = \frac{\Omega}{2}E_\theta[\cos(2\pi f_m \tau \cos\theta)]$$
$$= \frac{\Omega}{2}\int_{-\pi}^{\pi} \cos(2\pi f_m \tau \cos\theta) f_\theta(\theta) d\theta$$
$$= \frac{\Omega}{2}\frac{1}{\pi}\int_{0}^{\pi} \cos(2\pi f_m \tau \cos\theta) d\theta = \frac{\Omega}{2}J_0(2\pi f_m \tau) \quad (3.56)$$

In the above equation, $J_0(x)$ represents the zero-order Bessel function of the first kind [12]. In Fig. 3.25, $R_{h_I h_I}(\tau)$ is shown in terms of $f_m\tau$ with $\frac{\Omega}{2}$ being normalized to unity.

Similarly, the cross-correlation function is calculated as follows:

$$R_{h_I h_Q}(\tau) = E_{\tau,\theta}[h_I(t)h_Q(t+\tau)] = \frac{\Omega}{2}E_{\tau,\theta}[\sin(2\pi f_m \tau)\cos(\theta)] \quad (3.57)$$

which can be computed as follows:

$$\begin{aligned} R_{h_I h_Q}(\tau) &= \frac{\Omega}{2}E_{\tau,\theta}[\sin(2\pi f_m \tau)\cos(\theta)] \\ &= \frac{\Omega}{2}\frac{1}{2\pi}\int_{-\pi}^{\pi}\sin(2\pi f_m \tau)\cos(\theta)d\theta = 0 \end{aligned} \quad (3.58)$$

Substituting equations (3.56) and (3.58) into equation (3.51), the autocorrelation function of the received signal can be expressed as follows [6]:

$$R_{rr}(\tau) = \frac{\Omega}{2}J_0(2\pi f_m \tau)\cos(2\pi f_c \tau) \quad (3.59)$$

From equation (3.59), we can derive important metrics for the design of mobile communication systems. Many wireless mobile communication systems, including WLAN and cellular communication, transmit signals on a frame-by-frame basis as shown in Fig. 3.26. These frames are divided into pilot signals and data information signals. The pilot signals contain data information known to both the transmitter and the receiver, and by receiving these pilot signals, the receiver can detect the data information signals through time and frequency synchronization and channel prediction. To predict the channel using a limited number of pilot signals, it is preferable for the channel to exhibit minimal change over the frame duration. This notion of "minimal change in the channel" can be expressed as "high autocorrelation coefficient values of the channel." Therefore, if we define the frame length as T_f, and then it can be said that channel prediction is possible without performance degradation if the autocorrelation

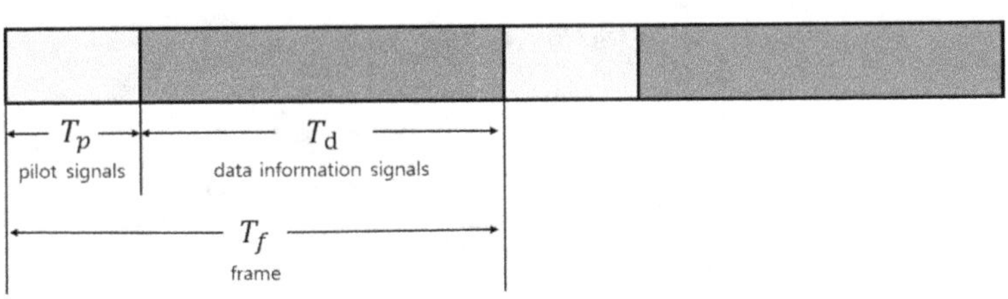

Fig. 3.26. Frame structure with pilot and data symbols.

coefficient attains the following relationship for a specific value ρ_{th}:

$$\frac{\Omega}{2} J_0(2\pi f_m T_f) > \rho_{th} \tag{3.60}$$

Upon revisiting equation (3.34), we can express it in terms of the maximum Doppler frequency value as follows:

$$T_f \geq \frac{1}{2\pi f_m} J_0^{-1}(\rho_{th}) = \frac{\lambda_c}{2\pi \nu} J_0^{-1}(\rho_{th}) \tag{3.61}$$

where $J_0^{-1}(\bullet)$ represents the inverse function of $J_0(\bullet)$. As evident from the equation, when users are moving at higher speeds, shorter frame lengths are required for communication. For instance, let's consider designing a system where communication is feasible even at 2 GHz carrier frequencies and speeds of $\nu = 500$ km/hr. Moreover, if $\rho_{th} \geq 0.5$, let's assume that channel prediction is feasible without performance degradation in such cases, $J_0^{-1}(0.5) = 1.521$ from Fig. 3.25 and then the frame length is calculated using equation (3.61):

$$T_f \geq \frac{\lambda_c}{2\pi \nu} J_0^{-1}(\rho_{th}) = 261.47\ [\mu s]$$

3.3.4 *Signal correlation in LoS channel*

In a channel environment with line-of-sight characteristics, as illustrated in Fig. 3.24(b), the channel consists of both line-of-sight and numerous non-line-of-sight propagations. In a channel with line-of-sight propagation, equation (3.50) can be expressed as follows [13]:

$$\begin{aligned} h_I(t) &= c_0 \cos \varphi_0(t) + \sum_{n=1}^{N} c_n \cos \varphi_n(t) \\ h_Q(t) &= -c_0 \sin \varphi_0(t) - \sum_{n=1}^{N} c_n \sin \varphi_n(t) \end{aligned} \tag{3.62}$$

where c_0 and $\varphi_0(t)$ are not random values but deterministic and represent line-of-sight propagation. $\varphi_n(t)$ is expressed in equation (3.47). Let's define the line-of-sight propagation part in equation (3.62) as follows:

$$m_I(t) = c_0 \cos \varphi_0(t)$$
$$m_Q(t) = -c_0 \sin \varphi_0(t) \tag{3.63}$$

If s_d^2 represents the power value of the channel affected by line-of-sight propagation, it can be expressed as follows:

$$s_d^2 = m_I^2(t) + m_Q^2(t) \tag{3.64}$$

On the other hand, equation (3.62) only represents the part of the non-line-of-sight propagation, when excluding the part representing line-of-sight propagation. The non-line-of-sight propagation part, as described in equation (3.50), can be modeled as independent and identically distributed Gaussian random processes due to the central limit theorem. In other words, if the non-line-of-sight component has a mean value of 0 and a variance of σ_{Non}^2, it can be represented as Gaussian random variables as follows:

$$\mathrm{X} = \sum_{n=1}^{N} c_n \cos \varphi_n(t) \sim N(0, \sigma_{\text{Non}}^2)$$
$$\mathrm{Y} = -\sum_{n=1}^{N} c_n \sin \varphi_n(t \sim N(0, \sigma_{\text{Non}}^2) \tag{3.65}$$

The channel gain Ω can also be represented as the sum of the powers of line-of-sight and non-line-of-sight components, as follows:

$$\Omega = s_d^2 + 2\sigma_{\text{Non}}^2 \tag{3.66}$$

Now let's define the ratio of the power of the line-of-sight component to the power of the non-line-of-sight component as K, as follows:

$$K = \frac{s_d^2}{2\sigma_{\text{Non}}^2} \tag{3.67}$$

The term K in equation (3.67) is known as the Rician K-factor, which is an important measure used to indicate the degree of line-of-sight propagation. From equations (3.66) and (3.67), s_d^2 and σ_{Non}^2 can be expressed in terms of K and Ω as follows:

$$s_d^2 = \frac{K\Omega}{K+1} \quad \text{and} \quad \sigma_{\text{Non}}^2 = \frac{\Omega}{K+1} \tag{3.68}$$

In a channel with line-of-sight propagation, the angle of arrival of the received signal can be expressed by the probability density function as follows [3]:

$$f_\theta(\theta) = \frac{1}{1+K} \cdot \frac{1}{2\pi} + \frac{K}{K+1}\delta(\theta - \theta_0) \tag{3.69}$$

Using the same approach as in equations (3.56) and (3.57) for a non-line-of-sight channel, we can derive the autocorrelation function for the line-of-sight channel:

$$\begin{aligned} R_{h_I h_I}(\tau) &= \frac{1}{K+1}\frac{\Omega}{2}J_0(2\pi f_m \tau) + \frac{K}{K+1}\frac{\Omega}{2}\cos(2\pi f_m \tau \cos\theta_0) \\ R_{h_I h_Q}(\tau) &= \frac{K}{K+1}\frac{\Omega}{2}\sin(2\pi f_m \tau \cos\theta_0) \end{aligned} \tag{3.70}$$

3.3.5 *Probability distribution of signal-to-noise ratio*

The probability distribution of the received signal-to-noise ratio (SNR), $\gamma(t)$, can be derived from equation (3.48) and Fig. 3.23 as follows:

$$\gamma(t) = \frac{|\tilde{h}(t)|^2 \Omega P_t}{N_0} \tag{3.71}$$

where $\tilde{h}(t)$ is the normalized as

$$\tilde{h}(t) = \frac{h(t)}{\sqrt{\Omega}} \tag{3.72}$$

Note that since the received signal power is a random process varying over time, the instantaneous SNR is a random process.

3.3.5.1 *Rayleigh channel*

In the case of a non-line-of-sight (NLoS) channel, according to equations (3.50) and (3.58), both $h_I(t)$ and $h_Q(t)$ can be modeled as independent Gaussian random processes due to the central limit theorem both with the mean 0 and the variance $\frac{\Omega}{2}$. Thus, we can express

$$h_I(t), h_Q(t) \sim N\left(0, \frac{\Omega}{2}\right) \tag{3.73}$$

By definition of equation (3.72), if we express $\tilde{h}(t) = \tilde{h}_I(t) + j\tilde{h}_Q(t)$, both $h_I(t)$ and $h_Q(t)$ can be represented as Gaussian random variables with mean zero and variance 1/2. Now, to understand the distribution of SNR in equation (3.71), let's first examine the distribution of $|\tilde{h}(t)|^2$. $|\tilde{h}(t)|^2$ can be represented as follows:

$$|\tilde{h}(t)|^2 = \tilde{h}_I^2(t) + \tilde{h}_Q^2(t) \tag{3.74}$$

In other words, $|\tilde{h}(t)|^2$ is the sum of squares of two independent Gaussian random variables. Then $|\tilde{h}(t)|^2$ follows a central chi-square distribution with two degrees of freedom. Therefore, through a variable transformation, $|\tilde{h}(t)|^2$ has the following probability density function:

$$f_\gamma(\gamma) = \frac{1}{\overline{\gamma}}\exp\left(-\frac{\gamma}{\overline{\gamma}}\right), \quad \gamma \geq 0 \tag{3.75}$$

In the above equation, $\overline{\gamma}$ represents the mean value of the received SNR, which can be calculated from equation (3.71) as follows:

$$\overline{\gamma} = E[\gamma] = \frac{\Omega P_t}{N_0} \tag{3.76}$$

Figure 3.27 illustrates equation (3.75) for various $\overline{\gamma}$ values. Through this figure, it can be observed that even when $\overline{\gamma}$ is high, there are many instances where the received SNR values are distributed at low levels.

The Cumulative Distribution Function (CDF) of SNR can be derived from equation (3.75) as follows:

$$F_\gamma(x) = \Pr[\gamma \leq x] = 1 - \exp\left(-\frac{x}{\overline{\gamma}}\right), \quad x \geq 0 \tag{3.77}$$

Using the above CDF, the outage probability is defined as the probability that the instantaneous SNR is less than a certain threshold γ_{th}, which can be expressed as follows:

$$P_{\text{out}}(\gamma_{\text{th}}) = \Pr[\gamma \leq \gamma_{\text{th}}] = F_\gamma(\gamma_{\text{th}}) = 1 - \exp\left(-\frac{\gamma_{\text{th}}}{\overline{\gamma}}\right) \tag{3.78}$$

Figure 3.28 represents the outage probability for various threshold values using equation (3.78). From the figure, it can be observed that for a mean

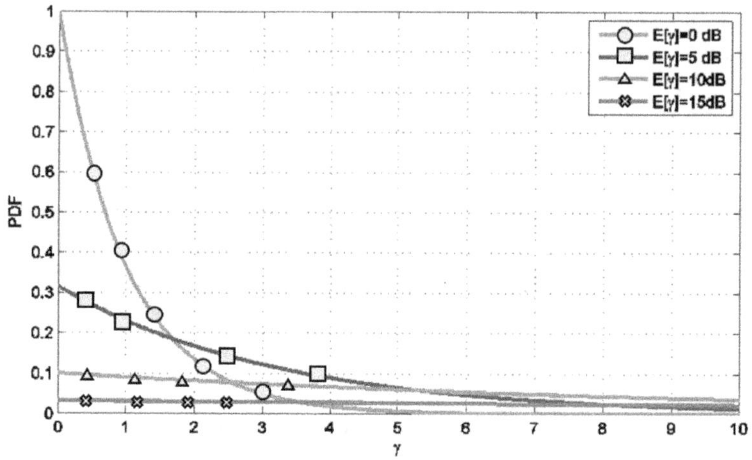

Fig. 3.27. Probability density function of SNR in NLoS.

SNR value of $\overline{\gamma} = 10$ dB, when the threshold value γ_{th} is set at 10 dB, the outage probability is approximately 63.2%. This indicates a relatively high probability of outage. Additionally, it can be noted that the outage probability increases when the average SNR value is lower than this threshold. Here we mean that the average SNR value is equal to the local mean power introduced in large-scale propagation models divided by the power spectral density of the noise.

3.3.5.2 *Rician channel*

The Rician channel refers to the probabilistic modeling of the SNR values of received signals in a channel with line-of-sight. From equations (3.62), (3.63), and (3.65) for a channel with line-of-sight, it can be observed that $h_I(t)$ and $h_Q(t)$ are Gaussian random processes with mean values of $m_I(t)$ and $m_Q(t)$, respectively. Therefore, the channel gain $|h(t)|^2 = h_I^2(t) + h_Q^2(t)$ is understood to be a non-central chi-square random variable with two degrees of freedom. Using equation (3.71) and variable transformation, the probability density function (PDF) of SNR can be derived as follows [6]:

$$f_\gamma(\gamma) = \frac{(K+1)}{\overline{\gamma}} \exp\left[-K - \frac{(K+1)\gamma}{\overline{\gamma}}\right] I_0\left(2\sqrt{\frac{K(K+1)\gamma}{\overline{\gamma}}}\right) \tag{3.79}$$

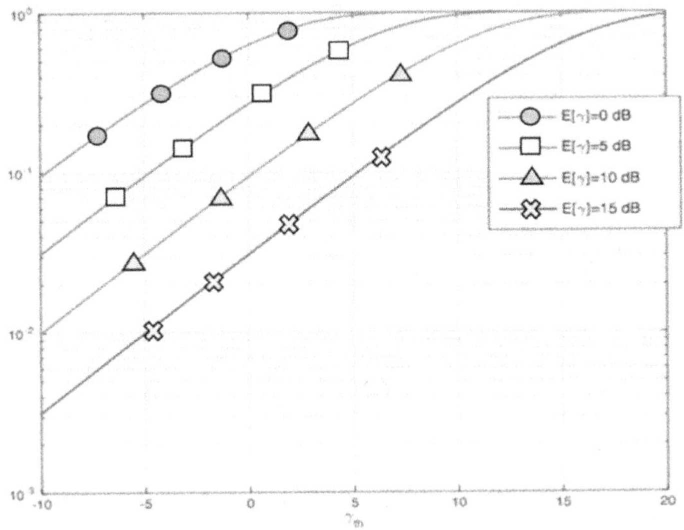

Fig. 3.28. Outage probability of SNR.

where $I_0(x)$ represents the modified Bessel function of the first kind and K is the Rician factor defined in equation (3.67). When $K = 0$, there is no line-of-sight component, and the PDF function becomes equivalent to equation (3.75).

Figure 3.29 depicts the PDF function for various values of K with $\overline{\gamma} = 1 (= 0 \text{ dB})$. As K approaches infinity, the channel converges to a white noise channel.

On the other hand, the cumulative distribution function (CDF) is given by

$$F_\gamma(x) = 1 - Q\left(\sqrt{2K}, \sqrt{\frac{\overline{\gamma} x}{2(K+1)}}\right) \tag{3.80}$$

In the above equation, $Q(a, b)$ is Marcum's Q-function defined as follows:

$$Q(a, b) = \int_b^\infty x \exp\left(-\frac{a^2 + b^2}{2}\right) I_0(ax) dx \tag{3.81}$$

When calculating the outage probability for a threshold value γ_{th} as in the Rayleigh channel, we can substitute γ_{th} for x in equation (3.80). Figure 3.30 illustrates the outage probability values for different γ_{th} when $\overline{\gamma} = 1$. As seen in the figure, it can be observed that better outage performance

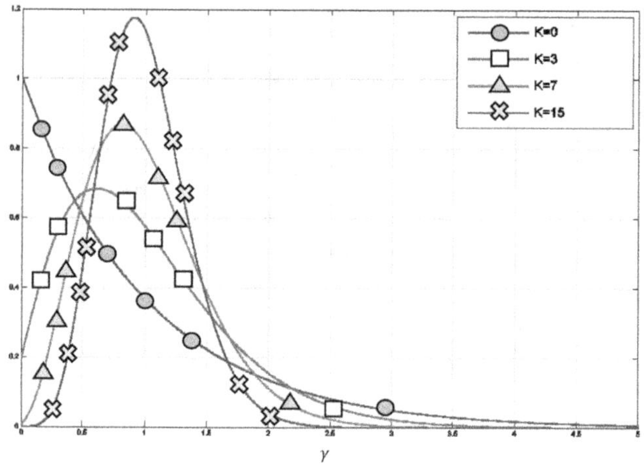

Fig. 3.29. PDF vs. γ [dB] for a Rician channel with $\overline{\gamma} = 1$.

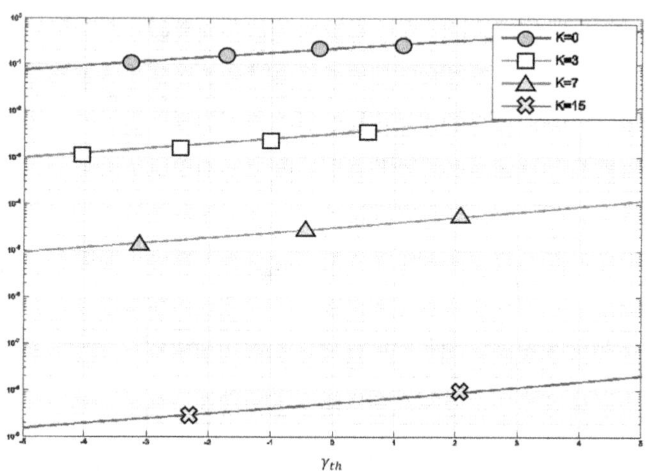

Fig. 3.30. Outage vs. γ_{th}[dB] for Rician channel with $\overline{\gamma} = 1$.

is achieved for higher K values compared to channel environments with smaller K values.

3.3.5.3 *Nakagami-m channel*

The Nakagami-m channel was introduced by Nakagami [14]. Nakagami modeled the received signals in wireless communications as probability density functions based on experimental data, allowing the characterization of fading environments based on the parameter m.

According to the Nakagami channel, the probability density function (PDF) of the received SNR is as follows:

$$f_\gamma(\gamma) = \left(\frac{m}{\overline{\gamma}}\right)^m \frac{\gamma^{m-1}}{\Gamma(m)} \exp\left[-\frac{m\gamma}{\overline{\gamma}}\right], \quad \gamma > 0, \quad m \geq \frac{1}{2} \tag{3.82}$$

where $\Gamma(q)$ is the gamma function defined as follows:

$$\Gamma(q) = \int_0^\infty t^{q-1} e^{-t} dt, \quad q > 0 \tag{3.83}$$

In equation (3.82), m is referred to as the Nakagami parameter. When $m = 1$, it reduces to the Rayleigh channel, and when $m = 0.5$, it represents the distribution of the square of a Gaussian random variable. Moreover, as m increases, the channel environment approaches a white noise channel. A rough equivalence between the Rician K-factor and the Nakagami parameter m is shown as follows [14]:

$$K = \frac{\sqrt{m^2 - m}}{m - \sqrt{m^2 - m}}, \quad m > 1 \tag{3.84}$$

where

$$m = \frac{(K+1)^2}{(2K+1)}$$

The cumulative distribution function for the Nakagami channel is given by

$$F_\gamma(x) = 1 - \frac{\Gamma\left(m, \frac{m}{\overline{\gamma}}x\right)}{\Gamma(m)} \tag{3.85}$$

where $\Gamma(a, x)$ is the incomplete gamma function defined as follows:

$$\Gamma(a, x) = \int_x^\infty t^{a-1} e^{-t} dt \tag{3.86}$$

3.4 Summary

This chapter aims to mathematically represent complex wireless channels to facilitate cost-effective validation of system designs via computer simulations. It distinguishes between large-scale and small-scale channel models, focusing on average received power and dynamically changing received power, respectively. Understanding and applying these models are essential for designing mobile communication systems efficiently.

References

[1] S. Haykin, *Communication Systems*, 4th edn. Wiley, New York, 2006.

[2] K. Feher, *Wireless Digital Communications*, Prentice-Hall, Englewood Cliffs, 1995.

[3] G. L. Stuber, *Principles of Mobile Communication*, 2nd edn. Kluwer Academic Publisher, New York, 2001.

[4] W. C. Y. Lee, *Mobile Communications Engineering*, McGaw Hill, New York, 1982.

[5] W. C. Jakes, *Microwave Mobile Communications*, IEEE Press, New York, 1993.

[6] A. Papoulis and S. Unnikrishna Pillai, *Probability, Random Variables, and Stochastic Process*, 4th edn. McGraw Hill, New York, 2002.

[7] K. Feher, *Advanced Digital Communications*, Prentice-Hall, Englewood Cliffs, 1987.

[8] T. S. Rappaport, *Wireless Communications — Principles and Practice*, 2nd edn. Prentice-Hall, Englewood Cliffs, 2001.

[9] A. Goldsmith, *Wireless Communications*, Cambridge University Press, Cambridge, England, 2005.

[10] J. G. Proakis, *Digital Communications*, 4th edn. McGraw Hill, New York, 2001.

[11] R. H. Clarke, A statistical theory of mobile radio reception, *Bell Syst. Tech. J.* 957–1000, July/August 1968.

[12] I. S. Gradshteyn and I. M. Ryzhik, *Table of Integrals, Series, and Products*, 5th edn. Academic, New York, 1994.

[13] S. O. Rice, Mathematical analysis of random noise. *Bell Syst. Tech. J.* 282–333, July 1944; 45–156, January 1945.

[14] M. Nakagami, The m-distribution — A general formula of intensity distribution of rapid fading, in *Statistical Methods in Radio Wave Propagation*, Pergamon Press, Oxford, 1960, pp. 3–36.

[15] K. S. Kwak and Y. C. Ko, *Fundamentals of Mobile Communications: Recent Trend and Outlook of Mobile Communication Systems*, Hanbit Academy, Inc. Korea, 2016.

Chapter 4

Fixed Orthogonal Multiple Access Systems

In wireless communication, multiplexing and multiple accesses occur across the dimensions of space, time, frequency, and code. The primary objective of multiplexing in wireless communication is to achieve optimal transmission efficiency over the medium while minimizing interference. This involves allocating space, time, frequency, and code to each communication channel, where a communication channel represents the transmission path between a transmitter and a receiver. The process of allocating the medium to multiple users through multiplexing techniques is known as medium access control (MAC).

In this context, "multiplexing" and "multiple accesses" both refer to the sharing of communication resources. However, there exists a subtle distinction between the two terms. Multiplexing typically involves fixed or slowly changing user requirements for communication resource sharing, commonly within a local site such as a circuit board. On the other hand, multiple access often involves remote resource sharing, as observed in satellite communications. In dynamically changing multiple access schemes, a system controller must be cognizant of each user's communication resource needs, which introduces overhead and sets an upper limit on resource utilization efficiency.

Within the ISO/OSI reference model, the second layer consists of the logical link control (LLC) and the MAC. The MAC layer encompasses mechanisms that regulate multiple access for numerous users by partitioning space, time, frequency, code, and other resources. Channel access methods fall under four main categories: fixed-assignment (orthogonal or non-orthogonal), random access, demand-assignment, adaptive-assignment protocols, and various combinations thereof. Fixed-assignment protocols involve dividing the total resource among users in a predetermined manner, where users occupy a portion of the channel capacity regardless of whether or not they have data to transmit. Examples of such protocols include time-division multiple access (TDMA), frequency-division multiple access (FDMA), code-division multiple access (CDMA), and space-division multiple access (SDMA). Random access, exemplified by the early ALOHA packet-switched system, involves transmitting a packet and then waiting to detect collisions with other packets on the system. While this approach offers inefficient throughput, it is operationally straightforward. Slotted ALOHA improves channel utilization by avoiding partial packet overlap. Similar concepts are found in LANs, such as CSMA, CSMA/CD, or CSMA/CA [1, 2].

4.1 Allocation of a Communication Resource

Enhancement of the throughput (the total data rate) of a communication resource can be achieved through three fundamental approaches. The first involves increasing the transmitter's effective isotropic radiated power (EIRP) or minimizing system losses to boost the received signal-to-noise power (E_b/N_0). The second method focuses on providing more channel bandwidth. The third and crucial approach is to optimize the allocation of the communication resource, a domain governed by multiple access methods. The primary ways of allocating communication resources include the following:

- **Frequency division (FD):** This involves allocating specified subbands of frequency.
- **Time division (TD):** This identifies periodically recurring time slots. Users may have fixed assignments in time with some systems, while others allow random access to the resource.
- **Code division (CD):** This allocates specified members of a set of orthogonal or nearly orthogonal spread-spectrum codes, each utilizing the full channel bandwidth.
- **Space division (SD):** This utilizes spot beam antennas to separate radio signals by pointing in different directions, enabling the reuse of the same frequency band.

- **Polarization division (PD):** This uses orthogonal polarizations to separate signals, facilitating the reuse of the same frequency band.

The essence of all multiplexing and multiple access schemes lies in enabling various signals to share a communication resource without causing unmanageable interference during the detection process. The permissible limit for such interference is that signals on one communication resource channel should not significantly increase the probability of error in another channel. The use of orthogonal signals on separate channels effectively avoids interference between users.

Signal waveforms, denoted as $s_i(t)$, are considered orthogonal if they can be expressed in the time domain as

$$\int_{-\infty}^{\infty} s_i(t)s_j(t)dt = \begin{cases} 0 & \text{for } i \neq j \\ c & \text{for } i = j \end{cases} \tag{4.1}$$

Similarly, signals are orthogonal in the frequency domain if they can be expressed as

$$\int_{-\infty}^{\infty} S_i(f)S_j(f)df = \begin{cases} 0 & \text{for } i \neq j \\ c & \text{for } i = j \end{cases} \tag{4.2}$$

Two coded signals, $c_i(t)$ and $c_j(t)$, are orthogonal in the code domain if they can be expressed as

$$\int_{-\infty}^{\infty} c_i(t)c_j(t)dt = \begin{cases} 0 & \text{for } i \neq j \\ c & \text{for } i = j \end{cases} \tag{4.3}$$

where c is a nonzero constant and $S_i(f)$ is the Fourier transform of the signal waveforms.

Channelization characterized by orthogonal waveforms, as in the time domain expression, is referred to as time-division multiplexing or time-division multiple access (TDM/TDMA). Conversely, channelization characterized by orthogonal spectra, as in the frequency domain expression, is known as frequency-division multiplexing or frequency-division multiple access (FDM/FDMA). Similarly, the orthogonal waveforms in the coded domain are called code-division multiplexing or code-division multiple access (CDM/CDMA). Figure 4.1 represents the multiple access techniques in time, frequency, and code domains for four-user cases (U_1, U_2, U_3, and U_4).

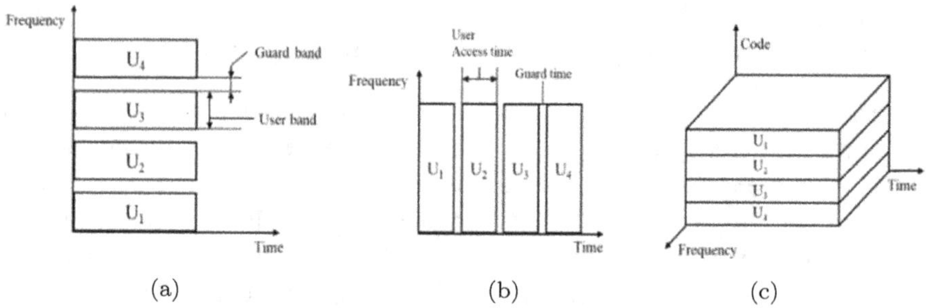

Fig. 4.1. Multiple access techniques: (a) FDMA, (b) TDMA, (c) CDMA.

4.2 Multiplexing and Multiple Accesses

4.2.1 *Duplexing techniques*

Duplexing allows the separation of the uplink and downlink communications. There are two primary techniques used for duplexing in wireless communication systems: Frequency Division Duplexing (FDD) and Time Division Duplexing (TDD). FDD separates uplink and downlink channels by using distinct frequencies, enabling simultaneous data transmission in both directions and continuous communication. On the other hand, TDD shares the same frequency for both uplink and downlink but alternates between them in time, with downlink typically accessed first, followed by uplink (see Fig. 4.2). The two main advantages of TDD are such that a greater simplicity of the RF part since it works only on one frequency at a time (against two for the FDD: one on the uplink and one on the downlink)

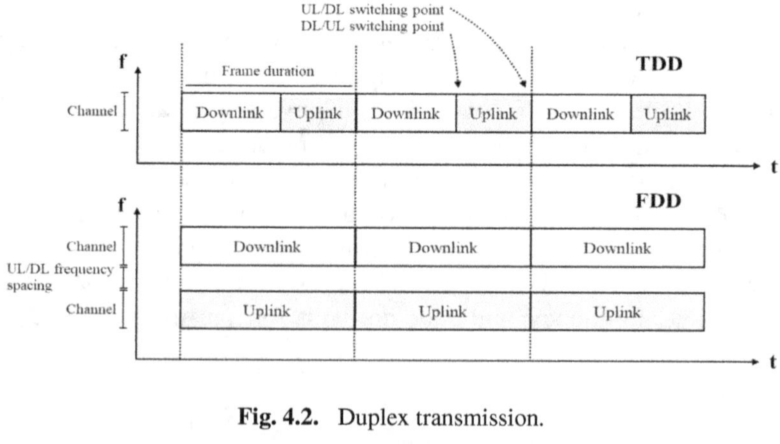

Fig. 4.2. Duplex transmission.

and the channel is reciprocal, i.e., the base station sees the same transmission channel as the terminal. Therefore, the power adjustment is simple. The power adaptation adapts its transmission power to have a good signal-to-noise ratio at the receiver.

4.2.2 *Frequency-division multiplexing and multiple access*

FDM, or FDMA, partitions the frequency domain into distinct and non-overlapping frequency bands, as illustrated in Fig. 4.1(a). Each channel is assigned its own dedicated frequency band, and a transmitter consistently utilizes its allocated band. To prevent frequency band overlap or adjacent channel interference, FDM incorporates guard bands, buffer zones, that mitigate interference between adjacent frequency channels.

This straightforward multiplexing technique finds common application in radio broadcasting stations operating within the same geographical area, where each station is assigned a unique frequency band. The simplicity of FDM eliminates the need for intricate coordination between transmitters and receivers. Receivers merely need to tune to the specific frequency band used by the transmitter.

In the modulation process, a baseband signal undergoes transformation to a higher frequency signal through a mixer, allowing it to occupy a higher frequency band. When two sinusoidal input signals with frequencies f_1 and f_2 are processed through a mixer, the resulting multiplication generates new frequencies at the sum $(f_1 + f_2)$ and the difference $(f_1 - f_2)$. This mixing effect is mathematically expressed by the trigonometric identity:

$$\cos(2\pi f_1)\cos(2\pi f_2) = \frac{1}{2}\{\cos 2\pi(f_1 + f_2) + \cos 2\pi(f_1 - f_2)\} \quad (4.4)$$

This identity, represented as equation (4.4), encapsulates the outcomes of the mixer. Illustrated in Fig. 4.3 is the mixing scenario involving a typical voice-grade telephone signal $x(t)$ (with a baseband frequency range of 300–3400 Hz) and a sinusoid from a 20-kHz oscillator. The baseband two-sided magnitude spectrum, $|X(f)|$, is depicted in Fig. 4.3(a). It's important to note that the output signal of a linear device will contain only the same component frequencies as the input signal, albeit with variations in amplitude and/or phase.

In Fig. 4.3(b), the one-sided magnitude spectrum at the mixer output is depicted. Following the mixing process described by equation (4.4), the output spectrum takes the form of a frequency-upshifted version of the baseband spectrum, centered at the oscillator frequency of 20 kHz.

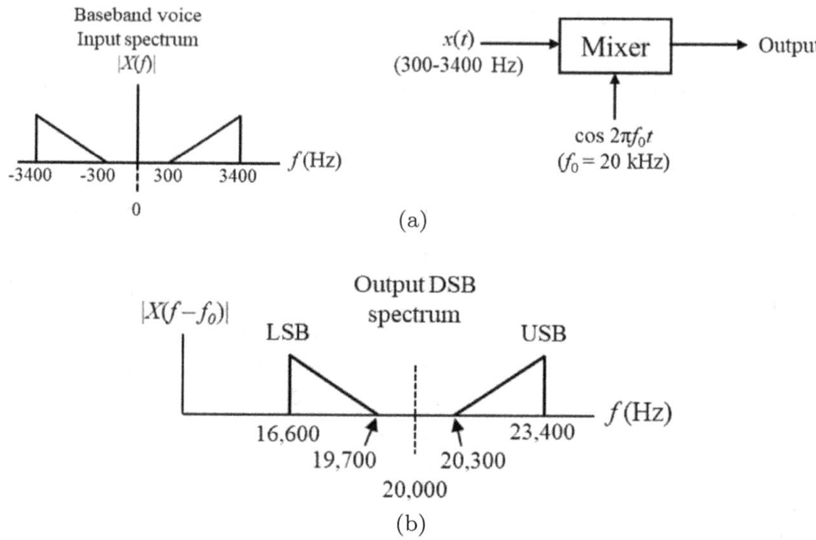

Fig. 4.3. Heterodyning (mixing): (a) mixing operation and (b) mixer output spectrum: lower sideband and upper sideband.

This type of spectrum is referred to as a double-sideband (DSB) spectrum, as the information is distributed across two distinct bands in the positive frequency domain. The lower sideband (LSB), spanning 16,600–19,700 Hz, is a product of filtering the DSB spectrum, and a similar filtering process can separate the upper sideband (USB), ranging from 20,300 to 23,400 Hz. Both sidebands of the DSB spectrum contain identical information, implying that either the USB or the LSB is sufficient for recovering the original baseband data.

As seen in Fig. 4.4, FDMA can be integrated with both FDD and TDD for the implementation of the uplink and downlink communications. It is crucial to recognize that comprehending FDMA is distinct from understanding FDD and TDD.

Nevertheless, FDM/FDMA presents drawbacks from a multiplexing standpoint. For instance, radio stations broadcast continuously for 24 hours, whereas in mobile communications, users typically engage with the service for only a few minutes at a time. If a frequency band is allocated each time a mobile phone call is initiated, it would result in the inefficient use of frequency resources. Additionally, the fixed allocation of frequencies to transmitters is undesirable as it imposes limitations on the number of available transmitters.

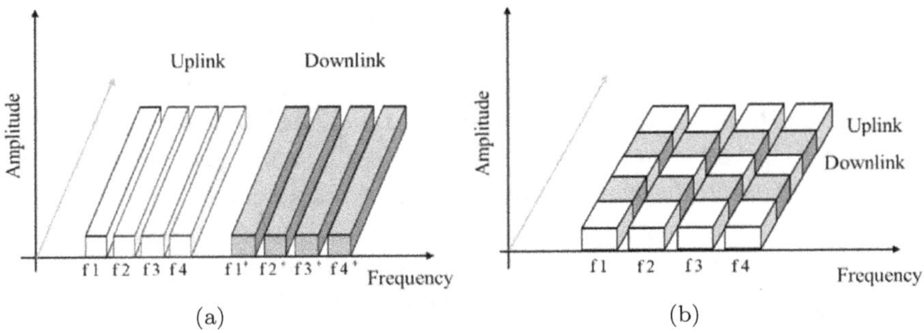

Fig. 4.4. Duplex transmission: (a) FDMA-FDD and (b) FDMA-TDD.

4.2.3 *Time-division multiplexing and multiple access*

A more adaptable multiplexing approach for mobile communication is TDM. Illustrated in Fig. 4.1(b), this technique shares the communication resource by allocating each of the multiple users the complete spectral occupancy of the system for brief durations known as time slots. Guard times, situated between slot assignments, remain unused and serve as buffer zones to mitigate interference, allowing for some time uncertainty between signals in adjacent time slots.

To avoid this type of interference, precise synchronization between communication transmitters is necessary. However, achieving precise synchronization by maintaining accurate clocks in all transmitters or distributing synchronization signals to all transmitters can be challenging. From the perspective of the receiving transmitter, this process involves not only adjusting the frequency but also precisely timing the transmission and reception of signals. However, TDM allows for flexible allocation of time, assigning more time to terminals with large amounts of data and less time to terminals with fewer data, making it a highly adaptable multiplexing method.

For TDMA, two allocation methods can be used: fixed allocation, where one channel is assigned to a specific time slot, and dynamic allocation, where time slot assignments can be adjusted as needed. Dynamic allocation is similar to the approach used in wired MAC technologies and may require a MAC address as an identification for the transmitter. In fixed allocation, the fixed positions of time slots can serve as unique identification information, eliminating the need for separate information for user identification. However, it may face challenges to adapt flexibly to changing environments.

FDM and TDMA can be combined for use together. In other words, a channel, can utilize a specific frequency for a particular period of time. However, guard intervals are required for both time and frequency domains. This method exhibits strong resistance to frequency-selective interference, where certain frequency components of a signal experience more interference than others. In essence, the wireless channel uses these narrow frequency bands for short durations. Moreover, in this method, the receiver needs to know the transmitter's frequency selection order in advance, providing some level of eavesdropping prevention.

4.2.4 *Fixed-assignment TDM/TDMA*

The TDD method, which supports both uplink and downlink using the same frequency, can be combined with TDMA. As seen in Fig. 4.5, a terminal uses one of the 12 time slots for the uplink, while the base station also uses one of the 12 time slots for the downlink. The uplink and downlink are separated by time. A maximum of 12 bidirectional transmissions can be supported without interference on the same frequency band. Each connection is assigned one pair of uplink and downlink time slots. For example, in DECT systems, this pattern repeats every 10 ms, so each time slot has a length of 417 μs (=10 ms/24). Terminals can access the medium every 10 ms based on this repeating pattern, and this access operates independently of other connections.

A fixed-assignment TDM/TDMA scheme is highly efficient when source requirements are predictable, and the traffic is consistently heavy (with time slots mostly always filled). However, for bursty or sporadic traffic, the fixed-assignment scheme becomes wasteful. Let's consider the simple example depicted in Fig. 4.6, where four time slots are allocated per

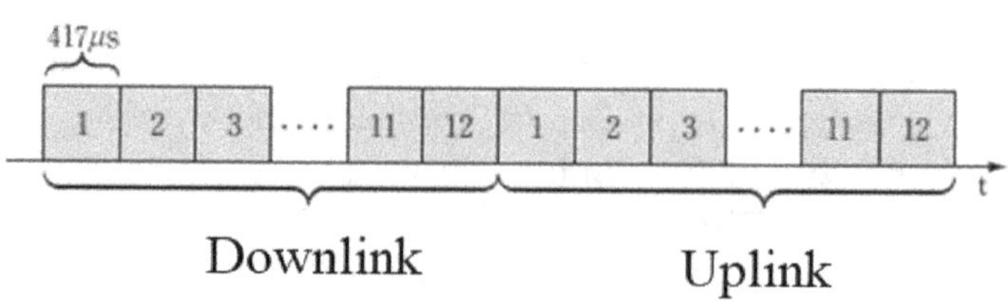

Fig. 4.5. TDM-TDMA.

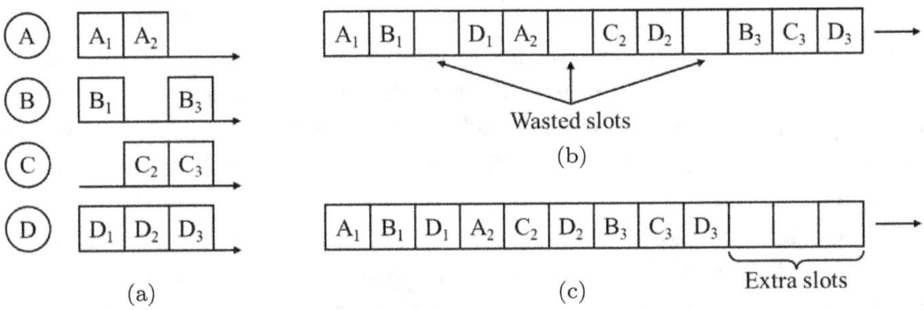

Fig. 4.6. Fixed-assignment TDM versus packet switching. (a) Data source, (b) TDM and (c) packet switching.

frame, with each slot preassigned to users A, B, C, and D, respectively. In Fig. 4.6(a), we observe a typical activity profile of the four users.

During the first frame time, user C has no data to transmit, and during the second frame time, user B is inactive, followed by user A during the third frame time. In a fixed-assignment TDMA scheme, all slots within a frame are preassigned. If the allocated user of a slot has no data to send during a particular frame, that slot is wasted. The data stream, depicted in Fig. 4.6(b), highlights the inefficiency of the fixed-assignment approach, showing the periods of wasted time slots.

For scenarios where source requirements are unpredictable, as illustrated in this example, more efficient schemes involve a dynamic slot assignment rather than a fixed assignment. These dynamic schemes, known as statistical multiplexing (Fig. 4.6(c)), utilize all slots in a frame efficiently, conserving capacity.

Fixed allocation methods may not be suitable for packet data with unpredictable and asymmetric characteristics. In situations involving packet data, fixed methods can lead to a substantial waste of wireless resources, making them unsuitable for modern Internet communication.

4.2.5 *Code-division multiplexing and multiple access*

Although initially primarily used in military communications due to its unique security features, CDM has also been applied in various commercial mobile communication systems as the availability of inexpensive digital circuits has increased.

The primary advantage of CDM in wireless transmission is its resilience to interference and eavesdropping. However, since different codes are assigned to each user and the code space is typically larger than the frequency space, there are usually no significant issues with code allocation. Codes assigned to each transmitter typically do not cause mutual interference. On the other hand, a drawback of CDM is that the receiver is relatively complex. The receiver can identify the user data channel from background noise, which is a combination of other signals and ambient noise, only if it knows the codes used by the transmitters. Furthermore, accurate synchronization between the transmitter and receiver is necessary for successful data recovery. All signals must reach the receiver with almost the same power. Otherwise, some signals may be overwhelmed by others. Therefore, precise power control is necessary for implementing CDM.

One method of implementing CDM/CDMA is through spread-spectrum techniques: direct-sequence spread spectrum (DSSS) and frequency-hopping (FH) spread spectrum. Figure 4.1(c) shows the CDMA method.

Figure 4.7 shows FH CDMA, where frequency bands are briefly assigned to different signal sources in succession. During each time slot, which is typically short, the frequency band assignments are rearranged. Each user utilizes a pseudo-noise (PN) code, orthogonal or nearly orthogonal to all other user codes, determining the FH band assignments. In this illustration, three signals (S_1, S_2, and S_3) are assigned according to a PN code, which shows a simplified representation of how the communication

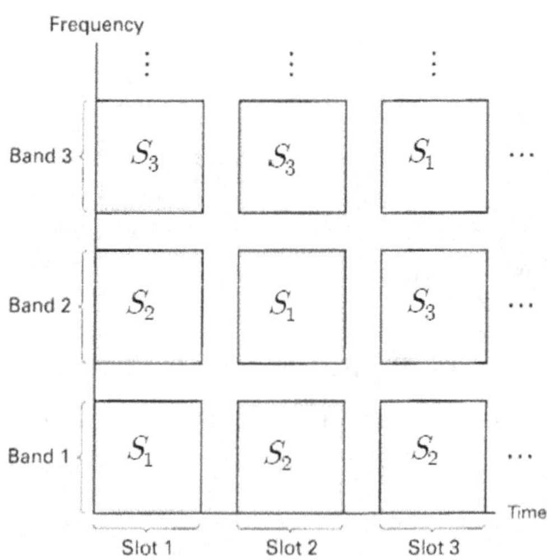

Fig. 4.7. Code-division multiplexing: frequency-hopping CDMA.

resource is shared in FH CDMA. However, the symmetry depicted implies that each FH signal is in time synchronism with every other signal, which is not the case. In fact, one advantage of CDMA over TDMA is the absence of synchronization requirements among user groups, necessitating synchronization only between a transmitter and a receiver within a group.

The FH modulation process, as depicted in Fig. 4.8, involves a two-step process of data modulation and FH modulation, though it can be implemented as a single step. In the single-step implementation, the modulator generates a transmission tone based on the outputs of the PN code and the data. During each frequency hop time, the PN generator supplies a code sequence to a device called a frequency hopper, which generates one of the allowable hop frequencies. Assuming that the data modulation follows an M-ary frequency shift keying (MFSK) format, the key difference between a conventional MFSK system and a FH MFSK system is that in the former, a data symbol modulates a carrier wave fixed in frequency, while in the latter, the data symbol modulates a carrier wave that hops across the entire communication resource bandwidth.

During a specific CDMA hop, the signal bandwidth aligns with the bandwidth of conventional MFSK, typically equal to the minimum bandwidth required for transmitting the MFSK symbol. However, across numerous time slots, the system undergoes FH over a band considerably broader than the data bandwidth, known as the spread spectrum.

In Table 4.1, compared to TDMA, CDMA significantly reduces the chance of collisions, except in cases where two messages are transmitted

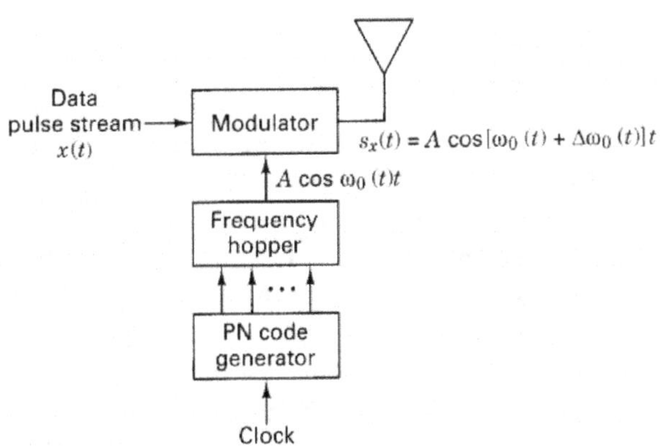

Fig. 4.8. CDMA frequency-hopping modulation process.

Table 4.1. Comparison of access methods.

Type of collision	TDMA	CDMA
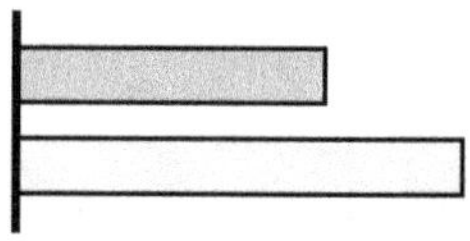	Collision	Collision (but not necessarily all lost)
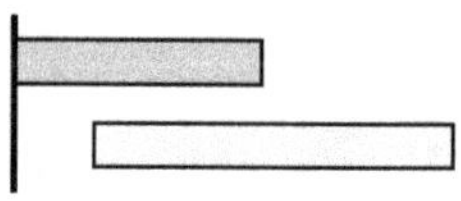	Collision	No collision

at exactly the same time, in which case there will be a collision and therefore a retransmission. So, there is more a concept of sharing of bandwidth (frequency or time) between users.

If a specific spectral segment experiences fading, signals within that frequency range undergo attenuation. In an FDMA scheme, a user assigned to the fading segment of a spectrum may encounter significantly degraded communications for the entire duration of the fading. However, in an FH-CDMA scheme, degradation is experienced by a user only when they hop into the affected spectral portion. Consequently, with CDMA, such degradation is distributed among all users.

The primary advantage of CDMA schemes over TDMA lies in the absence of the need for precise time coordination among simultaneous transmitters. The orthogonality between user transmissions on different codes remains unaffected by variations in transmission times.

4.2.5.1 *Principles of code-division multiple access*

In CDM, codes assigned to users must exhibit the property of having a correlation close to or equal to zero for them to be used in data transmission. In CDMA systems, each terminal uses its unique code in the code space to distinguish users and access a shared medium without significant interference. A crucial challenge in CDMA is how to find good codes and how to separate one's signal from the noise generated by the surroundings.

Orthogonality in the code space is analogous to orthogonality in a mathematical three-dimensional space, minimizing correlation with other codes.

Let's consider a three-dimensional vector system with coordinates starting from the origin. If the inner product of two vectors $\vec{a} = (a_1, a_2, \ldots, a_n)$ and $\vec{b} = (b_1, b_2, \ldots, b_n)$ is 0, then they are orthogonal, and otherwise, not orthogonal. Here, the inner product is defined by the following formula:

$$\vec{a} * \vec{b} = \sum_{i=1}^{n} a_i b_i \tag{4.5}$$

For example, the inner product of vectors $(2, 5, 0)$ and $(0, 0, 17)$ can be computed as $(2, 5, 0) * (0, 0, 17) = 0 + 0 + 0 = 0$ and since this value is 0, we can say that these two vectors are orthogonal. Similarly, vectors $(3, -2, 4)$ and $(-2, 3, 3)$ are orthogonal because their inner product is 0. On the other hand, vectors $(1, 2, 3)$ and $(4, 2, -6)$ cannot be considered orthogonal because their inner product is not 0 but -10. Vectors $(1, 2, 3)$ and $(4, 2, -3)$ have an inner product of -1, so they can be considered approximately orthogonal. Although "approximate orthogonality" does not strictly imply mathematical orthogonality, it's a commonly used characteristic in implementations because ensuring perfect orthogonality can be challenging due to various factors in practice.

Now, let's explain what it means for a code to have good autocorrelation in the code space. For example, codes known as Barker codes $(+1, -1, +1, +1, -1, +1, +1, -1, -1, -1)$ exhibit good autocorrelation. These codes are used in ISDN and IEEE 802.11, and their inner product with themselves yields 11. However, if a Barker code is shifted by 1 chip, the correlation value becomes 1 in absolute value. Barker codes maintain a correlation value of 1 until they complete a full cycle shift. This property can help synchronize the receiver to the data stream. Even if noise slightly distorts the original signal, this property aids in synchronization. Through the matching process, the receiver can discover the maximum value and align itself accordingly. This synchronization allows the receiver to precisely recover the original data.

Now, let's explain the basic operation of CDMA using the following example:

- Consider the case where users A and B want to transmit data in the CDMA system. In CDMA, for each transmission, an orthogonal key sequence is assigned. The key for user A is $A_k = 010011$, and the key for user B is $B_k = 110101$. User A will transmit $A_d = 1$, and B will transmit $B_d = 0$. Assuming coding for wireless transmission, binary 0 is coded as -1 and binary 1 is coded as $+1$.
- Both users spread their data bits using their key as a chipping sequence. Here, "spreading" means multiplying the entire chipping sequence of

the data bits. In the case of using a very long chipping sequence, only a portion may be used for spreading. After the spreading process, A's signal becomes $A_s = A_d * A_k = +1 * (-1, +1, -1, -1, +1, +1) = (-1, +1, -1, -1, +1, +1)$, and B's signal becomes $B_s = B_d * B_k = -1 * (+1, +1, -1, +1, -1, +1) = (-1, -1, +1, -1, +1, -1)$.

- Since the A and B signals are simultaneously transmitted using the same frequency band, the signals overlap in the transmission medium (ignoring analog modulation). Assuming that signals from other users and the surrounding noise are ignored, the two signals are received with the same intensity. Now, the combined signal C in the transmission medium will be

$$\mathrm{C} = A_s + B_s = (-2, 0, 0, -2, +2, 0).$$

- Now, let's assume that the receiving user receives the signal transmitted by user A. The receiving user A reconstructs the original signal through the de-spreading process using the same code that A used. Therefore, $\mathrm{C} * A_k = (-2, 0, 0, -2, +2, 0) * (-1, +1, -1, -1, +1, +1) = 2 + 0 + 0 + 2 + 2 + 0 = 6$. Since this value is significantly greater than 0, the receiving terminal determines that the received bit is 1. The same process can be applied when the receiving user B is receiving the signal from user B. Using B's code, $\mathrm{C} * B_k = (-2, 0, 0, -2, +2, 0) * (+1, +1, -1, +1, -1, +1) = -2+0+0-2-2+0 = -6$. Since this result is much less than 0, the receiving terminal determines the received bit as 0.

The basic functions of CDMA explained simplified the overall process of the CDMA system. In reality, noise or interference signals will be added during transmission, and this noise needs to be included in the signals. Therefore, the result could be different from −6 or +6, and in some cases, it may be close to 0. In such cases, determining whether the received information is exactly 0 or 1 becomes challenging. Additionally, we assumed that the A and B signals were added in exactly the same ratio during transmission, and the receiving terminal also assumed the same strength for both signals.

However, what if the B signal is much stronger? Assuming that the intensity of the B signal is five times greater than that of the A signal, C′ would be calculated as

$$\begin{aligned}\mathrm{C}' = A_s + 5B_s &= (-1, +1, -1, -1, +1, +1)\\ &\quad + (-5, -5, +5, -5, +5, -5) = (-6, -4, +4, -6, +6, -4).\end{aligned}$$

If the receiving terminal is receiving the bit from B, then $C' * B_k = -6 - 4 - 4 - 6 - 6 - 4 = -30$ is obtained, and the receiving terminal would naturally determine it as 0. If the receiving terminal wants to receive the bit from A, then

$$C' * A_k = 6 - 4 - 4 + 6 + 6 - 4 = 6.$$

However, it would be challenging for the receiving terminal to confidently decide that it received exactly 1 because, in terms of the absolute magnitude, 30 and 6 show a significant difference. In other words, the receiving terminal may find it difficult to distinguish whether it received 0 or 1 in such a situation.

4.2.5.2 *Direct-sequence spread spectrum*

DSSS systems spread the user bit stream with a chipping sequence using exclusive-OR (XOR) operations. In the example shown in Fig. 4.9, when the user bit is 0, it is spread to the sequence 0110101, and when the bit is 1, it is spread to the sequence 1001010. The period of each user bit

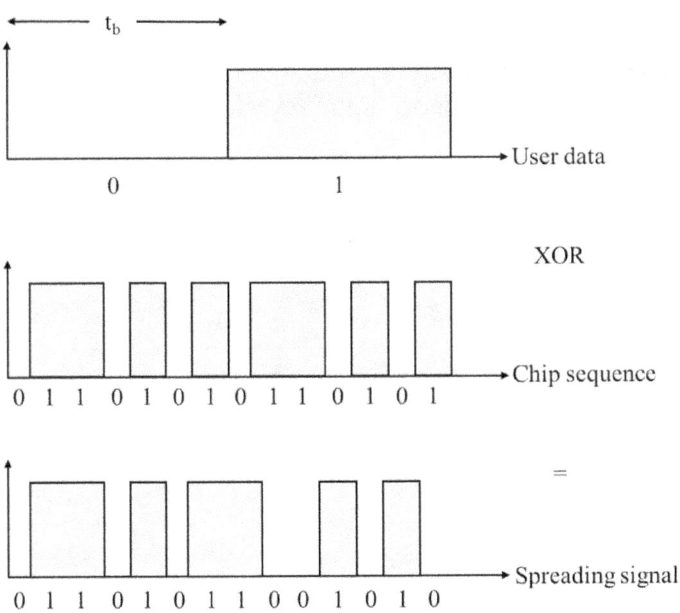

Fig. 4.9. Frequency hopping using DSSS.

is t_b, whereas the period of the chipping sequence is t_c. The spreading signal, the chipping sequence, consists of small pulses called chips. In fact, the chipping sequence often appears as pseudo-random noise rather than a typical signal, so it is sometimes referred to as a PN sequence. The spreading factor determines the bandwidth after spreading. In the simple example above, the spreading factor is 7, but commercial systems typically use spreading factors between 10 and 100, while military communications may use 10,000 or more. For example, if a wireless LAN following the IEEE 802.11 standard is implemented with DSSS, it usually uses a sequence called a Barker code, such as the sequence 10110111000. Barker codes are highly resistant to interference and not very sensitive to multipath delay. Other Barker codes with different lengths include 11, 110, 1110, 11101, 1110010, and 1111100110101 [3].

So far, we have described the simple spreading process, but transmitters and receivers using DSSS require additional components, as shown in Figs. 4.10 and 4.11. First, the DSSS transmitter spreads the user data using a chipping sequence, as shown in Fig. 4.10. Then, the spread signal is digitally modulated. Consider a user signal with a bandwidth of 1 MHz. If the signal is spread with a Barker code consisting of 11 chips, it will have a bandwidth of 11 MHz. Then, the radio carrier forwards this signal to the carrier frequency band and transmits it through the antenna.

The DSSS receiver is slightly more complex than the transmitter. However, the receiver can perform the reverse process of spreading after

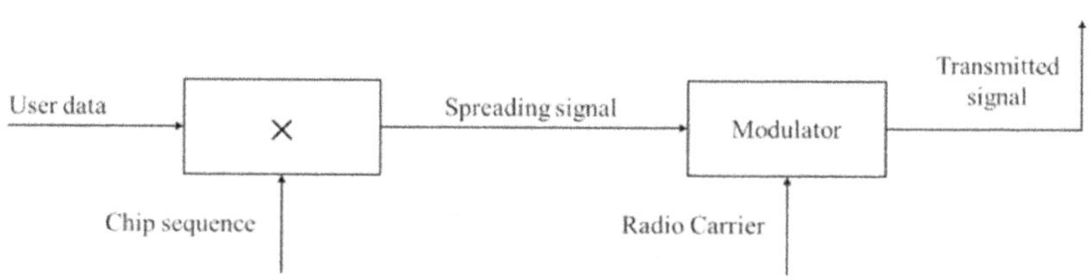

Fig. 4.10. DSSS transmitter.

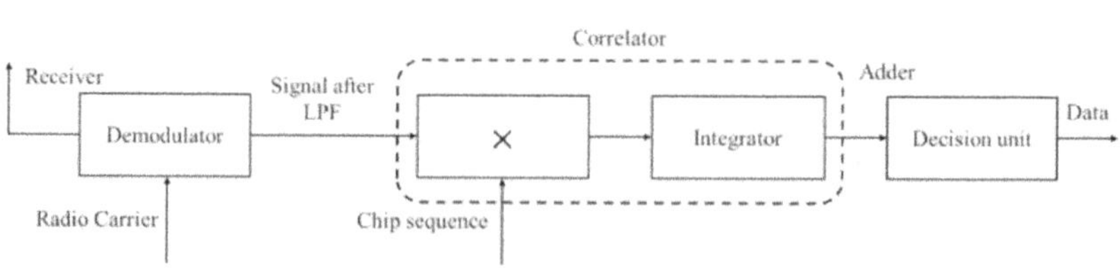

Fig. 4.11. DSSS receiver.

modulation by using the same two-stage modulation function used in the transmitter, namely by using chip sequences. In this process, additional techniques must be added to overcome noise and multipath propagation to restore the original data.

First of all, the receiver can obtain a signal with approximately the same bandwidth as the spread signal while performing the demodulation process using the same carrier wave as the transmitter. Additionally, to exclude signals outside this frequency band, the signal can be passed through a filter to obtain the required signal.

The next step corresponds to the actual DSSS receiver. Through this process, the complexity of the receiver increases. The receiver knows the chip sequence it uses, which means the receiver generates the same PN sequence used in the transmitter. The sequences of the transmitter and receiver must be perfectly synchronized, and the receiver calculates by multiplying the input signal and chip sequence. This process can be performed through XOR operations. Subsequently, the signal output during one-bit period is integrated through an integrator, adding up to measure the correlation between the two signals. The device that measures the correlation between the two signals is called a correlator. In the final step, based on the sum generated by the integrator during each bit period, the decision unit makes the final decision (1 or 0).

If the transmitter and receiver are perfectly synchronized and the signal is not distorted by noise or multipath signals, DSSS will function well according to the process described earlier. For example, if the user data to be transmitted is 01, and the Barker code 10110111000 consisting of 11 chips is applied to this data, the spread "signal" becomes 1011011100001001000111. At the receiving end, after demodulating the received signal, the receiver performs XOR operations with the same Barker code. Summing the values obtained chip by chip results in 0 for the first bit and 11 for the second bit. The decision unit determines the first bit as binary 0 and the second bit as 1.

However, in real situations, it becomes more complex. Let's assume that the decoded signal has some distortion, such as 1010010100001101000111. The integrator output for the first bit is 2, and for the second bit, it is 10. The decision unit will represent outputs less than 4 as 0 and outputs greater than 7 as 1. However, synchronization with the transmitter's signal is still crucial. What if multipath delays occur? Let's assume that multiple paths exist between the transmitter and receiver, each experiencing unique path losses. In this case, using a rake receiver can enhance the quality of the received signal. The rake receiver utilizes *n* correlators for *n* multipath signals. Each correlator is synchronized considering the delay for a specific path. In the rake receiver, if a stronger signal is detected than the existing weak signal, that correlator is assigned to the new strong signal. The outputs of each correlator are summed and then sent to the decision unit. The rake receiver has advantages in a multipath propagation environment since it can combine multiple signals [4].

4.2.5.3 *Frequency-hopping spread spectrum*

Frequency-hopping spread spectrum (FHSS) is a communication technique that involves dividing the entire bandwidth into small channels and hopping between them at specific time intervals. In this process, a channel is used for a certain period, and then after a short duration, the system switches to another channel. Therefore, FHSS can be implemented using FDM and TDM. The pattern of frequencies used for hopping is called a hopping sequence, and the time spent on a specific frequency band is referred to as dwell time. FHSS can be categorized into slow hopping and fast hopping, as depicted in Fig. 4.12.

A transmitter using the slow hopping method employs one frequency during the transmission of multiple bits, and the number of frequencies hopped during the one-bit period defines the hopping frequency. Figure 4.12 illustrates the bit periods t_b for five users. The transmitter uses a frequency f_2 for the first three bits during the dwell time t_d and hops to another

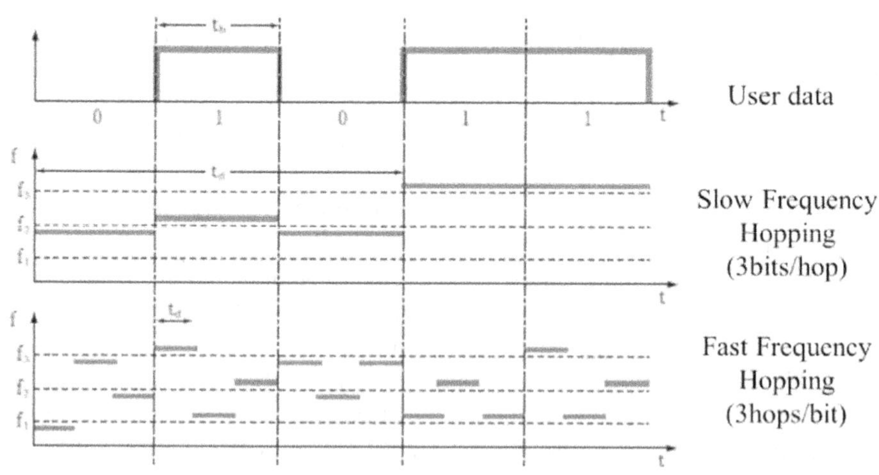

Fig. 4.12. Slow frequency hopping and fast frequency hopping.

frequency f_3 for the subsequent bits. This approach is cost-effective and less stringent on accuracy, but weaker against narrowband interference compared to fast hopping. In GSM, the slow FH method can be optionally utilized.

On the other hand, a transmitter using the fast hopping method changes frequencies multiple times to transmit a single bit. In the example in Fig. 4.12, the transmitter undergoes three frequency hops within the one-bit period. Fast hopping systems are more complex to implement because precise synchronization between the transmitter and receiver is essential for accurate hopping in a short time. However, this system is less susceptible to narrowband interference or frequency-selective fading since it uses each frequency for a short duration.

An example of a system utilizing FHSS is Bluetooth. Bluetooth changes frequencies 1,600 times per second and operates in the ISM band of a 2.4 GHz spectrum, using 79 separate channels spaced 1 MHz apart.

Figures 4.13 and 4.14 depict a simple circuit diagram for the transmitter and receiver of FHSS. The FHSS transmitter modulates user data using a modulation method such as FSK or BPSK to generate a spread-spectrum signal. If FSK is used, frequency f_0 represents 0, and frequency f_1 represents 1. Subsequently, FH is performed based on the hopping sequence. A frequency synthesizer is used to generate the carrier frequencies f_c according to the hopping sequence.

In the second modulation, a new spread-spectrum signal is generated using the modulated spread-spectrum signal and the carrier frequency. To transmit 0, $f_c + f_0$ is used, and to transmit 1, $f_c + f_1$ is used. To eliminate mutual interference between signals transmitted by multiple transmitters, the hopping sequences used by FHSS transmitters are configured not to overlap. In other words, coordination between the hopping sequences used by all transmitters is required to ensure that multiple transmitters do not use the same frequency f_c at the same time.

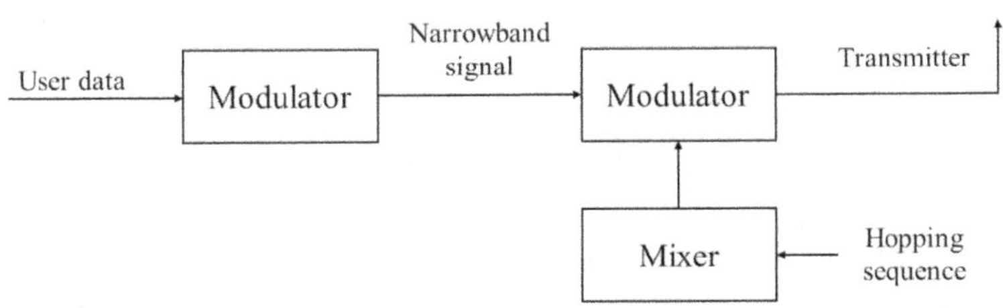

Fig. 4.13. FHSS transmitter.

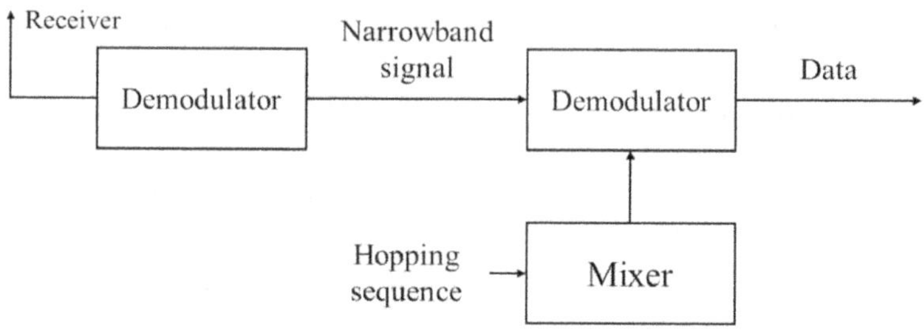

Fig. 4.14. FHSS receiver.

In DSSS systems, a pseudo-random hopping sequence can be used without any coordination. Such hopping sequences need to satisfy only a few characteristics to minimize interference. Although multiple transmitters may simultaneously select the same frequency, interference is minimized due to the very short dwell time caused by fast FH.

The receiver of an FHSS system needs to be aware of the hopping sequence in advance and must maintain synchronization with the transmitter simultaneously. The receiver reverses the modulation process performed by the transmitter to restore user data. Although not explicitly shown in Fig. 4.14, multiple filters may be additionally used in the process.

Compared to DSSS, FHSS is simpler to implement. FHSS systems use only one frequency band at a specific time, whereas DSSS systems always utilize the entire frequency band. However, DSSS systems have advantageous features in fading environments where the signal strength varies over time and in the multipath transmission of signals. While, DSSS signals, with much longer spreading codes compared to FHSS, make it extremely challenging, and even infeasible, to recover the original signal without the knowledge of the spreading code. If each transmitter uses a pseudo-random number sequence for signal spreading (DSSS or FHSS), the system performs CDM. For an in-depth understanding of spread-spectrum applications and theoretical background, please refer to the literature [4–7].

4.3 Performance of FDMA and TDMA

4.3.1 *Bit rate performance*

Figure 4.15 illustrates the fundamental differences between an FDMA system and a TDMA system operating within a communication resource

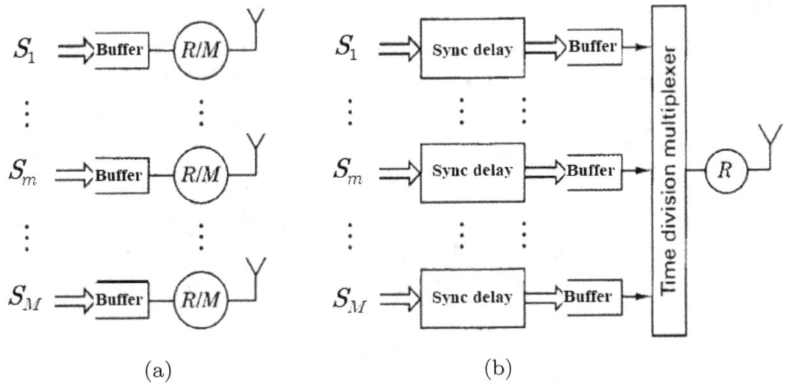

Fig. 4.15. (a) FDMA with M orthogonal frequency bands and (b) TDMA with M orthogonal time frames.

capable of supporting a total of R bits/s with the same bandwidth. In Fig. 4.15(a), the FDMA system's bandwidth is divided into M orthogonal frequency bands. As a result, each of the sources, S_m for $m = 1, 2, \ldots, M$, can simultaneously transmit at a bit rate of R/M bits/s.

In Fig. 4.15(b), representing TDMA, the frame is divided into M orthogonal time slots. Consequently, each of the sources bursts its transmission at R bits/s, M times faster than the equivalent FDMA user. In both cases, the source S_m transmits information at an average rate of R/M bits/s.

Let the information generated by each of the sources in Fig. 4.15 be organized into a length of b-bit packets. In the case of FDMA, the b-bit packets are transmitted in T seconds over each of the M disjoint channels. Therefore, the total bit rate is $R_{FD} = (Mb)/T$ bits/s. In the case of TDMA, the b bits are transmitted in T/M seconds from each source. Therefore, the bit rate is $R_{TD} = b/(T/M)$ bits/s. Hence, we can conclude that

$$R = R_{FD} = R_{TD} = \frac{Mb}{T} \text{ bits/s} \tag{4.6}$$

Thus, both systems require the full communication resource with the same data rate, R bits/s.

4.3.2 *Message-delay performance*

In the preceding sections, one might infer that the duality between FDMA and TDMA would yield equivalent performance. However, when considering the average packet delay as the performance metric, this assumption does not hold true. Studies [8, 9] have demonstrated that TDMA exhibits inherent superiority over FDMA, resulting in lower average packet delays.

As previously described, in FDMA, the system bandwidth is partitioned into orthogonal frequency bands, whereas in TDMA, the frame is divided into orthogonal time frames. For the analysis of message delay, the simplest scenario involves deterministic data sources. It is assumed that the communication resource is fully utilized, implying that all frequency bands in FDMA or all time slots in TDMA are occupied by data packets. To maintain simplicity, overhead costs such as guard bands or guard times are disregarded.

The message delay can be defined as

$$D = \omega + \tau \tag{4.7}$$

where ω is the average packet waiting time (prior to transmission) and τ is the packet transmission time. In the FDMA case, each packet is sent over a T-second interval, so the packet transmission time for FDMA is simply

$$\tau_{FD} = T \tag{4.8}$$

In the TDMA case, each packet is sent in slots of T/M seconds. We can thus write the TDMA packet transmission time with the use of equation (4.6) as

$$\tau_{TD} = T/M = b/R \tag{4.9}$$

Since the FDMA channel is continuously available and packets are sent as soon as they are generated see (Fig. 4.16(b)), the waiting time for FDMA is

$$\omega_{FD} = 0 \tag{4.10}$$

FDMA and TDMA bit streams are compared in Fig. 4.16. For TDMA, Fig. 4.16(a) illustrates that each user's slot begins at a different point in the T-second frame, that is, the kth packet of source m, S_{mk}, will start at $(m - 1)T/M$ seconds $(1 \leq m \leq M)$ after the packet generation instant. Therefore, the average waiting time for a TDMA packet before transmission begins is as follows:

$$\omega_{TD} = \frac{1}{M}\sum_{m=1}^{M}(m-1)\frac{T}{M} = \frac{T}{M}\sum_{n=1}^{M-1} n = \frac{T}{2}\left(1 - \frac{1}{M}\right) \tag{4.11}$$

Note that the minimum and the maximum waiting times before the transmission of a packet are 0 and $(M - 1)T/M$ seconds, respectively.

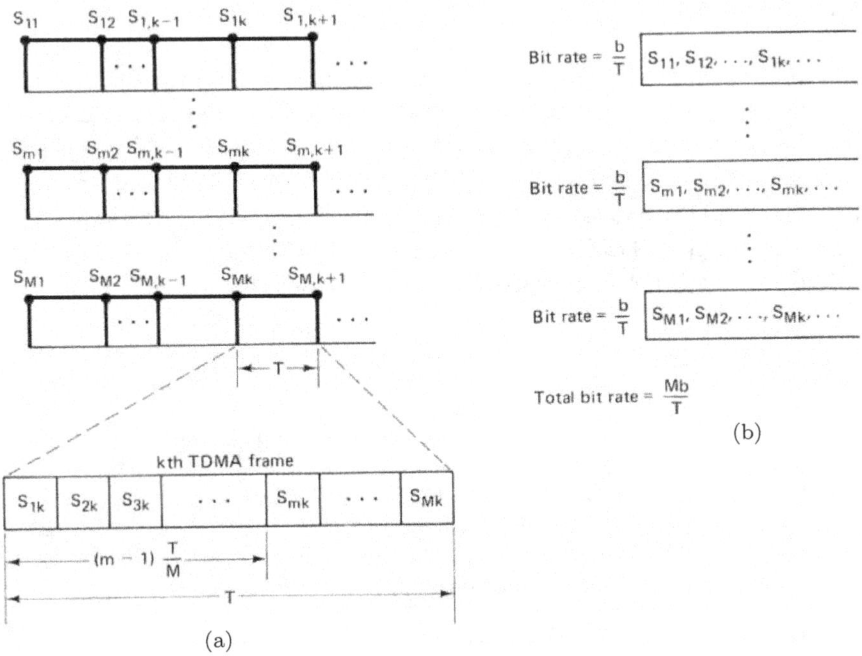

Fig. 4.16. (a) TDMA and (b) FDMA channelization.

So, on an average, a packet will wait $T(1 - 1/M)/2$ seconds, which is given by equation (4.11)

The average delay times can be compared for FDMA and TDMA, resulting in

$$D_{FD} = \omega_{FD} + \tau_{FD} = T \tag{4.12}$$

$$D_{TD} = \omega_{TD} + \tau_{TD} = \frac{T}{2}\left(1 - \frac{1}{M}\right) + \frac{T}{M} = D_{FD} - \frac{T}{2}\left(1 - \frac{1}{M}\right) \tag{4.13}$$

Using equation (4.6), equation (4.13) can be written as

$$D_{TD} = D_{FD} - \frac{b}{2R}(M - 1) \tag{4.14}$$

The result indicates that TDMA is inherently superior to FDMA from a message-delay perspective. Although equation (4.14) assumes that the data source is deterministic, the smaller average message delays for TDMA schemes hold true for any independent message arrival process [8, 9].

4.4 Summary

This chapter delves into fixed orthogonal multiple access systems, covering various aspects of resource allocation, multiplexing, and multiple access techniques. It discusses duplexing techniques, such as FDM and TDM, along with fixed-assignment FDM/FDMA and TDM/TDMA. Additionally, it explores CDM and CDMA, including principles and spread-spectrum techniques such as DSSS and FHSS. Finally, it evaluates the performance of FDMA and TDMA in terms of bit rate and message delay.

References

[1] B. Sklar, *Digital Communications, Fundamentals and Applications*, Prentice Hall, 1988.
[2] K. S. Kwak and Y. C. Ko, *Fundamentals of Mobile Communications: Recent Trend and Outlook of Mobile Communication Systems*, Hanbit Academy, Inc. Korea, 2016.
[3] W. Stallings, *Wireless Communications and Networks*, Prentice Hall, 2002.
[4] A. Viterbi, *CDMA: Principles of Spread Spectrum Communication*, Addison-Wesley Longman, 1995.
[5] T. Ojanpera and R. Prasad, *Wideband CDMA for Third Generation Mobile Communications*, Artech House, 1998.
[6] R. Ziemer, R. Peterson and D. Borth, *Introduction to Spread Spectrum Communications*, Prentice Hall, 1995.
[7] R. Dixon, *Spread Spectrum Systems with Commercial Application*, John Wiley, 1994.
[8] I. Rubin, Message delays in FDMA and TDMA communication channels, *IEEE Trans. Commun.*, 27(5), 769–777, May 1979.
[9] L. M. Nirenberg and I. Rubin, *Multiple Access System Engineering – A Tutorial*, IEEE WESCON/78, Los Angeles, September 13, 1978.

Chapter 5

Random Multiple Access Systems

5.1 Fundamentals of Random Multiple Access

5.1.1 *Multiple access in wireless LAN*

A fundamental inquiry in wireless communications concerns the applicability of advanced multiple access control (MAC) techniques [1], such as carrier-sense multiple access with collision detection (CSMA/CD), which were prevalent in early IEEE 802.3 networks like Ethernet, to wireless networks. To begin this exploration, let's delve into the operation of CSMA/CD. To address this, it's important to first examine the workings of CSMA/CD.

In CSMA/CD, the transmitter first senses whether the medium is available. Note that in wired communication, the medium refers to a wire or a coaxial cable but more broadly can be referred to as a channel. If the medium is busy, the transmitter waits until it becomes available. Once the medium is available, the transmitter starts transmitting data and continues to monitor the medium. If the transmitter detects a collision during data transmission, it immediately stops transmitting and sends a jamming signal to indicate the collision.

CSMA/CD is more concerned with collisions occurring at the receiver than those originating from the transmitter. While it is expected that the transmission signal should reach the receiver without collisions, the responsibility of detecting such collisions falls on the transmitter. Specifically, in a wired environment where the length of the wire is within a certain range, signals with nearly the same strength are received throughout the wire. Therefore, if a collision occurs on the wire, everyone is aware of it. As the transmitter monitors the medium during data reception, CSMA/CD works appropriately in a wired environment.

This situation can be applied differently in wireless networks. The strength of wireless signals decreases, due to the near–far effect, proportionally to the square of the distance between the transmitter and the receiver in a vacuum, and in real scenarios where obstacles are present, the signal strength is further attenuated. While a transmitter may determine transmission feasibility through carrier sensing and proceed with data transmission, collisions may still occur at the receiver due to another transmitter that was not detected by the initial transmitter. Additionally, although the transmitter expects data to be transmitted without errors, the receiver may not successfully receive the data. Particularly from the transmitter's perspective, detecting collisions in a wireless environment is challenging because the transmission power is much greater than the received power. For these reasons, conventional MAC methods used in wired environments may not be suitable for wireless situations. The hidden node problem is even more severe in wireless LAN systems compared to cellular communications.

5.1.1.1 *Hidden node and exposed node problems*

Let's consider a scenario with three terminals using wireless LANs, as depicted in Fig. 5.1. The transmission range of A includes B, but it does not reach C. Similarly, the transmission range of C includes B, but not A. In contrast, the transmission range of B includes both A and C. To summarize, A and C are in a situation where they are unaware of each other's presence.

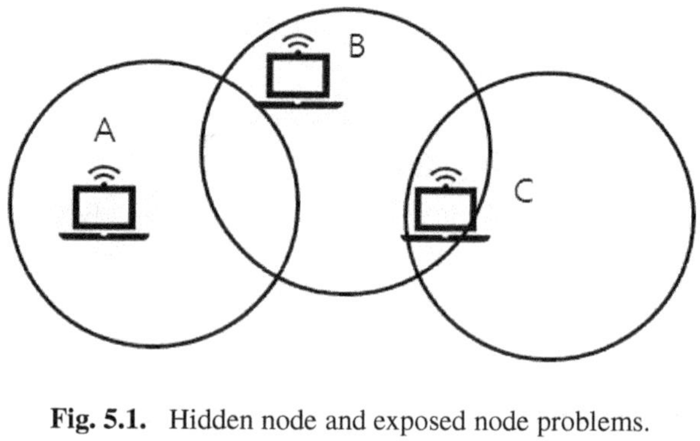

Fig. 5.1. Hidden node and exposed node problems.

When A starts transmitting to B, C cannot receive this signal. If C wants to transmit data to B, it first checks the availability of the wireless channel. Even if A is transmitting data to B, C perceives the medium as available, and C starts transmitting data to B. In this case, both A and C are transmitting data to B simultaneously, resulting in B being unable to receive any data.

However, since A and C are unaware of B's situation, they continue to transmit data persistently. In this situation, A and C become hidden nodes to each other.

While the hidden node problem causes transmission errors due to collisions, the exposed node problem leads to unnecessary transmission delays. For instance, if B is sending something to A and C plans to transmit data to another mobile device, not within the transmission range of A and B, C detects that the medium is busy since B is transmitting data and delays its transmission until the medium becomes available. However, since A and C are not within each other's transmission range, C does not need to delay its transmission. In this scenario, C is said to be exposed to B.

5.1.1.2 *Near–far effect*

Let's consider a situation depicted in Fig. 5.2. A and B are transmitting data with the same transmission power. Due to the signal strength decreasing proportionally to the square (or more) of the distance, A's signal interferes with B's signal, ultimately preventing C from receiving the data transmitted by A.

Now, let's assume that C performs the role of a base station controlling access to the transmission medium and granting authorization related to information transmission. In this case, just like before, C can only receive information from terminal B since A's signal interferes with B's signal, making it difficult to allocate resources for fair data transmission. This phenomenon, where the signal strength received is much higher for terminals closer in distance compared to those farther away, is called the near–far effect. In other words, it is the phenomenon where terminals closer to the transmitter receive signals with much higher intensity than those farther away, considering the same transmission power as a reference. The near–far

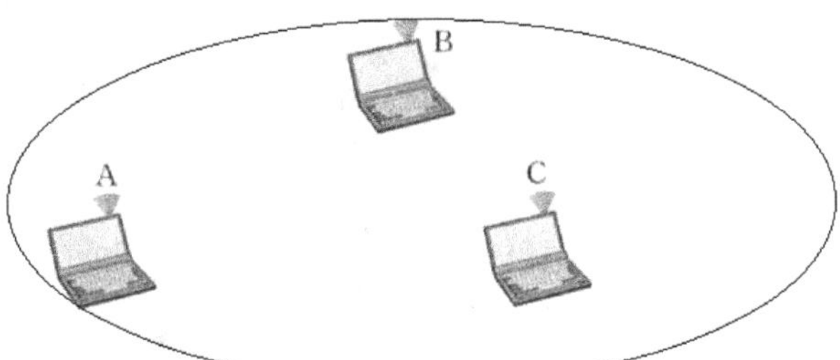

Fig. 5.2. Near–far effect of received signals.

effect is a crucial issue in CDMA-based 2G and 3G cellular communications and sensor-to-sensor wireless networks. All signals generally need to arrive at the receiver with approximately the same strength. Even if different codes are used, signals transmitted by a transmitter located nearby may not be detected due to interference with other signals. Therefore, precise power control is necessary for the receiver to capture all transmitted signals at the same intensity.

5.1.2 *Unslotted ALOHA multiple access*

This technology was proposed at the University of Hawaii and applied to ALOHAnet, which supported wireless connections between multiple terminals [2–4]. ALOHA does not control medium access, and the role of the MAC layer is very simple. Instead, each terminal can access the medium at any time, as illustrated in Fig. 5.3. In this figure, slots 1 and 2 overlap between terminals A and B, and slots 4 and 5 overlap between terminals C and D. However, slot 3 is occupied solely by terminal C, resulting in just one slot being fully utilized among the five available slots. This method does not require interference or coordination from a centrally controlled base station, and there is no need for temporal synchronization between terminals. If two terminals access the medium simultaneously, a collision occurs, and the transmitted data are not received correctly. The retransmission function to resolve data loss due to collisions is performed at a layer above the MAC layer.

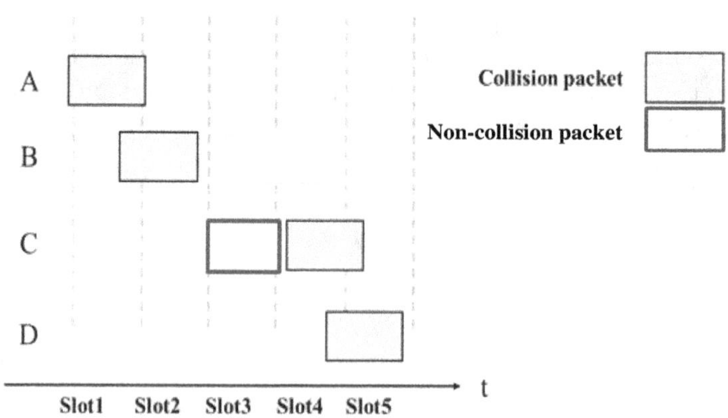

Fig. 5.3. Unslotted ALOHA multiple access.

The protocol concept was extremely simple, consisting of the following modes:

(i) **Transmission mode:** Users (nodes or terminals) transmit at any time they desire, encoding their transmissions with an error detection code.
(ii) **Listening mode:** After a message transmission, a user listens for an acknowledgment (ACK) from the receiver. Transmissions from different users will sometimes overlap in time, causing reception errors in the data in each of the contending messages. It is said that the messages have collided. In such cases, the errors are detected, and the users receive a negative acknowledgment (NAK).
(iii) **Retransmission mode:** When an NAK is received, the messages are simply retransmitted. If the colliding users were to retransmit immediately, they would collide again. Therefore, the users retransmit after a random delay.
(iv) **Timeout mode:** If, after a transmission, the user does not receive either an ACK or an NAK within a specified time, the user retransmits the message.

Let us find the utilization of a pure ALOHA system. Assume that the system demand requires a mean packet (message) arrival rate of λ_a accepted per second. Due to collisions, some of the messages will be unsuccessful or rejected. Therefore, the total traffic arrival rate λ_t is the sum of the acceptance rate λ_a and the rejection rate λ_r, which is $\lambda_t = \lambda_a + \lambda_r$.

Let us denote the length of each message or packet as L_b bits. Then the average amount of successful traffic or throughput, $\overline{\rho}$, on the channel in units of bits per second, is $\overline{\rho} = \lambda_a L_b$.

Let us define the total traffic, $\overline{G}$, on the channel, in units of bits per second, as $\overline{G} = \lambda_t L_b$.

With the channel capacity (maximum bit rate) designated as C bits per second, let us further define a normalized throughput to be $\rho = \overline{\rho}/C = \lambda_a L_b/C$ and a normalized total traffic to be

$$G = \overline{G}/C = \lambda_t L_b/C.$$

The normalized throughput, ρ, expresses the throughput as a fraction ($0 \leq \rho \leq 1$) of the channel capacity, and the normalized total traffic, G, expresses the total traffic as a fraction ($0 \leq G < \infty$) of the channel capacity. Let us also define the transmission time of each packet to be $\tau = L_b/C$.

Combining ρ, G, and τ, it is shown that

$$\rho = \lambda_a \tau \quad \text{and} \quad G = \lambda_t \tau$$

A user can successfully transmit a message as long as no other user began one within the previous τ seconds or starts one within the next τ seconds. If another user begins a message within the next τ seconds, it will collide with the tail end of the current message. Thus, a time interval of 2τ seconds is required for each message to be successfully transmitted.

The message arrival statistics for independent users is generally modeled as a Poisson process, $X(t)$. The probability of having k new messages arriving during a time interval of τ seconds is given by the Poisson distribution [5, 6] as

$$P(X(t) = k) = \frac{(\lambda\tau)^k e^{-\lambda\tau}}{k!} \quad \text{for } k \geq 0 \tag{5.1}$$

where λ is the average message arrival rate. Since the users transmit without regard for each other in the ALOHA system, this expression is useful for calculating the probability that exactly $k = 0$ other messages are transmitted during the time interval 2τ. This is the probability, P_s, that a user's message transmission is successful (no collisions). To compute P_s, assuming that total traffic is Poisson, using λ_t and 2τ in (5.1), thus,

$$P_s = P(X(t) = 0) = \frac{(2\lambda_t\tau)^0 e^{-2\lambda_t\tau}}{0!} = e^{-2\lambda_t\tau} \tag{5.2}$$

Since the total traffic arrival rate λ_t is $\lambda_t = \lambda_a + \lambda_r$, then, by definition, the portion of a successful packet, called the probability of a successful packet, can be expressed as

$$P_s = \frac{\lambda_a}{\lambda_t} \tag{5.3}$$

With (5.2)and (5.3), one has

$$\lambda_a = \lambda_t P_s = \lambda_t e^{-2\tau\lambda_t} \tag{5.4}$$

By combining (5.4) with ρ, G, and τ, finally it is shown that

$$\rho = G\, e^{-2G}$$

This result relates the normalized throughput, ρ, to the normalized total traffic, G, on the channel for the pure ALOHA system. As shown in Fig. 5.4, as G increases, ρ increases until a point is reached where further traffic

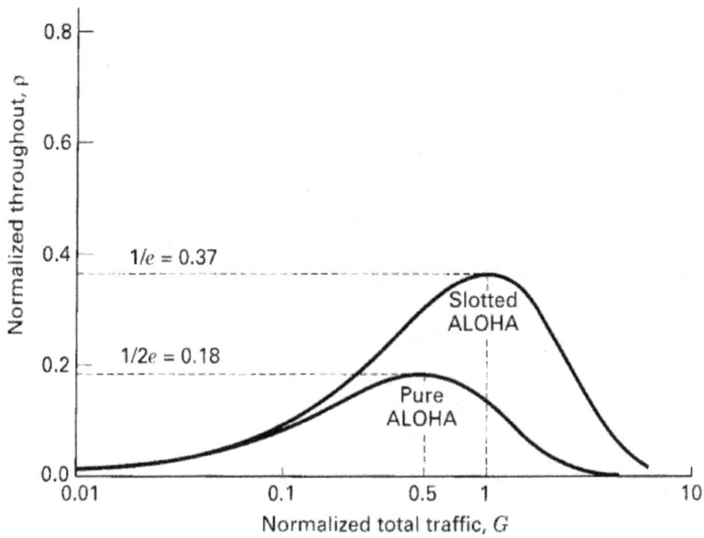

Fig. 5.4. Throughput in ALOHA channels.

increases create a large enough collision rate to cause a reduction in the throughput. The maximum ρ, equal to $0.5\,e^{-1} = 0.18$, occurs at a value of $G = 0.5$. Therefore, for a pure ALOHA channel, only 18% of the communication resource can be utilized [6]. The simplicity of control is achieved at the expense of the channel capacity [4, 7].

5.1.3 *Slotted ALOHA*

To overcome the low transmission efficiency of traditional ALOHA, slotted ALOHA was introduced, which utilizes time synchronization based on time slots. For slotted ALOHA to be supported, synchronization must be established between all transmitters, and transmission can only begin at the start of a time slot. However, like pure ALOHA, slotted ALOHA does not include a medium access control function. As with pure ALOHA, packet lengths are constant. Messages are required to be sent in the slot time between synchronization pulses, which can be started only at the beginning of a time slot. This simple change reduces the rate of collisions by half, since only messages transmitted in the same slot can interfere with one another. Collisions can happen if one or more packets are fully overlapped in a slot (see Fig. 5.5 of the four-user case).

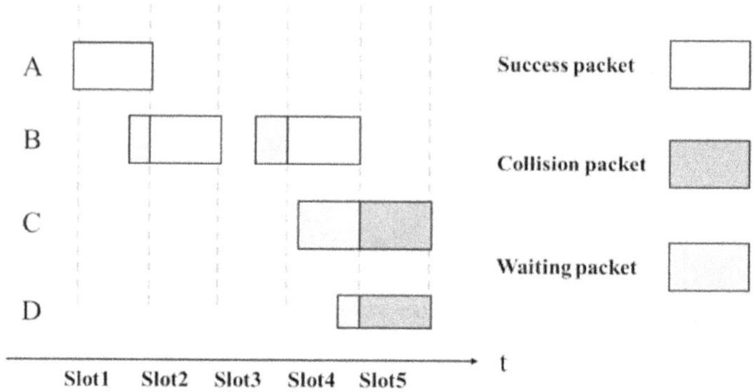

Fig. 5.5. Slotted ALOHA multiple access.

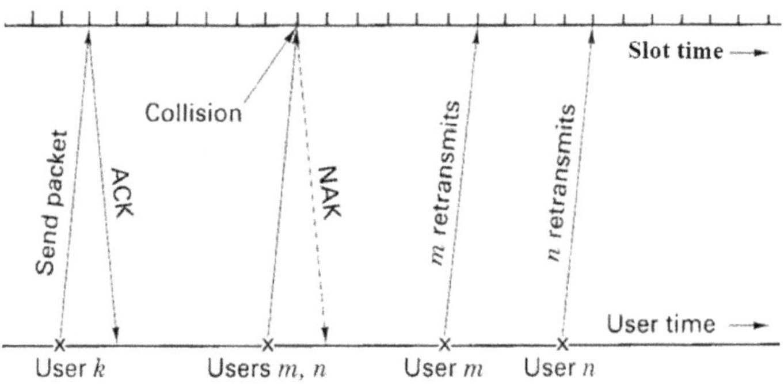

Fig. 5.6. Slotted ALOHA operation.

Figure 5.6 illustrates the slotted ALOHA operation. A packet of data bits is shown to be transmitted by user k followed by the ACK. Also when users m and n simultaneously transmit packets, which results in a collision, an NAK is returned. Each node employs a random number generator to select its retransmission time. Users m and n retransmit data at their randomly selected times. Of course, there is some probability that users m and n will collide again. However, in that case, they simply repeat the retransmission, using another random delay.

For slotted ALOHA [6], the reduction in the collision window from 2τ to τ, using the normalized throughput ρ and the normalized total traffic G, results in the following:

$$\rho = Ge^{-G} \tag{5.5}$$

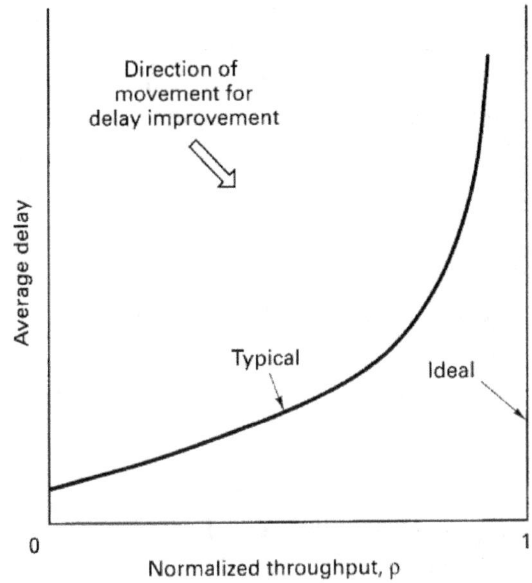

Fig. 5.7. Delay-throughput characteristic curve.

The plot of equation (5.5) is shown in Fig. 5.4. Here, the maximum value of ρ is $e^{-1} = 0.37$, or an improvement of twice the efficiency compared to the pure Aloha protocol.

The basic quality measure of a digital modulation scheme is the bit error probability which is expressed as P_b versus E_b/N_0 (per-bit energy to noise spectral density) curve. This measure is particularly useful because E_b/N_0 is a normalized signal-to-noise ratio, and being normalized, the curves allow us to compare the performance of various modulation schemes. There is a similar performance measure for multiple access schemes. Here, the average delay versus the normalized throughput is the key performance criterion. Figure 5.7 illustrates such a curve and shows a typical delay-throughput characteristic curve and the direction in which the curve will move as the delay performance improves. For normalized throughput values of $0 \leq \rho < 1$, the delay starts from a minimum point and increases without bound [5, 6].

While ALOHA systems (pure and slotted) perform well under low traffic load, they do not guarantee criteria, such as the maximum delay time from accessing the medium to completing the transmission or the minimum transmission efficiency. If the traffic increases beyond a certain limit, the delay increases without bound, which is called unstable. To address these shortcomings, additional mechanisms can be incorporated. For example, a combination of control mechanisms with ALOHA methods can be employed. In 3G mobile communication systems, slotted ALOHA is often used for initial connection setup or transmitting small amounts of data to supplement the communication process.

5.1.4 *Reservation ALOHA*

A significant improvement was made to the ALOHA system with the introduction of the reservation ALOHA (R-ALOHA) [8] scheme. The R-ALOHA system has two basic modes: an unreserved mode and a reserved mode, each of which is described as follows:

- **Unreserved mode**
  - (i) A time frame is established and divided into a number of small reservation sub-slots.
  - (ii) Users use these small sub-slots to reserve message slots.
  - (iii) After requesting a reservation, the user listens for an ACK and a slot assignment.
- **Reserved mode**
  - (i) The time frame is divided into $M + 1$ slots whenever a reservation is made.
  - (ii) The first M slots are used for message transmissions.
  - (iii) The last one slot is subdivided into sub-slots to be used for reservation/requests.
  - (iv) Users send messages only to their assigned portions of the M slots.

Consider the R-ALOHA example shown in Fig. 5.8. In the quiescent state, with no reservations, time is partitioned into short sub-slots for making reservations. Once a reservation is made, the system is configured in such a way that, in this example, $M = 5$ message slots followed by $V = 6$ reservation sub-slots. In this example, the station seeks to reserve three message slots. The reservation ACK advises the user where to locate their first data packet. Since the control is distributed so that all participants receive the downlink transmissions and are thus aware of the reservations, the ACK need not disclose any more than the location of the first slot. As shown in Fig. 5.8, the user sends their second packet in the slot following the first packet. The user further knows that the next slot comprises six

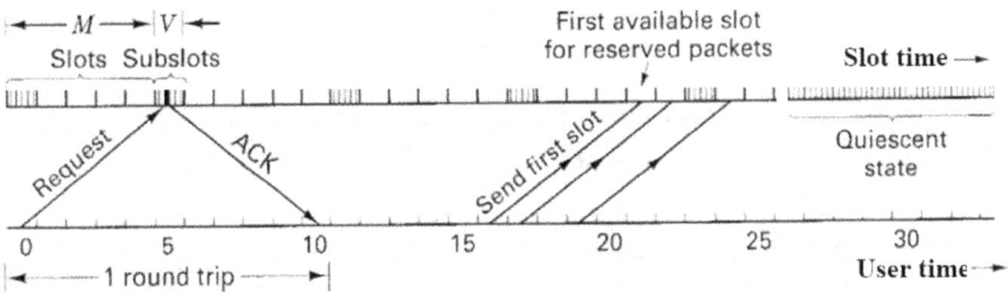

Fig. 5.8. Example of reservation ALOHA.

sub-slots for reservations, so no packets are transmitted during this time. The third and final packet is sent in the following slot. When there are no reservations taking place, the system reverts back to its quiescent state of sub-slots only. Since the control is distributed, all the participants are made aware of the quiescent state by receiving appropriate synchronizing pulses on the downlink.

5.1.5 *Spread ALOHA multiple access*

The CDMA technique [9, 10], which uses unique codes to spread data, can be considered a highly useful multiple access technology. However, the transmitters and receivers for CDMA communication are not as simple as in practice. The fact that multiple nodes are communicating implies that the receiver must have the ability to decode multiple codes. Fortunately, in typical mobile systems, the complexity required to receive multiple CDMA signals is handled by the base station. However, in cases where CDMA technology is employed to support sporadic traffic in packet communications, it may appear to have too much overhead.

In architectures like *ad-hoc* networks, there's little desire for a program managing multiple types of spreading codes. Conversely, ALOHA technology, being straightforward, grapples with reduced transmission efficiency from data collisions. So, what if we merge CDMA spreading technology with ALOHA's medium access approach? Can we support CDMA with just one code instead of many? The resultant technology born from this inquiry is termed spread ALOHA multiple access (SAMA) [11], which combines CDMA and TDMA. In SAMA, all transmitters use the same spreading code. In the example shown in Fig. 5.9, all terminals are using the code 110101.

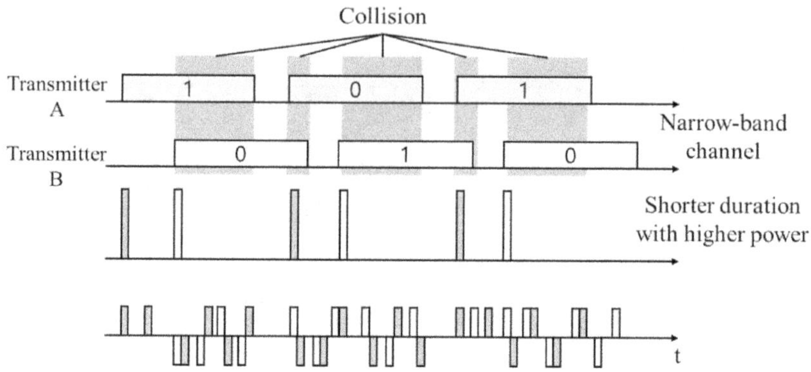

Fig. 5.9. Operating principle of SAMA.

The common issue encountered in ALOHA is described in the upper part of Fig. 5.9, where terminals A and B simultaneously access the narrow-band frequency spectrum, resulting in a collision of three bits.

Another approach, as shown in the middle part of Fig. 5.9, involves sending the same data at a shorter interval with higher power. Let's switch the approach to now considering transmission through frequency spreading. In this method, the bandwidth increases, and in this example, the spreading factor is 6. Both signals are spread by the same code, but the code's positions are slightly different. If a specific receiver synchronizes with terminal A and another receiver synchronizes with terminal B, the signals from A and B can be separated at each receiver. Signals from terminals not synchronized with the receiver appear as simple noise, and this approach is influenced by the ambient noise level. The most critical issue when applying SAMA is finding codes with good characteristics. These codes should have good autocorrelation properties and low correlation values when slightly out of sync. From the perspective of maximum transmission rate, the efficiency is around 18%, similar to ALOHA, but this method uses frequency spreading technology. This implies that it is robust to narrow-band interference and can coexist with other systems using the same frequency band.

5.1.6 *Carrier-sense multiple access*

The main defect of ALOHA is its lack of spectral efficiency. In ALOHA, users do not take into account the activity of their peers, and there is no mechanism to prevent collisions. A simple way to avoid collisions is to simply listen to the channel before transmitting a packet. If the channel is used, the terminal will wait before transmitting. The protocols that use this principle are called carrier-sense multiple access (CSMA) and Listen Before Talk (LBT) [12, 13].

Detecting the carrier before accessing the medium reduces the probability of collision. However, the hidden terminal problem, as mentioned earlier, can still occur. If a hidden terminal starts transmitting when another transmitter is already sending data, the receiver may fail to receive the data due to collisions. In CSMA, a backoff algorithm can be used for data retransmission. When a terminal detects a collision for the data it has transmitted, it generates a random number and attempts retransmission during the slot corresponding to that random number, reducing data loss due to collisions.

Figure 5.10 illustrates the operation of CSMA: User 1, after checking that the channel is free, sends two frames. User 2 finds that the channel is being used and delays its emissions for a certain period of time. At the time when it sends its packet, users 1 and 3 want to transmit a packet, find that the channel is used, and also delay their transmission with a random delay. The delays are such that there is still collision between the packets and that a retransmission is necessary. The CSMA protocol reduces the collision probability significantly compared to ALOHA; however, these collisions may occur. In particular, if the propagation time between terminals is large, the probability of collision will be high. Therefore, CSMA is typically used in local networks, whereas ALOHA is preferred for larger networks.

CSMA has several versions [14]. In its most basic form, known as non-persistent CSMA, a terminal immediately transmits its data if it senses that the carrier is idle. If the medium is busy, the terminal defers its transmission

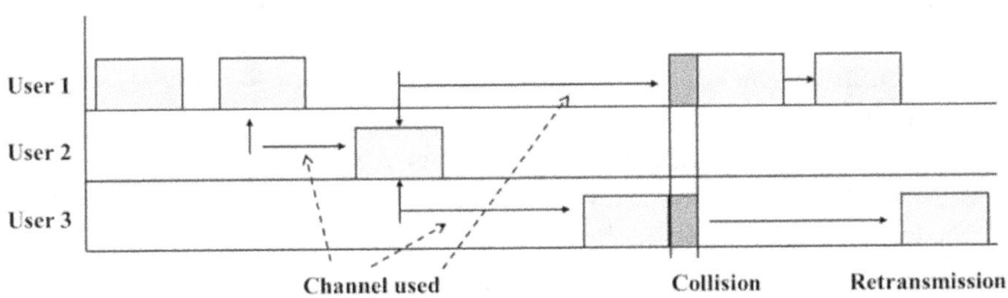

Fig. 5.10. Basic principle of CSMA.

and continues monitoring the medium's availability. In p-persistent CSMA systems, even if the medium is sensed as idle, the terminal does not immediately transmit data; instead, it decides whether to transmit with a probability of p. In 1-persistent CSMA systems, all terminals attempt to transmit simultaneously. If many terminals have data to transmit, mutual interference increases, resulting in numerous collisions.

The advanced methods from the Ethernet ALOHA and two varieties have been defined to deal with different scenarios. Among them, CSMA/CD and carrier-sense multiple access with collision avoidance (CSMA/CA) are widely deployed in many networks, such as Ethernet. Carrier sense defined here is the situation where the physical layer listens to the Ethernet wire before transmitting data across the network [13].

CSMA/CA is one of the access techniques used in wireless LANs based on the IEEE 802.11 standard. In this approach, collision avoidance is combined with the backoff algorithm when the medium is already in use to ensure a certain level of fairness among the competing terminals. Another example is Elimination-Yield-Non-Preemptive Multiple Access used in HiperLAN. This method adds stages of medium sensing and medium access to resolve collisions until the final winner is determined among the competing terminals to access the medium. Priority-based transmission methods may also be included to ensure preferential transmission for terminals with more critical data.

The Ethernet scheme is based on the assumption that each local machine can sense the state of a common broadcast channel before attempting to use it. This technique is known as carrier-sense multiple access with collision detection (CSMA/CD). Figure 5.11(a) illustrates the bit-field format for the Ethernet specification, with details listed as follows:

- The maximum packet size is 1526 bytes, where a byte is 8 bits.
- The packet breakdown is 8-byte preamble + 14-byte header +1500-byte data + 4-byte parity.
- The minimum packet size is 72 bytes, consisting of 8-byte preamble + 4-byte header + 46-byte data + 4-byte parity.
- The minimum spacing between packets is 9.6 μ s.
- The preamble contains a 64-bit synchronization pattern of alternating ones and zeros, ending with two consecutive ones: (1 0 1 0 1 0 ... 1 0 1 0 11).
- The receiving station examines a destination address field in the header to see if it should accept a particular packet.

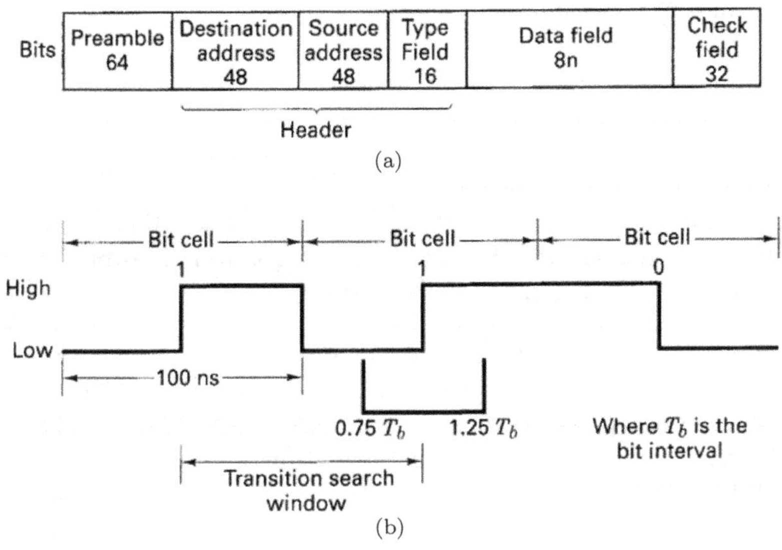

Fig. 5.11. Ethernet bit field and PCM format: (a) Ethernet frame format and (b) Manchester PCM format.

- The first bit indicates the type of address: 0 denotes single address, 1 denotes group address, and an entire field of ones denotes an all-station broadcast.
- The source address is the unique address of the transmitting machine.
- The type field determines how the data field is to be interpreted. For example, bits in the type field can be used to describe features, such as data encoding, encryption, and message priority. The data field is an integer number of bytes from a minimum of 46 to a maximum of 1500.
- The parity check field houses the parity bits which are generated by the following generating polynomial:

$$X^{32} + X^{26} + X^{23} + X^{22} + X^{14} + X^{12}$$
$$+ X^{11} + X^{10} + X^{8} + X^{7} + X^{5} + X^{4} + X^{2} + X + 1$$

The Ethernet multiple access algorithm defines the following user action or response [12, 13]:

(i) **Defer:** The user must not transmit when the carrier is present or within the minimum packet spacing time.
(ii) **Transmit:** The user may transmit unless deferring until the end of the packet or detecting a collision.

(iii) **Abort:** If a collision is detected, the user must terminate packet transmission and transmit a short jamming signal to ensure that all collision participants are aware of the collision.
(iv) **Retransmit:** The user must wait a random delay time (similar to the ALOHA system) and then attempt retransmission.
(v) **Backoff:** The delay before the nth attempt is a uniformly distributed random number from 0 to $2^n - 1$, for $0 < n \leq 10$. For $n > 0$, the interval remains 0–1023. The unit of time for the retransmission delay is 512 bits (51.2 μ s).

Figure 5.11(b) shows a 10 Mbps data stream with Manchester PCM formatting from the Ethernet specification. Note that with such formatting, each bit cell or bit position contains a transition. A binary one is characterized by transitioning from a low level to a high level, while a binary zero has the opposite transition. Therefore, the presence of data transitions denotes to all listeners that the carrier is present. If a transition is not seen between 0.75 and 1.25 bit times since the last transition, the carrier has been lost, indicating the end of a packet.

5.1.7 *Demand assigned multiple access*

One way to improve the ALOHA access system is through a reservation-based approach. This method involves a reservation step before transmitting data, allowing the reservation of time slots for future use. With this method, collisions only occur during the reservation process, and data transmission can occur without collisions. While reservation-based transmission methods typically suffer from longer delays even at low traffic loads, they inherently have fewer collisions, resulting in higher transmission efficiency.

One basic reservation method is demand assigned multiple access (DAMA), also known as reserved ALOHA. This method is commonly used in satellite systems. During the contention phase driven by slotted ALOHA technology, as shown in Fig. 5.12, all terminals attempt to reserve slots for future use. For example, consider multiple ground terminals attempting to reserve time slots for satellite communication. Collisions during the reservation phase mean that short requests for data transmission fail rather than collisions occurring during data transmission. Upon successful reservation, the reserved time slot is allocated for future use, preventing other terminals from transmitting data during that time slot. The satellite collects all successfully received reservation requests from terminals and then broadcasts

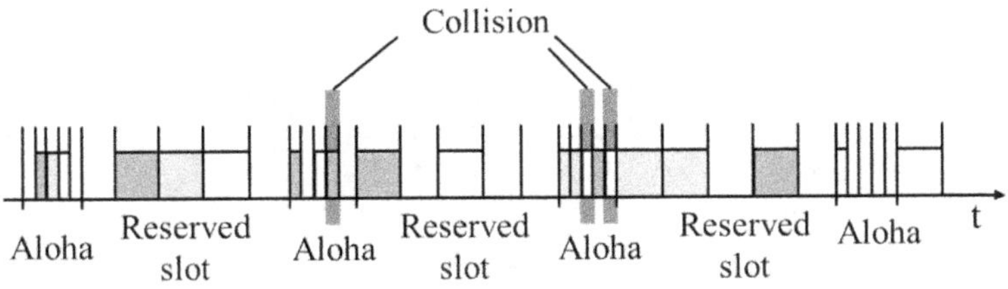

Fig. 5.12. DAMA with explicit reservation.

a reservation list to grant access rights for future slot usage. All ground terminals must adhere to the reservation information provided in this list.

To maintain the TDM pattern related to reservation and transmission, all terminals occasionally undergo a synchronization process. DAMA falls into the category of explicit reservation since it involves the explicit exchange of information regarding resource reservation and receiving confirmation for the reservation. In other words, the reservation of slots for data transmission requires a prior explicit information exchange process.

5.1.8 *Packet reservation multiple access*

A representative example of implicit reservation, which is the counterpart of explicit reservation, is packet reservation multiple access (PRMA). In PRMA, time slots are implicitly reserved through a process without attempting reservations or exchanging information for resource allocation. PRMA organizes multiple time slots into one frame, as illustrated in Fig. 5.13, where the example shows eight time slots per frame. This frame structure continues to repeat, and once a terminal reserves a specific time slot, it transmits data from the same time slot position in each subsequent frame.

The PRMA process proceeds as follows:

(i) The base station supporting PRMA broadcasts the reservation status of time slots for each frame to all terminals.
(ii) This information includes details about the occupancy status of each time slot, allowing all terminals to determine the availability of each time slot.
(iii) If data transmitted by a terminal is successfully delivered, the base station broadcasts information indicating that the terminal has successfully transmitted data and simultaneously reserved the respective time slot.

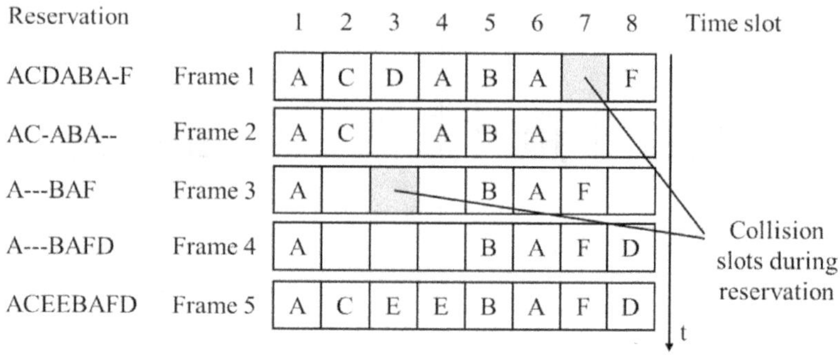

Fig. 5.13. PRMA with implicit reservation.

In Fig. 5.13, the base station notifies the reservation status by the terminal names, and one can see the broadcast of "ACDABA-F." This information indicates that among the eight time slots, slots 1–6 and 8 are already occupied, but slot 7 is empty. All terminals wishing to transmit use the information provided by the base station to attempt transmission via ALOHA in the empty time slots. Terminals cannot transmit in already occupied time slots, and if multiple terminals attempt to access an empty time slot simultaneously, data collision occurs. In our example, since multiple terminals attempt transmission simultaneously in slot 7, the reservation status of the next frame will remain the same as before, indicated by "ACDABA-F." Once again, terminals competing for transmission attempt to occupy the empty slot 7. If slot 7 is successfully occupied, the respective terminal can continue to use that time slot. Particularly, terminals D and F stop transmission in slots 3 and 8, respectively, in the first frame. The base station considers these slots unused if data are not received and thus announces before the start of the third frame that slots 3 and 8 are empty.

PRMA differs from previous methods in that it allocates fixed time slots simultaneously with reservation. In other words, once the data transmitted by a terminal successfully reaches the base station, the reservation is considered successful, and the terminal continues to use the same time slot in subsequent frames.

Terminals continuously use the reserved time slot as long as data are present, and when there is no more data to send, they simply stop transmission. If the base station does not receive data in the reserved time slot, it assumes that there is no more data to send and cancels the reservation,

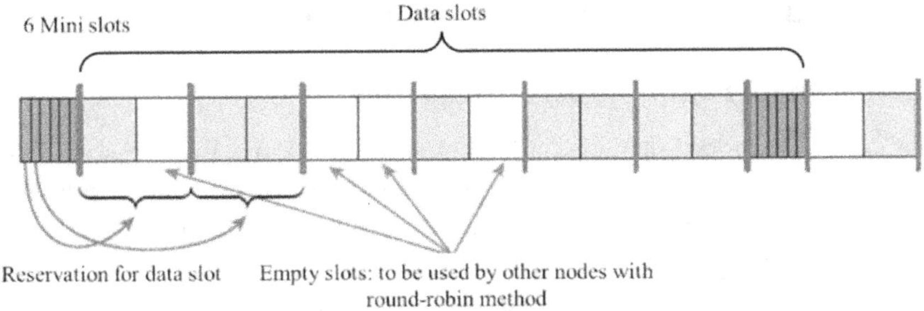

Fig. 5.14. Access using reservation TDMA.

notifying this information in the next frame. It's important to note that slotted ALOHA, used for time slot reservation, is only employed in empty slots. It is noted that PRMA ensures a certain level of guaranteed data transfer rate.

5.1.9 *Reservation TDMA*

Reservation TDMA, as shown in Fig. 5.14, supports transmission through random access while maintaining a somewhat fixed pattern. In this TDM method, one frame consists of N mini-slots and $N \cdot k$ data slots, and this frame structure repeats continuously ($N = 6$ and $k = 2$ in this example). Each terminal is assigned its own mini-slot, which it can use to reserve k data slots. This method ensures a consistent level of transmission speed and latency for each terminal. Other terminals can transmit data in unused data slots. These vacant slots can be accessed through a simple round-robin or ALOHA method. Ultimately, this approach facilitates a combination of traffic with fixed transmission speeds and best-effort traffic with no guarantees.

5.1.10 *Summary*

This section provides an in-depth exploration of the fundamental principles underlying random multiple access techniques in wireless settings. It covers a range of methods, including unslotted ALOHA, slotted ALOHA, reservation ALOHA, and spread ALOHA multiple access. Additionally, it discusses CSMA, DAMA, PRMA, and reservation TDMA while addressing key challenges, such as hidden node and exposed node problems as well as the near–far effect.

5.2 Stabilized Random Multiple Access Systems

5.2.1 *Random multiple access modeling*

The idealized model provides a framework for addressing the contention that arises when multiple nodes attempt to use the channel simultaneously [15, 16]. Conceptually, a satellite system is depicted in Fig. 5.15, comprising many transmitting nodes and one receiver.

This model can also be applied to scenarios where nodes can hear each other, with some considerations for propagation delay. First, the general assumptions of the model are outlined with their implications.

The model is a slotted system for which all transmitted packets have the same length, and each packet (a message or a frame) requires one time unit (a slot) for transmission. All transmitters are synchronized so that each packet's reception begins at an integer time and ends before the next integer time. This assumption introduces two key effects. First, it transforms the system into a discrete-time framework, simplifying the analysis. Second, synchronizing transmitters for slotted arrival at the receiver can be achieved with relatively stable clocks, a small amount of feedback from the receiver, and the incorporation of a consistent guard time between the end of packet transmission and the beginning of the next slot.

Packet arrivals for transmission follow Poisson processes independent from each of m transmitting nodes. Let λ be the overall arrival rate to the system and λ/m be the arrival rate at each transmitting node. This assumption serves the purpose of providing approximate performance criteria.

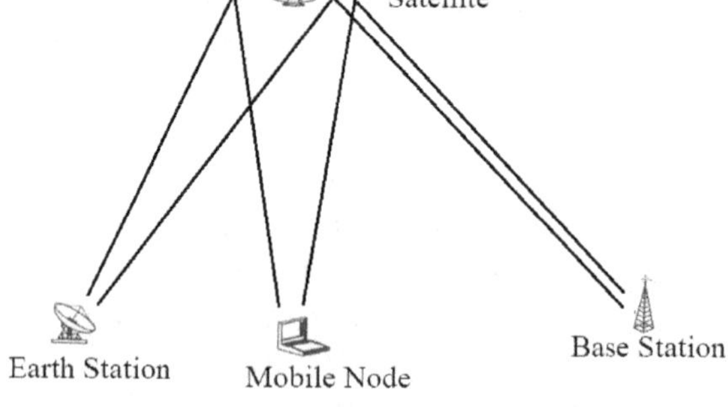

Fig. 5.15. Multiple access model in satellite communications.

If two or more nodes send a packet in a given time slot, a collision occurs, and the receiver obtains no information about the contents or the source of the transmitted packets. If only one node sends a packet in a given slot, the packet is successfully received. This assumption overlooks the potential impact of errors resulting from noise and does not account for capture techniques, wherein a receiver can sometimes capture one transmission among multiple simultaneous transmissions.

At the end of each slot, each node receives immediate feedback from the receiver, indicating whether 0 packet (empty), 1 packet (successful), or more than one packet (collision) were transmitted in that slot. This assumption is notably unrealistic but is made to simplify the analysis.

Each packet involved in a collision must be retransmitted in a later slot, with further retransmissions until the packet is successfully received. A node with a packet requiring retransmission is considered backlogged. This assumption is deemed reasonable for ensuring reliable communication.

The system can be configured to have either a finite set of nodes or an infinite set of nodes. For the finite population with no buffering, if one packet at a node is waiting for transmission or colliding with another packet during transmission, new arrivals at that node are discarded and not transmitted. It implies that new arrivals to backlogged nodes are discarded without any consequence, while packets that have been transmitted must be retransmitted until successful.

Our investigation focuses on multiple access channels characterized by a large number of nodes, relatively low arrival rates, and conditions where time division multiplexing (TDM) is impractical. In such scenarios, the proportion of backlogged nodes is typically small, and new arrivals at backlogged nodes are negligible. Consequently, the delay in a system without buffering should closely resemble that of a system with buffering. Moreover, the delay experienced in an unbuffered system sets a lower bound for the delay observed across a wide array of buffered systems.

For systems featuring an infinite number of nodes without buffering, the delay represents an upper limit on the achievable delay with a finite number of nodes. Specifically, in any multiple access algorithm, each node within a finite set can be viewed as a collection of virtual nodes, with one corresponding to each arriving packet. By independently applying the designated algorithm to each packet, the scenario becomes analogous to one featuring an infinite set of nodes [15, 16].

5.2.2 Slotted ALOHA revisited

For ALOHA systems, each node transmits a newly arriving packet in the first slot right after the packet's arrival, risking occasional collisions but achieving a very small delay if collisions are rare. Slotted ALOHA transmits packets almost immediately with occasional collisions, while TDM avoids collisions at the expense of large delays.

The behavior of slotted ALOHA can be described as a discrete-time Markov chain. Assuming that we have a finite number of users with m, let n be the number of backlogged nodes at the beginning of a given slot. Each of these nodes will transmit a packet in the given slot, independently of each other, with a probability q_r. Each of the $(m - n)$ other nodes will transmit a packet in the given slot with a probability q_a if one or more such packets arrived during the previous slot. Let $P_r(i, n)$ be the probability that i among n backlogged nodes transmit, and $P_a(i, n)$ be the probability that i among $(m - n)$ unbacklogged nodes transmit. We can have

$$P_r(i, n) = \binom{n}{i} (1 - q_r)^{n-i} {q_r}^{i} \tag{5.6}$$

$$P_a(i, n) = \binom{m-n}{i} (1 - q_a)^{m-n-i} {q_a}^{i} \tag{5.7}$$

Between consecutive slots, the state — measured as the number of backlogged packets — increments by the count of new arrivals transmitted by unbacklogged nodes. It is reduced by one if there is no new arrival and one backlogged packet is transmitted. The state remains unchanged if either one new arrival and no backlogged packet or no new arrival and more than one backlogged packet are transmitted.

Consequently, the transition probability from state n to $n + 1$ is expressed as

$$p_{n,n+1} = \left\{ \begin{array}{ll} P_a(0, n) P_r(1, n) & i = -1 \\ P_a(1, n) P_r(0, n) + P_a(0, n) & \\ \quad \cdot [1 - P_r(1, n)] & i = 0 \\ P_a(1, n) [1 - P_r(0, n)] & i = 1 \\ P_a(i, n) & 2 \leq i \leq (m - n) \end{array} \right\} \tag{5.8}$$

It's noteworthy that the state (backlog) can only decrease by a maximum of 1 in a single transition, simplifying the iterative computation of steady-state probabilities, p_n. This process involves calculating p_n for each progressively larger n in terms of p_0 and deriving p_0 as a normalizing constant. Subsequently, the expected number of backlogged nodes can be obtained, and employing Little's theorem, the average delay can be computed.

Unfortunately, this system exhibits peculiar properties, especially with a large number of nodes, and the steady-state results can be quite misleading. To gain some intuitive understanding, consider the desire to choose q_r, a moderately large retransmission probability to avoid substantial delays following collisions. When the arrival rate is small, and only a few packets are involved in collisions, this approach works effectively, and retransmissions tend to be successful. However, in scenarios where the system faces an unfavorable event, causing the number of backlogged packets n to become large enough to satisfy certain conditions of $nq_r \gg 1$, collisions become prevalent in almost every successive slot. As a result, the system remains heavily backlogged for an extended period.

To quantify this phenomenon, let's introduce the drift $D(n)$ in state n as the expected change in backlog over one slot time, starting in state n. Thus, $D(n)$ represents the expected number of new arrivals accepted into the system (i.e., $(m - n)q_a$) minus the number of successful transmissions in the next slot. If P_{suc} is denoted as the expected number of successful transmissions, then the drift is

$$D(n) = (m - n)q_a - P_{\text{suc}} \tag{5.9}$$

where

$$P_{\text{suc}} = P_a(1, n)P_r(0, n) + P_a(0, n)P_r(1, n) \tag{5.10}$$

Define $G(n)$ as the expected number of attempted transmissions in a slot when the system is in state n, called the total arrival rate:

$$G(n) = (m - n)q_a + nq_r$$

if q_a and q_r are small, P_{suc} is approximately as follows:

$$P_{\text{suc}} \simeq G(n)\, e^{-G(n)} \tag{5.11}$$

This approximation can be derived directly from (5.10), using the approximation $(1 - x)^y \simeq e^{-xy}$ for small x in the expressions for P_a and P_r. Similarly, the probability of an idle slot is approximately $e^{-G(n)}$.

Hence, the number of packets in a slot can be well approximated as a Poisson random variable, but the $G(n)$ parameter varies with the state.

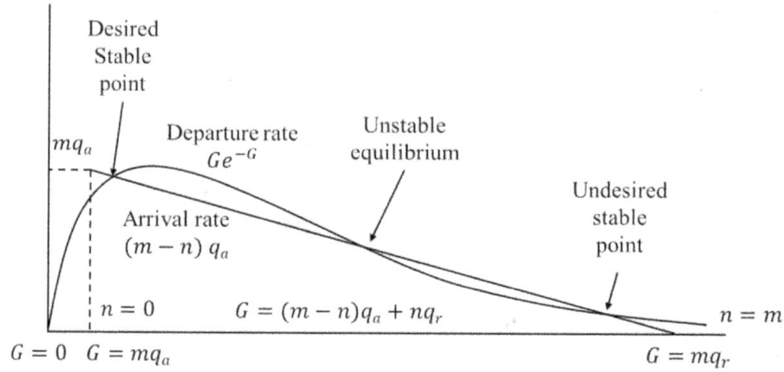

Fig. 5.16. Instability of slotted ALOHA.

Figure 5.16 illustrates (5.9) and (5.11) for the case $q_r > q_a$. The drift is the difference between the curve and the straight line. Since the drift represents the expected change in state from one slot to the next, the system tends to move in the direction of the drift. Consequently, it tends to cluster around the two stable points with rare excursions between the two. The horizontal axis denotes the state n and the attempt rate G, which are related by the linear equation $G = (m - n)\ q_a + nq_r$ with $q_r > q_a$. For n to the left of the unstable point, $D < 0$ and n drifts toward the desired stable point. For n to the right of the unstable point, $D \geq 0$ and n drifts toward the undesired stable point. Two significant conclusions can be drawn from this figure. First, the departure rate P_{suc} is at most $1/e$ for large m. Second, the departure rate is almost 0 for long periods whenever the system jumps to the undesirable stable point.

If we replace the no-buffering assumption with the infinite node assumption, then the attempt rate $G(n)$ becomes $\lambda + nq_r$, and the straight line representing arrivals in Fig. 5.16 becomes horizontal. In this case, once the state of the system passes the unstable equilibrium, it tends to increase without bound. In this case, the corresponding infinite-state Markov chain has no steady-state distribution, and the expected backlog increases without bound as the system continues to run.

From a practical standpoint, if the arrival rate λ is much smaller than $1/e$, and if q_r is moderate, then the system could be expected to remain in the desired stable state for very long periods. In the rare occurrence of a transition to an undesired stable point, the system could be reset with the loss of backlogged packets. However, rather than continuing to analyze this rather flawed system, attention shifts to modifications of slotted ALOHA aimed at resolving this stability issue.

5.2.3 *Stabilization algorithms of slotted ALOHA*

One straightforward approach to achieve stability becomes apparent. The parameter P_{suc} is approximately $G(n)e^{-G(n)}$ with $G = (m - n)q_a + nq_r$, which is maximized at $G = 1$. Therefore, it is desirable to dynamically adjust q_r to maintain the attempt rate at 1. However, the challenge lies in the fact that n is unknown to the nodes and can only be estimated from the feedback. Various strategies exist for estimating n or determining the appropriate value of q_r [17, 22, 23]. Essentially, all of these strategies involve increasing q_r when an idle slot occurs and decreasing q_r when a collision occurs.

Under the no-buffering assumption, the system exhibits a well-defined steady-state behavior for all arrival rates. In the absence of buffering, the system discards many arriving packets, resulting in a very large yet finite delay. Conversely, with the infinite-node assumption, no steady-state distribution exists, leading to an unbounded growth in the expected delay as the system operates over time. Under the infinite-node assumption, no arrivals are discarded, but the delay becomes infinite.

It's important to note that if a system is stable, then for a sufficiently large but finite number of nodes m, the system (viewing each arrival as corresponding to a new virtual node) achieves a smaller expected delay than TDM. This is significant because the delay of TDM, for a fixed overall arrival rate, increases linearly with m.

In the context of the infinite-node model, a multiple access system is considered stable for a given arrival rate if the expected delay per packet (measured as a time average or an ensemble average in the limit of increasing time) is finite. The maximum stable throughput of a multiple access system is defined as the least upper bound of arrival rates for a stable system. For instance, the maximum stable throughput of an ordinary slotted ALOHA is zero. With these definitions, multiple access algorithms that maintain small delays (for a given λ) independently of m do not require the knowledge of the number of nodes.

Returning to the stabilization of slotted ALOHA, note that when the estimate of the backlog is perfect, and $G(n)$ is maintained at the optimal value of 1, then (according to the Poisson approximation) idles occur with a probability $1/e \simeq 0.368$, successes occur with a probability $1/e$, and collisions occur with a probability $1 - 2/e \simeq 0.264$. Thus, the rule for changing q_r should allow fewer collisions than idles. The maximum stable throughput of such a system is at most $1/e$. To see this, note that when the backlog is large, the Poisson approximation becomes more accurate, the success rate is then limited to $1/e$, and thus, the drift is positive for $\lambda > 1/e$.

5.2.4 *Controlled ALOHA with pseudo-Bayesian algorithm*

The pseudo-Bayesian algorithm, introduced independently by researchers [18, 19], provides a clear and effective strategy for stabilizing ALOHA. Unlike slotted ALOHA, this algorithm considers new arrivals as backlogged immediately upon arrival. Instead of being transmitted with certainty in the next slot, these packets are sent with a probability q_r similar to packets involved in prior collisions.

The algorithm operates by maintaining an estimate $\hat{n}$ of the backlog at the beginning of each slot. Each backlogged packet is then transmitted independently with a probability $q_r(\hat{n}) = \min\{1, 1/\hat{n}\}$. Thc minimum operation limits q_r to at most 1 and, subject to this limitation, attempts to achieve an attempt rate $G(n) = nq_r$ of 1. For each k, the estimated backlog at the beginning of slot $k + 1$ is updated from the estimated backlog and feedback for slot k according to the following rule:

$$\hat{n}_{k+1} = \begin{cases} \max[\lambda, \ \hat{n}_k + \lambda - 1] & \text{for idle or success} \\ \hat{n}_k + \lambda + (e-2)^{-1} & \text{for collision} \end{cases} \tag{5.12}$$

The addition of λ to the previous backlog accounts for new arrivals, and the maximum operation ensures that the estimate is never less than the contribution from new arrivals. On successful transmissions, 1 is subtracted from the previous backlog to account for the successful departure. Finally, subtracting 1 from the previous backlog on idle slots and adding $(e-2)^{-1}$ on collisions has the effect of decreasing $\hat{n}$ when too many idles occur and increasing $\hat{n}$ when too many collisions occur. For large backlogs, if $\hat{n} = n$, each of the n backlogged packets is independently transmitted in a slot with a probability $q_r = 1/n$. Thus, $G(n)$ is 1 and, by the Poisson approximation, idles occur with a probability $1/e$ and collisions with a probability $(e-2)/e$ so that decreasing $\hat{n}$ by 1 on idles and increasing $\hat{n}$ by $(e-2)^{-1}$ on collisions maintains the balance between n and $\hat{n}$ on the average.

It is noted that if the *a priori* probability distribution on $\hat{n}$ is Poisson with a parameter $\hat{n} > 1$, then, given an idle or a successful transmission

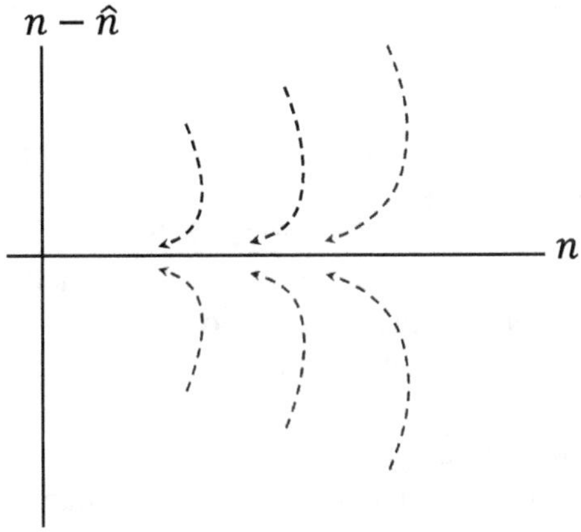

Fig. 5.17. Drift of n and $n - \hat{n}$ for pseudo-Bayesian stabilization.

in slot k, the probability distribution of n_{k+1} is Poisson with a parameter $\hat{n}_k + \lambda - 1$. Given a collision, the resulting distribution of n_{k+1} is not quite Poisson but is reasonably approximated as Poisson with a parameter $\hat{n}_{k+1}$. For this reason, the algorithm is called pseudo-Bayesian.

Figure 5.17 provides some heuristic understanding of why the algorithm is stable for all $\lambda < 1/e$. The state of the system is characterized by n and $\hat{n}$, and the figure shows the expected drift in these variables from one slot to the next. If n and $\hat{n}$ are large and equal, then the expected drift of each is $\lambda - 1/e$, which is negative. On the other hand, if the absolute value of $n - \hat{n}$ is large, then the expected drift of n is positive, but the expected reduction in $|n - \hat{n}|$ is considerably larger. Thus, if the system starts at some arbitrary point in the $(n, \hat{n})$ plane, n might increase on the average for a number of slots, but eventually n and $\hat{n}$ will come closer together and then n will decrease on the average. When the absolute value of $n - \hat{n}$ is large, it approaches 0 faster than n increases.

In applications, the arrival rate λ is typically unknown and slowly varying. Thus, the algorithm must either estimate λ from the time average rate of successful transmissions or set its value within the algorithm to some fixed value. It has been shown [20, 21] that if the fixed value $1/e$ is used within the algorithm, then stability is achieved for all actual $\lambda < 1/e$. Nothing has been proven about the behavior of the algorithm when a dynamic estimate of λ is used within the algorithm.

In some systems, a node receives feedback only about whether or not its own packets were successfully transmitted; it receives no feedback about slots in which it does not transmit. Such limited feedback is sufficient for slotted ALOHA, but it is insufficient for the backlog estimation of the pseudo-Bayesian strategy. The binary exponential backoff algorithm [22, 23] is a stabilization strategy used in Ethernet that employs only this more limited form of feedback.

This strategy is very simple; if a packet has been transmitted unsuccessfully k times, then the probability of transmission in successive slots is set at $q_r = 2^{-k}$ (or is uniformly distributed over the next 2^k slots after the kth failure). When a packet initially arrives in the system, it is transmitted immediately in the next stot.

Unfortunately, this strategy is unstable (i.e., has an infinite expected delay for $m = \infty$) for every arrival rate λ greater than 0 [22].

5.2.5 *Approximate delay for the pseudo-Bayesian algorithm*

An exact analysis of the expected delay for this algorithm, though λ is known, appears to be very difficult, but it is instructive to analyze an approximate model [16, 17]. Assume that λ is known and that the probability of a successful transmission $P_{\text{suc}} = 1/e$ whenever the backlog n is 2 or more, and that $P_{\text{suc}} = 1$ for $n = 1$. This is a reasonable model for very small λ since very few collisions occur and q_r is typically 1. It is also reasonable for large $\lambda < 1/e$ since typically n is large and $\hat{n} \simeq n$.

Let W_i be the delay from the arrival of the ith packet until the beginning of the ith successful transmission. Note that if the system were first-come first-serve, then W_i would be the queueing time of the ith arrival. It is assumed that the time average of W_i for ith arrival is equal to the expected queueing delay W. Let n_i be the number of backlogged packets at the instant before i's arrival; n_i does not include any packet currently being successfully transmitted but does include current unsuccessful transmission. Therefore, W_i can be expressed as

$$W_i = R_i + \sum_{j=1}^{n_i} t_j + y_i \tag{5.13}$$

where R_i is the residual time to the beginning of the next slot, and t_1 (for $n_i > 0$) is the subsequent interval until the next successful transmission is completed. Similarly, t_j, $0 < j \leq n_i$, is the interval from the end of the $(j-1)$th subsequent success to the end of the jth subsequent success. After those n_i successes, y_i is the remaining interval until the beginning of the next successful transmission, i.e., the ith transmission overall. Let p_n denote the probability of n backlogs at a slot boundary.

Observe that for each interval t_j, the backlog is at least two, counting the new ith arrival and the n_i packets already backlogged. Thus, each slot is successful with a probability $1/e$, and the expected value of each t_j is e. Next, observe that Little's formula relates the expected values of W_i and n_i (i.e., W_i is the wait not counting the successful transmission, and n_i is the number in the system not counting successful transmissions in the process). Finally, the expected value of R_i is $1/2$ (the counting time in slots). Thus, taking the expected value of (5.13) and averaging over i, we get

$$W = 1/2 + \lambda e W + E[y] \tag{5.14}$$

$$E[y] = \frac{(e-1)(\lambda - p_1)}{\lambda} \tag{5.15}$$

Finally, we must determine p_1. From the argument above, we see that $\lambda = p_1 + (1 - p_0 - p_1)/e$. Also, state 0 can be entered at a slot boundary only if no arrivals occurred in the previous slot and the previous state was 0 or 1. Thus, $p_0 = (p_0 + p_1)e^{-\lambda}$. Solving for p_1,

$$p_1 = \frac{(1 - \lambda e)(e^{\lambda} - 1)}{[1 - (e-1)(e^{\lambda} - 1)]} \tag{5.16}$$

Combining (5.14) and (5.16) yields

$$W = \frac{e - 1/2}{1 - \lambda e} - \frac{(e^{\lambda} - 1)(e - 1)}{\lambda[1 - (e-1)(e^{\lambda} - 1)]} \tag{5.17}$$

Figure 5.18 plots W for this approximate model and compares it with the queueing delay for TDM with $m = 8$ and 16 nodes. W is the expected waiting time from arrival until the beginning of a successful transmission. It is quite surprising that the delay is so small even for stabilization strategies of this type to be stable; the argument is that the feedback delay will make the estimate $\hat{n}$ less accurate, but $n/\hat{n}$ will still stay close to 1 for large n. For small arrival rates, the delay of stabilized ALOHA is little more than waiting for the next slot, whereas as the arrival rate approaches $1/e$, the delay becomes unbounded.

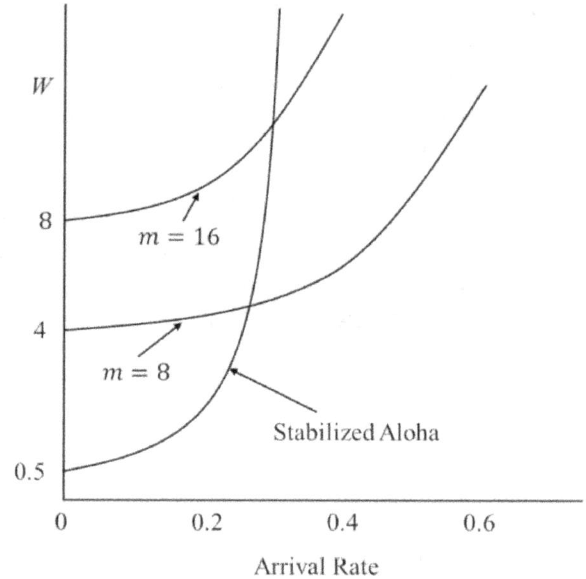

Fig. 5.18. Comparison of expected waiting time W in slots.

5.2.6 *Summary*

This section focuses on exploring the various aspects of random multiple access systems, particularly in the context of slotted ALOHA. It delves into the modeling of random multiple access, revisits the principles of slotted ALOHA, and examines stabilization algorithms designed to validate its performance. Additionally, it discusses the implementation of controlled ALOHA with a pseudo-Bayesian algorithm and provides insights into approximating delays associated with this algorithm.

5.3 Capture-enabled ALOHA Systems

In a capture channel, variations in the levels of received power and the times of arrival allow the receiver to successfully receive a packet even when two or more are transmitted simultaneously. The slotted ALOHA system operating over a capture channel is recognized to be unstable without the use of a control policy [34]. System stabilization is achieved through the implementation of a transmission control algorithm, and the throughput is derived as a function of the capture parameter.

5.3.1 *Introduction*

In conventional ALOHA channels, simultaneous transmission by two or more users leads to a collision, necessitating transmission of the unsuccessful packets according to a transmission algorithm. However, in a radio environment, users are often at different distances from the central receiver, resulting in substantially different power levels for their received signals [24]. The packet arriving with the highest energy now has a good chance of being detected accurately. Similarly, in some spread-spectrum random access systems, the earliest arriving packet has a high probability of being detected [25, 26]. A model that encompasses these various capture phenomena is presented, along with the formulation of a stable control policy. As anticipated, the capture mechanism enhances the overall system throughput of the standard ALOHA system.

The throughput analysis conducted in certain studies [24, 26] relies heavily on the composite Poisson traffic assumption. However, since the actual composite traffic does not adhere to a Poisson distribution, these systems frequently exhibit instability. Previous research has investigated the impact of different types of errors, including erasures, captures, and noise, on the slotted controlled ALOHA system [27, 28, 36]. Additionally, the capture effect within the framework of collision resolution algorithms has been examined in related studies [29].

In Ref. [30], the conditions leading to instability of the infinite user capture ALOHA model were identified in the absence of transmission control. It was subsequently noted that other researchers also investigated these and related questions concerning both uncontrolled and controlled ALOHA systems with multiple reception capability [31]. This note emphasizes the stabilization of the capture ALOHA system using a decentralized multiplicative control algorithm and investigates the stability region as a function of the capture parameter.

In the following, a Markov model is introduced to describe the fluctuation of the backlog. The stability issue is then addressed. Subsequently, the transmission control policy is formulated, and the throughput is analyzed as a function of the capture parameter, with comparisons to results from Ref. [32]. The analysis demonstrates that the transmission control policy ensures a stable system with a finite backlog. Finally, the findings are summarized.

5.3.2 *System model*

5.3.2.1 *User and channel model*

Assuming an infinite number of independent and identical users in the system, each user is limited to having at most one packet waiting to be transmitted at any given time, and there are no queues associated with the users. The cumulative incoming traffic in slot t, denoted as Y_t, follows a Poisson distribution with a mean rate of λ packets per slot.

The channel is assumed to be noiseless. The time axis is divided into constant intervals called slots. The messages are transmitted in packets whose lengths are equal to or less than the length of a slot. The channel is of capture type in the sense that a receiver can, with some positive probability, decode one of several simultaneous signals on the channel. This is a broad definition in the sense that it includes capture effects due to power variations (random or fixed) and due to different times of arrivals in spread-spectrum communication environments. Propagation delays are negligible or ignored. All users in the system will be informed of the outcome of the feedback signal, Z_t, occurring immediately at the end of a time slot. Additionally, it is assumed that the identity of a captured packet is known to all the users. One way to arrange for this is to have the central receiver broadcast the identity of all successfully received packets.

5.3.2.2 *Channel access model*

This is the procedure executed by each user to schedule the transmissions of new arrival packets. We will consider two protocols, namely immediate first-transmission (IFT) and delayed first-transmission (DFT), which are common to ALOHA systems. Under the IFT protocol, new packets that arrive during slot $t-1$ are first transmitted in slot t with a probability of one. By slot t, the time segment is $(t, t+1]$. Under the DFT protocol, the new packets join the backlog before their first transmissions are attempted, and each packet is independently transmitted in slot t with a probability f_t.

During each slot, each user with a backlogged packet must decide whether or not to transmit the packet in that slot. Let Z_t be the feedback random variable during slot $(t, t+1]$. The user is informed via the common feedback at the end of slot t that slot t was either the **empty** slot ($Z_t = 0$), the **capture** slot ($Z_t = 1$), or the **collision** slot ($Z_t = e$). In addition, at the end of slot t, each user learns whose packet is captured. When two or more users transmit a packet but do not successfully capture the channel, each of these users joins the backlog and awaits another transmission opportunity at a later time.

In our model, we define the capture probability, C_n, to be the probability that one packet is successfully received from among n simultaneous packets on the channel. C_n, therefore, depends on the number of users transmitted in the same slot.

5.3.2.3 *Markov chain model*

In the ALOHA systems with a capture, it is important to determine the statistics of the capture. In spread-spectrum random access systems, the capture probability may be defined to be the probability that one packet is correctly decoded, given several users in a slot, or to be the probability that the receiver captures the first arriving packet out of many contending ones, each arriving at a random time instant. With these definitions, a capture is dependent on the coding, processing gain, and jamming margin of the given system.

We know that $C_0 = 0$ and $C_1 = 1$ by definition. If $C_n = 1$ for $n > 1$, it is called a perfect capture channel and otherwise a non-perfect capture one. Different systems may yield different statistics of capture [27, 29, 30] in general:

$$C_n = ab^n \text{ for some } a \text{ and } b$$

The channel transition probabilities, which describe the dynamics of the user backlogs, are derived and provided in Appendix A.5.3, where $\{N_t\}$ denotes the number of backlogged packets at time t.

Let $S(n, f)$ be the probability of a successful packet in a time slot given that $(N_t = n, f_t = f)$, i.e.,

$$S(n, f) = P[Z_t = 1 | N_t = n, f_t = f]$$

It is shown that from the transition probabilities,

$$S(n, f) = \begin{cases} \sum_{j=0}^{\infty} \sum_{m=0}^{n} \frac{\lambda^j e^{-\lambda}}{j!} \binom{n}{m} f^m (1-f)^{n-m} C_{j+m}, & j+m \geq 2, n \geq 1, \\ & \text{for IFT} \\ \sum_{m=1}^{n} \binom{n}{m} f^m (1-f)^{n-m} C_m \text{ for DFT} & \end{cases}$$

Note that if $f = f(n)$ depends on n, then

$$\lim_{n\to\infty} nf = \mu > 0$$

and from the Poisson approximation to binomial, one has

$$\lim_{n\to\infty} \binom{n}{m} f^m (1-f)^{n-m} = \frac{\mu^m e^{-\mu}}{m!} \quad \text{for } m = 0, 1, 2, \ldots$$

This Poisson approximation to binomial is possible only if f is controlled according to the variation of n. With this assumption, one may show that with $\mu = nf$,

$$S(\mu) = \begin{cases} \sum_{l=0}^{\infty} \sum_{m=0}^{\infty} \frac{\lambda^l \mu^m e^{-(\lambda+\mu)}}{l!m!} C_{l+m}, & l+m \geq 2 \quad \text{for IFT} \\ \sum_{k=1}^{\infty} \frac{\mu^k e^{-\mu}}{k!} C_k & \text{for DFT} \end{cases}$$

Note that $S(\mu)$ for the DFT protocol follows from the standard Poisson assumption first introduced by Abramson [35], but this is not valid at all unless f is controlled, depending upon n [37]. In fact, the channel backlog in each slot is fluctuating because of the randomness of the arrival process. Hence, at least intuitively, the ALOHA system with capture is not stable for an infinite user model if we cannot maintain $\mu = nf$ at a nearly constant level.

In the following section, we will prove that the Markov chain $\{N_t\}$, for both protocols, is not ergodic, and thus, the system is not stable. Accordingly, neither $S(n, f)$ nor $S(\mu)$ is an achievable channel throughput, and so, the throughput in a steady state is zero.

5.3.3 *Stability issue*

ALOHA systems (pure or slotted) in IFT or DFT protocols are unstable, so the throughput is not achievable. It is shown that the capture-enabled ALOHA system is also unstable unless $\lim_{n\to\infty} C_n > 0$. We will make use of results due to Kaplan [38].

Proposition. *For any* f, $0 < f < 1$ *and any* C_n, $0 < C_n < 1$ *such that* $\lim_{n\to\infty} C_n = 0$, *the Markov chain* $\{N_t\}$ *of backlog dynamics with state space* $\{0, 1, 2, \ldots\}$ *is not ergodic.*

Proof. Note that the downward transitions of the chain $\{N_t\}$ are uniformly bounded since

$$P[N_{t+1} = k|N_t = n, f_t = f] = 0 \quad \text{whenever} \quad k < n - 1.$$

Now, define the drift function

$$D(n, f) = E[N_{t+1} - N_t|N_t = n, f_t = f]$$

Hence, it suffices to show that there are some positive constants ϵ and N such that

$$D(n, f) \geq \epsilon \text{ for all } n \geq N$$

It is easy to see that with the result of $S(n, f)$,

$$D(n, f) = \lambda - S(n, f)$$

So, it suffices to show that there are some positive constants ϵ and N such that

$$S(nf) \leq \lambda - \epsilon \quad \text{for all} \quad n \geq N$$

For an IFT model, it can be expressed as,

$$\begin{aligned} S(n, f) &= e^{-\lambda} nf(1-f)^{n-1} + \lambda e^{-\lambda}(1-f)^n \\ &+ e^{-\lambda} \sum_{m=2}^{n} \binom{n}{m} f^m (1-f)^{n-m} C_m \\ &+ \lambda e^{-\lambda} \sum_{m=1}^{n} \binom{n}{m} f^m (1-f)^{n-m} C_{m+1} + \sum_{j=2}^{\infty} \frac{\lambda^j e^{-\lambda}}{j!} \\ &\times \sum_{m=0}^{n} \binom{n}{m} f^m (1-f)^{n-m} C_{m+j} \end{aligned}$$

and for a DFT model,

$$S(n, f) = nf(1-f)^{n-1} + \sum_{m=2}^{n} \binom{n}{m} f^m (1-f)^{n-m} C_m$$

Now, let

$$\sum_{m=m_0}^{n} \binom{n}{m} f^m (1-f)^{n-m} C_m$$

$$= \sum_{m=m_0}^{M} \binom{n}{m} f^m (1-f)^{n-m} C_m$$

$$+ \sum_{m=M+1}^{n} \binom{n}{m} f^m (1-f)^{n-m} C_m$$

But there exists some M such that $m > M$ implies C_m is made arbitrarily small, and $C_m \leq c$ for any $c > 0$:

$$\sum_{m=M+1}^{n} \binom{n}{m} f^m (1-f)^{n-m} C_m \leq c$$

Assume that n is sufficiently large and $m_0 \leq M < \infty$, using $\frac{1}{n^m}\binom{n}{m} = \frac{1}{m!}$ and $\lim_{n\to\infty} (nf)^m (1-f)^{n-m} = 0$ for $0 < f < 1$, then

$$\sum_{m=m_0}^{M} \binom{n}{m} f^m (1-f)^{n-m} C_m$$

$$= \sum_{m=m_0}^{M} \frac{1}{m!} (nf)^m (1-f)^{n-m} C_m \to 0$$

Thus,

$$\sum_{m=m_0}^{n} \binom{n}{m} f^m (1-f)^{n-m} C_m \leq c$$

Note also that

$$e^{-\lambda} \leq 1, \quad \lambda e^{-\lambda} \leq e^{-1} \quad \text{and} \quad \sum_{j=2}^{\infty} \frac{\lambda^j e^{-\lambda}}{j!} \leq 1$$

Hence, for any fixed $0 < f < 1$ and $\lambda > 0$, there exists $\epsilon > 0$ such that

$$S(n, f) \leq \lambda - \epsilon$$

Therefore, the Markov chain is non-ergodic and so the system is unstable. □

5.3.4 *Stabilization control algorithm*

5.3.4.1 *Geometric capture model*

The channel access protocol is executed independently by each user to schedule the transmissions of newly arrived packets. The IFT model is only considered in the following, but the DFT model is simply modified with $G = \mu$ instead of $G = \lambda + \mu$. Any new packets that arrive during slot $t - 1$ are immediately transmitted in slot t with a probability one. Slot t represents the time segment $(t, t + 1]$.

Recall C_n as the probability of successfully receiving one packet from among n simultaneous packets on the channel. One commonly employed capture model is a form of geometric capture probability:

$$C_n = Q^n, 0 \leq Q \leq 1, \text{ for } n \geq 2 \tag{5.18}$$

where Q, the capture parameter, depends on the system model [25, 26].

5.3.4.2 *Transmission control algorithm*

The decentralized procedure is executed by the backlogged users in the system to schedule the transmissions. Remember that f_t is the transmission probability for the backlogged users at time t and Z_t is the channel feedback information given to all users in the system at time $t + 1$, then f_t must be a function of the channel output history $\{Z(s)\text{: } s < t\}$ for each t, and $0 \leq f_t \leq 1$. Consider the transmission control algorithm for the backlogged user at time $t + 1$, and the transmission probability is defined as

$$f_{t+1} = \{a^{\gamma}(Z_t) f_t \wedge \beta\} \tag{5.19}$$

for some positive constants γ, β, a(0), a(1), and a(e). The notations, $(x \wedge y)$ and $(x \vee y)$, denote $\min(x, y)$ and $\max(x, y)$, respectively. The goal of the control policy is to steer the transmission traffic in the direction of the optimal point. Consequently, the choice of γ, β, and a(0), a(1), and a(e) is closely associated with the dynamics of the traffic intensity.

Consider the number of backlogged users at time t, N_t, then the process $\{N_t,\ f_t\}$ is a two-dimensional Markov process with state space $\{[0, 1, 2, 3...] \times (0, 1)\}$.

In the previous section, it is shown that the Markov chain obtained, using the number of backlogged users at each slot as the state, is non-ergodic.

The goal is to identify the stability region when a decentralized transmission control policy is employed. However, it is not obvious how the transmission control will change the capacity region.

Define for $j = 0, 1, e$,

$$P_j = P[Z_t = j | N_t = n,\ f_t = f] \tag{5.20}$$

Then it can be shown that with $\theta = 1 - Q$,

$$P_0 = e^{-\lambda}(1 - f)^n \tag{5.21}$$

$$P_1 = \theta e^{-\lambda} n f (1 - f)^{n-1} + \theta \lambda e^{-\lambda}(1 - f)^n + (1 - \theta f)^n e^{-\lambda\theta} - e^{-\lambda}(1 - f)^n \tag{5.22}$$

$$P_e = 1 - [\theta e^{-\lambda} n f (1 - f)^{n-1} + \theta \lambda e^{-\lambda}(1 - f)^n + (1 - \theta f)^n e^{-\lambda\theta}] \tag{5.23}$$

Let $\psi = \ln[(n \vee 1) f]$ with $\mu = nf$.

Now, the approximate version, $\widehat{P_j}$ of P_j, can be expressed as follows after using the local Poisson approximation [28] (see Appendix B.5.3)

$$\hat{P}_0 = e^{-G} \tag{5.24}$$

$$\hat{P}_1 = (\theta G - 1)\, e^{-G} + e^{-\theta G} \tag{5.25}$$

$$\hat{P}_e = 1 - [\theta G e^{-G} + e^{-\theta G}] \tag{5.26}$$

where $G = \lambda + \mu = \lambda + e^{\psi}$.

The parameters $a(0)$, $a(1)$, and $a(e)$ are chosen such that the dynamics of the traffic intensity will be maintained at the level that yields the maximum throughput.

With $\psi_t = \ln\{(N_t \vee 1) f_t\}$ and (5.24–5.26), the limiting mean drift function can be obtained:

$$\lim_{\substack{n\to\infty \\ \text{or} \\ f\to 0}} E[\psi_{t+1} - \psi_t | N_t = n,\ f_t = f] = \gamma \hat{m}(\psi) \tag{5.27}$$

where

$$\hat{m}(\psi) = c_0 \hat{P}_0 + c_1 \hat{P}_1 + c_e \hat{P}_e \tag{5.28}$$

and $c_0 = \ln[a(0)]$, $c_1 = \ln[a(1)]$, and $c_e = \ln[a(e)]$.

In order for ψ_t to move an optimally stable point ψ^*, c_i for $i = 0, 1, e$ should be chosen to have the following conditions:

$$\hat{m}(\psi) = 0 \quad \text{for} \quad \psi = \psi^*$$
$$\hat{m}(\psi) < 0 \quad \text{for} \quad \psi > \psi^*$$
$$\hat{m}(\psi) > 0 \quad \text{for} \quad \psi < \psi^*$$

This objective is achieved with the following choice of parameters:

$$c_0 = \frac{1 - (G^* e^{-G^*} + e^{-G^*})}{1 + e^{-G^*} - (\theta G^* e^{-G^*} + e^{-G^*})} \tag{5.29}$$

$$c_1 = 0 \tag{5.30}$$

$$c_e = \frac{e^{-G^*}}{1 + e^{-G^*} - (\theta G^* e^{-G^*} + e^{-G^*})} \tag{5.31}$$

Table 5.1 gives the system parameters in relation to the capture parameter Q. When no capture exists ($Q = 0$), the parameters align with those of the traditional slotted ALOHA. Notably, the parameters are insensitive to Q for $Q \leq 0.3$, but they are very sensitive for $Q \geq 0.6$.

The choice of β depends on the capture parameter Q; following the development in Ref. [39], β is chosen to be

$$\beta(\theta) = \frac{[1 + (1 - \lambda)\theta]e^{-\lambda} - \theta e^{-\lambda\theta}}{[1 + (2 - \lambda)\theta]e^{-\lambda} - \theta^2 e^{-\lambda\theta}} \tag{5.32}$$

Note $\beta(0) = 1$ and $\beta(1) = \frac{1-\lambda}{2-\lambda} \leq \frac{1}{2}$. It is expected that the choice of γ will affect the rate of convergence and delay. Experimental results suggest that values between 0.1 and 0.7 yield reasonable performance. For the proof of stability, it is sufficient that $\gamma > 0$. With these choices, it follows from the definition that

$$\hat{m}(\psi) = c_0 e^{-G} + c_e[1 - (\theta G e^{-G} + e^{-\theta G})] \tag{5.33}$$

with $G = \lambda + e^{\psi}$. $\hat{m}(\psi)$ has the following properties.

Table 5.1. System parameter, optimal traffic, and capacity with Q parameter.

Q	G^*	S^*
0.00	1.000200	0.3678795
0.10	1.005400	0.3697870
0.20	1.023098	0.3758491
0.30	1.055849	0.3867780
0.40	1.108671	0.4036872
0.50	1.188268	0.4283534
0.60	1.308150	0.4637134
0.70	1.491668	0.5149133
0.80	1.765286	0.5918868
0.90	2.392178	0.7176850
0.91	2.488173	0.7349055
0.92	2.597041	0.7533827
0.93	2.722147	0.7732963
0.94	2.866582	0.7948757
0.95	3.039557	0.8184243
0.96	3.254815	0.8443622
0.97	3.533480	0.8733132
0.98	3.929426	0.9063099
0.99	4.613717	0.9454535

Lemma 1.

(a) *For any* λ *and any* $0 < \theta \leq 1$, *the function* $\hat{m}(\psi)$ *is strictly decreasing in* ψ.

(b) *For any* $0 < \theta \leq 1$, *there exists a unique* $\psi^* \in R = \{$*set of real numbers*$\}$, *such that* $\hat{m}(\psi^*) = 0$.

Proof. (a) This is simply implied by the inequality

$$\frac{\partial}{\partial \psi}\hat{m}(\psi) = G'\{-c_0 e^{-G} + c_e(\theta G e^{-G} + \theta(e^{-\theta G} - e^{-G}))\} < 0 \quad \forall \psi$$

because $G' = e^{\psi} > 0$ and $e^{-\theta G} - e^{-G} \geq 0$ for $0 < \theta \leq 1$ and $c_0 > 0$ and $c_e < 0$.

(b) The existence of a solution in R follows from the continuity of $\hat{m}(\psi)$ and because $\hat{m}(+\infty) < 0$ and $\hat{m}(-\infty) > 0$. Uniqueness comes from the strict monotonicity of $\hat{m}(\psi)$. □

5.3.5 *Stability analysis*

In this section, the Markov process $X(t) = (N_t, f_t)$ is confirmed to be stable using the technique established by Mikhailov [40] (see Appendix C.5.3). Mikhailov's result established a simple and intuitive technique to prove the stability of the multidimensional Markov process. The main difficulty is in choosing a suitable Lyapunov function. If the stochastic Lyapunov function is radially linear and sufficiently smooth, then the stability criterion is determined by the inner product of the limiting mean drift vector function and the gradient of the Lyapunov function.

Theorem 1. *If $\lambda < S^*$ and $0 < \gamma$, then the Markov process $X(t) = (N_t, f_t)$ is stable.*

Note that (N_t, ψ_t) has an equivalent representation (N_t, f_t) since it is an instantaneous function of both N_t and f_t. The state space for the Markov process $X(t) = (N_t, \psi_t)$ is now $S = \{x_1 \geq 0$ and $-\infty < x_2 < +\infty\}$.

Proof. We use Mikhailov's result. First, we show that the Markov process $X(t)$ satisfies (i) and (ii) and then we present a Lyapunov function that satisfies (a)–(d) in Appendix C.5.3.

Condition (i) is established in Lemma 1, which, along with Lemma 2, validates conditions (a)–(c), while Lemma 3 verifies condition (d), and thus Proposition 1 completes the theorem. □

Lemma 1.

$$E\left[(N_{t+1} - N_t)^2 \,\middle|\, N_t = n,\, f_t = f\right] < 2E\left[Y_t^2\right] + 2$$

$$E\left[(\psi_{t+1} - \psi_t)^2 \,\middle|\, N_t = n,\, f_t = f\right] < 4E\left[Y_t^2\right] + 4 + 2\gamma^2 \max_i c_i^2$$

Proof. Use the fact that for $a > 0$ and $b > 0$, $(a + b)^2 \leq 2(a^2 + b^2)$.

Define

$$m_1(n, f) = E[N_{t+1} - N_t | N_t = n, f_t = f]$$

and

$$m_2(n, f) = E[\psi_{t+1} - \psi_t(t) | N_t = n, f_t = f]$$

Then the mean drift vector function is with $X = (n, f)$

$$M(X) = (m_1(X), m_2(X))$$

The following lemma shows that the process also satisfies the condition (ii) in Appendix C.5.3. □

Lemma 2. *With* $X = (n, \psi)$

$$\hat{m}_1(\psi) = \lim_{\|X\|\to\infty} m_1(X) = \lambda - \hat{S}(\psi)$$

and

$$\hat{m}_2(\psi) = \lim_{\|X\|\to\infty} m_2(X) = \gamma\,\hat{m}(\psi)$$

Proof. See Proposition B.2 in Appendix B.5.3 and (5.27).

Then the limiting mean drift vector function is

$$\hat{M}(X) = (\hat{m}_1(\psi),\ \hat{m}_2(\psi))$$

Introduce a test function with $X = (x_1, x_2)$

$$V(X) = x_1 + \alpha \ln \cosh(x_2 - \psi^*) \text{ for some } \alpha > 0$$

where

$$\psi^* = \ln(G^* - \lambda)$$

and α will be specified below.

Using power series expansions, we can establish the following facts:

$$\ln \cosh(x) = \frac{x^2}{2} + O(x^4) \text{ for } |x| < 1 \tag{5.34}$$

$$\ln \cosh(x) = |x| - \ln 2 + O(e^{-2|x|}) \text{ for } |x| > 1 \tag{5.35}$$

It is obvious that $V(X)$ satisfies the conditions (a)–(c). The following lemma verifies that $V(X)$ also satisfies (d) in Appendix C.5.3. □

Lemma 3. *With* $X = (n, \psi)$,

$$LV(X) \to \hat{M}(X) \text{ grad } V(X) \quad \text{for } \|X\| \to \infty$$

Proof. Let $\Delta V(t) = V(X(t+1)) - V(X(t))$ and $\Delta X(t) = X(t+1) - X(t)$. Then with an arbitrary r_1 we have

$$\begin{aligned} LV(X) = {} & E\,[\Delta V(t) I(\Delta X(t) \le r_1) \mid X(t) = X] \\ & + E[\Delta V(t) I(\Delta X(t) > r_1) | X(t) = X] \end{aligned} \tag{5.36}$$

□

Since $V(X)$ satisfies the Lipschitz condition and $X(t)$ satisfies (i) and (ii), it may be shown that

$$E\left[\Delta V(t) I(\Delta X(t) > r_1) \mid X(t) = X\right] = O(r_1^{-1}) \tag{5.37}$$

Since $\frac{\partial^2 V(x_1,x_2)}{\partial x_2^2} = O(e^{-2|x_2|})$ for all x_2, we can show that

$$E\left[\Delta V(t) I\left(\Delta X(t) \le r_1\right) \mid X(t) = X\right] = M(X) \text{grad} V(X) + O(r_1^{-1}) \tag{5.38}$$

Hence, using (5.35)–(5.38) and noting that $M(X) \to \hat{M}(X)$ for $\|X\| \to \infty$, we arrive at the result.

Note that

$$\hat{M}(X) \text{grad } V(X) = \lambda - \hat{S}(\psi) + \gamma\, \alpha \tanh(\psi - \psi^*) \widehat{m}(\psi) \tag{5.39}$$

Proposition 1. *If* $\lambda < S^* - \epsilon$ *for any arbitrarily small* $\epsilon > 0$ *and* $\gamma > 0$, *then*

$$\widehat{M}(X) \text{ grad } V(X) < -\epsilon/2 \text{ for } \forall X \in S \tag{5.40}$$

Let

$$D(\psi) = \lambda - \hat{S}(\psi) + \gamma\, \alpha \tanh\left(\psi - \psi^*\right) \widehat{m}(\psi) \tag{5.41}$$

By the continuity of $\hat{S}(\psi)$, it follows that there exists a δ such that if $|\psi - \psi^*| < \delta$, then $S^* - \hat{S}(\psi) < \epsilon/2$. Now, consider two cases: $|\psi - \psi^*| < \delta$ and $|\psi - \psi^*| \ge \delta$. Note that

$$\tanh(\psi - \psi^*) \hat{m}(\psi) \le 0\ \forall \psi$$

Therefore, if $|\psi - \psi^*| < \delta$, then

$$D(\psi) \le \lambda - \hat{S}(\psi) \le \lambda - S^* + \epsilon/2 = -\epsilon/2 \tag{5.42}$$

For $|\psi - \psi^*| \ge \delta$, we can show that

$$\left|\tanh(\psi - \psi^*)\right| \ge \delta > 0$$

and that there exists a $\sigma > 0$ such that

$$\left|\hat{m}(\psi)\right| > \sigma > 0$$

Hence, by choosing $\alpha = \lambda + 0.5\epsilon/\gamma\, \delta\sigma$, we have

$$D(\psi) < -\epsilon/2 \quad \text{for } |\psi - \psi^*| \ge \delta$$

Combining the above result with (5.42), Proposition 1 is proved.

Now, according to Mikhailov's result, the Markov process.

$X(t) = (N_t, \psi_t)$ or $X(t) = (N_t, f_t)$ is stable, so Theorem 1 is proved.

If the choice of a^γ is such that, for some relatively prime integer $i, j, a^\gamma = (v^i, 1, v^j)$ where $v = v(\theta)$ such that $0 < v < 1$ and $0 < \theta \leq 1$, then the process $X(t) = (N_t, f_t)$ is a Markov chain on a denumerable state space [39]; $\{(n, \beta v^k): n, k$ nonnegative integers$\}$. It is then easily shown that if $\lambda < S^*$ and $0 < \gamma$, then the Markov chain $X(t) = (N_t, f_t)$ is ergodic. The positive recurrence follows from Mikhailov's theorem, and the ergodicity follows from the uniform bound on the mean drift of the chain.

5.3.6 *Throughput performance*

Having specified the control policy, let us find the achievable maximum throughput. Let $S(n, f)$ be the probability of a successful packet in a time slot given that $(N_t = n, f_t = f)$, i.e.,

$$S(n, f) = P\left[Z_t = 1 | N_t = n, f_t = f\right]$$

Then it is shown that

$$S(n, f) = \sum_{j=0}^{\infty} \sum_{m=0}^{n} \frac{\lambda^j e^{-\lambda}}{j!} \binom{n}{m} f^m (1 - f)^{n-m} C_{j+m} \tag{5.43}$$

Using Poisson approximations to binomial, we may approximate $S(n, f)$ by

$$\hat{S}(\psi) = \overline{S}(G) = \theta G e^{-G} + e^{-\theta G} - e^{-G} \tag{5.44}$$

where $G = \lambda + e^{\psi}$. Since $\overline{S}(G)$ in (5.44) is achievable, we find the maximum throughput. Letting $(\partial \overline{S}(G)/\partial G) = 0$, one can get

$$G = 1 + \theta^{-1} - e^{(1-\theta)G} \tag{5.45}$$

Note that the left-hand side is increasing, while the right-hand side is decreasing in G. Since $S''(G^*) < 0$, for $0 < \theta \leq 1$, then $G^* = 1 + \theta^{-1} - e^{(1-\theta)G^*}$ is the unique solution for $\overline{S}(G)$. Thus, with G^*, the maximum achievable throughput for the slotted ALOHA system with capture is found to be

$$S^* \equiv \hat{S}(\psi^*) = \left(\theta G^* - 1\right) e^{-G^*} + e^{-\theta G^*} \tag{5.46}$$

Figure 5.19 shows the trend of the throughput versus traffic intensity for some specific capture parameters. Table 5.1 provides the

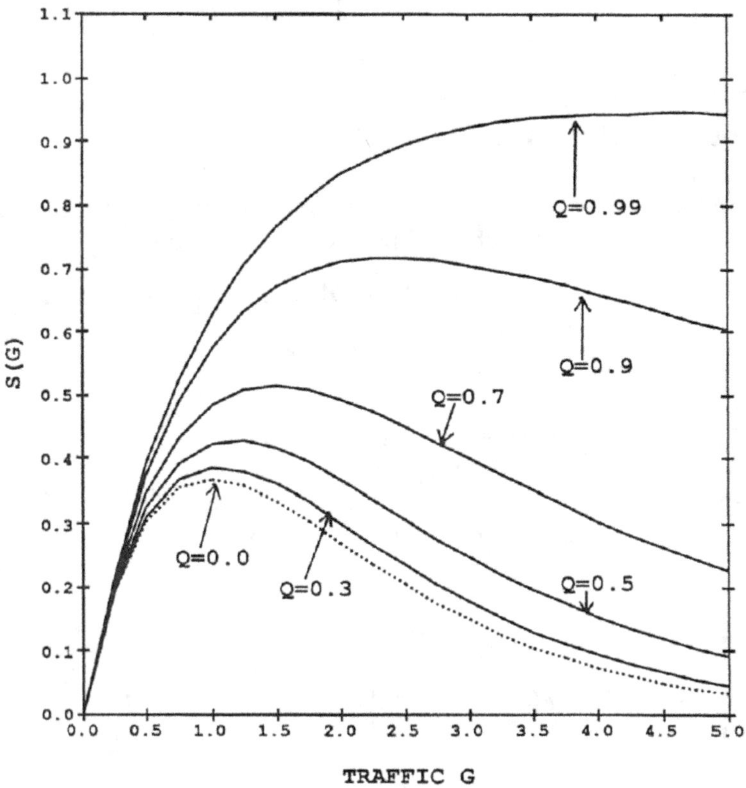

Fig. 5.19. Throughput versus total traffic intensity for different Q.

optimal channel traffic intensity and capacity as Q varies. As Q approaches unity, both G^* and S^* increase toward 5 and 1, respectively. At $Q = 0$, $G^* = 1$, resulting in $S^* = e^{-1}$, representing the maximum throughput for the slotted ALOHA system with no capture. The maximum achievable throughput in this scheme can be arbitrarily close to 1, which is expected.

5.3.7 *Summary*

The capture-enabled ALOHA system has been thoroughly examined using a Markovian model. Addressing the system's instability for a broad range of capture probabilities necessitated the implementation of transmission control policies. The achievable channel capacity was determined as a function of the capture parameter, showcasing substantial improvements even when the capture level was low.

Appendix A.5.3. Markov Transition Probability

The two-dimensional channel transition probabilities $\{N_t, f_t\}$ that characterize the dynamics of the channel backlog are as follows:

(a) **An IFT model with the general capture probability of C_n:**

$$P[N_{t+1} = k, Z_t = 0 | N_t = n, f_t = f] = e^{-\lambda}(1-f)^n \quad \text{for } k = n. \tag{A.1}$$

$$P[N_{t+1} = k, Z_t = 0 | N_t = n, f_t = f]$$
$$= \begin{cases} e^{-\lambda} n f (1-f)^{n-1} + e^{-\lambda} \cdot & \text{for } k = n-1, \\ \quad \sum_{m=2}^{n} I(n \geq 2) \binom{n}{m} f^m (1-f)^{n-m} C_m & \\ \lambda e^{-\lambda}(1-f)^n + \lambda e^{-\lambda} \cdot & \text{for } k = n \\ \quad \sum_{m=1}^{n} I(n \geq 1) \binom{n}{m} f^m (1-f)^{n-m} C_{m+1} & \\ \frac{\lambda^j e^{-\lambda}}{j!} \sum_{m=0}^{n} I(n \geq 0) \cdot & \text{for } k = n-1+j, \\ & j \geq 2. \\ \quad \binom{n}{m} f^m (1-f)^{n-m} C_{m+j} & \end{cases} \tag{A.2}$$

$$P\left[N_{t+1} = k, Z_t = 0 | N_t = n, f_t = f\right]$$
$$= \begin{cases} e^{-\lambda} \sum_{m=2}^{n} I(n \geq 2) \binom{n}{m} \cdot & \text{for } k = n, \\ \quad f^m (1-f)^{n-m} (1 - C_m) & \\ \lambda e^{-\lambda} \sum_{m=1}^{n} I(n \geq 1) \binom{n}{m} \cdot & \text{for } k = n+1 \\ \quad f^m (1-f)^{n-m} (1 - C_{m+1}) & \\ \frac{\lambda^j e^{-\lambda}}{j!} \sum_{m=0}^{n} I(n \geq 0) \binom{n}{m} \cdot & \text{for } k = n+j, \quad j \geq 2. \\ \quad f^m (1-f)^{n-m} (1 - C_{m+j}) & \end{cases} \tag{A.3}$$

$$P[N_{t+1} = k, Z_t = i | N_t = n, f_t = f] = 0 \quad \text{for } k < 0 \quad \forall i \quad \text{and} \quad n \geq 0 \tag{A.4}$$

(b) **A DFT model with the general capture probability of C_n:**

$$P\left[N_{t+1}=k, Z_t=e|N_t=n, f_t=f\right]$$
$$=\begin{cases} e^{-\lambda}\sum_{m=2}^{n}\binom{n}{m}f^m(1-f)^{n-m}(1-C_m) & \text{for } k=n,\\ \lambda e^{-\lambda}\sum_{m=1}^{n}\binom{n}{m}f^m(1-f)^{n-m}(1-C_{m+1}) & \text{for } k=n+1,\\ \frac{\lambda^j e^{-\lambda}}{j!}\sum_{m=0}^{n}\binom{n}{m}f^m(1-f)^{n-m}(1-C_{m+j}) & \text{for } k=n+j,\\ & j\geq 2. \end{cases} \tag{A.5}$$

Appendix B.5.3

In this appendix, Propositions B.1–B.3 are stated upon Lemma B.1 and can be validated upon Refs. [28, 29, 33].

Lemma B.1. *For any* $0<\theta\leq 1$ *and* $0\leq f\leq 1$

$$0\leq e^{-\theta nf}-(1-\theta f)^n\leq \frac{8e^{-2}}{n}$$

$$0\leq e^{-\theta nf}-(1-\theta f)^n\leq 2\theta e^{-1}f$$

Proof. Use the fact that $\theta f\leq \theta\beta \leq 1/2$ for $0<\theta\leq 1$.

With P_0, P_1, P_e, $\hat{P}_0$, $\widehat{P_1}$, and $\widehat{P_e}$ as defined in Section 5.3.4.2, the following three propositions may be established. □

Proposition B.1.

$$|P_0-\hat{P}_0|\leq \frac{8e^{-2}}{n} \quad \text{and} \quad |P_0-\hat{P}_0|\leq 2e^{-1}f$$

Proposition B.2.

$$|P_1-\hat{P}_1|\leq \frac{1}{n}[16e^{-2}+62\theta e^{-3}] \quad \text{and}$$

$$|P_1-\hat{P}_1|\leq f\left[10e^{-2}+2(1+\theta)e^{-1}\right]$$

Proposition B.3.

$$|P_e - \hat{P}_e| \leq \frac{1}{n}[24e^{-2} + 62\theta e^{-3}] \quad \text{and}$$

$$|P_e - \hat{P}_e| \leq f[10e^{-2} + 2(2+\theta)e^{-2}]$$

Appendix C.5.3. Mikhailov's Result

Define a vector of n-tuple in n-dimensional Euclidean space R^n

$$X = (x_1, x_2, \ldots, x_n)$$

and the metric

$$\|X - Y\| = \sqrt{\sum_{i=1}^{n} (x_i - y_i)^2}$$

Let $X(t)$, $t = 1, 2, \ldots$, be an n-dimensional Markov process with the state space $S \subset R^n$. Then the mean drift vector function of the process is defined as

$$M(X) = E[X(t+1) - X(t)|X(t) = X]$$

and the limiting mean drift vector function of the process (if it exists and converges uniformly) is defined as

$$\widehat{M}(X) = \lim_{\|X\| \to \infty} M(X)$$

Let us assume that a homogeneous Markov process $X(t)$ satisfies the following conditions:

(i) There exists a constant $C < \infty$ such that

$$E\left[\|X(t+1) - X(t)\|^2 | X(t) = X\right] < C \quad \forall\, X \in S$$

(ii) The limiting mean drift vector function of the process exists and converges uniformly. Also, we assume that there exists a scalar function $V(.)$, which satisfies the following conditions:

(a) $V(X) \geq 0 \quad \forall\, X \in R^n$
(b) $V(X) \leq C_1 \|X\| + C_2 \quad \forall\, X \in R^n$ and for some finite C_1 and C_2
(c) $\inf_{\|X\|>r} V(X) \to \infty$ for $r \to \infty$

(d) For every ϵ_0, there exists r_0 such that for all X: $\|X\| > r_0$

$$\left| LV(X) - \hat{M}(X) \text{ grad } V(X) \right| < \epsilon_0$$

where $LV(X) = E[V(X(t+1)) - V(X(t))|X(t) = X]$

A scalar function $V(.)$ that satisfies (a)–(d) is called the stochastic Lyapunov function.

Mikhailov has established that if $X(t)$ is a homogeneous Markov process which satisfies (i) and (ii) and if there exists a scalar function $V(.)$ which satisfies (a)–(d), and if $\hat{M}(X)$ grad $V(X) < -\epsilon$ is satisfied for some $\epsilon > 0$ and for all $X \in S$, then the Markov process is stable.

5.4 Capture Model in Random Access Systems

Fading is an inevitable phenomenon in various communication systems, including urban and mobile radio networks. While fading typically leads to increased error rates in data transmission, the random fluctuation of the received signal strength caused by fading can result in varying power levels. This variation, known as capture effects, can actually enhance the performance of ALOHA networks. In this section, Nakagami's general fading model for the radio channel is explored. The exact capture statistics probabilities are derived, and the channel capacity of the ALOHA system operating over a capture channel is determined.

5.4.1 *Introduction*

In the idealized channel model in ALOHA systems, the major source of capacity reduction comes from the mutual interference of messages from other users. A successful transmission occurs only if a single packet is transmitted during a slot. In ALOHA channels, fading also causes capture as in FM radio communication systems [41]. If signals sent to a common radio receiver arrive from different distances and with independent fading levels, it is no longer true that all colliding packets will always be destroyed by others. In fact, the random variation of the natural fading level divides the received signals into many different power levels, a method known to improve the utilization of the ALOHA network [42, 44, 45]. Thus, we anticipate a higher probability of successful reception in a real radio channel, leading to a higher throughput than the idealized ALOHA channels.

The objective of this section is to calculate the capture probabilities for the radio communication channel. Previous studies have investigated the basic ALOHA system operating over a Rayleigh fading channel [42, 43]. The model presented in this work extends the previous research by exploring various fading models, including the Rayleigh type. Detailed results are provided regarding the exact form of capture probabilities along with numerical data for a wide range of fading parameters.

In Section 5.4.2, a comprehensive survey of existing literature on frequently modeled fading statistics is conducted, focusing particularly on the Nakagami fading model. This model is recognized as a general representation of fading channels, encompassing the Rayleigh, half-Gaussian, Rician, and lognormal fading models. Section 5.4.3 involves analytical computations and comparisons of capture probabilities. The computational results are presented to compare fading models based on different statistical characteristics. Section 5.4.4 demonstrates that capture probabilities can be effectively modeled using a semi-exponential approach, enabling the derivation of the channel capacity of a controlled ALOHA random access system. Conclusions drawn from the analysis are presented in Section 5.4.5.

5.4.2 *Capture model in fading channel*

In the analysis, the noise level is regarded as very low and thus negligible compared to the signal level. It's assumed that the power level of all packets follows an independent Nakagami fading. The Nakagami distribution with parameters (p_0, m) is as follows [46]: For $x \geq 0$ and $m \geq 1/2$,

$$f_{A_i}(x) = \frac{2}{\Gamma(m_i)} \left(\frac{m_i}{p_0}\right)^{m_i} x^{2m_i-1} e^{(-m_i/p_0)x^2} \tag{5.47}$$

where $\Gamma(x)$ is the Gamma function and A_i is the fading envelope of the ith received signal, with $p_0 = E[A^2]$ and $m_i = {p_0}^2/var[{A_i}^2]$.

This distribution is also called m-distribution and was introduced by Nakagami [46] as a general model for fading channels. The theoretical model for deriving the Nakagami distribution assumes that the received signal is a sum of vectors with random amplitudes and random phases, which leads to more flexibility in matching experimental data than the use of Rayleigh or Rice distributions [46, 47].

For Rayleigh fading [47], the fading envelope A has the following distribution:

$$f_A(x) = \frac{2x}{p_0} e^{-x^2/p_0} \quad \text{for } x \geq 0$$

which has a single parameter p_0, mean power, and is frequently modeled for short-term and local-area fading. The theoretical model for deriving this distribution models the received signal as a sum of vectors with almost equal envelopes and independently, uniformly distributed phases [47]. The experimental results [48] support this model.

In Rician fading, we have a two-parameter distribution, (σ, α), for the fading envelope A:

$$f_A(x) = \frac{2x}{\sigma^2} e^{(x^2+\alpha^2)/\sigma^2} I_0\left(\frac{2\alpha}{\sigma^2} x\right) \quad \text{for } x \geq 0$$

where σ is the standard deviation. This is a statistical description of the strength of a signal received via a propagation medium in which a path consists of a strong stable component, α, the so-called specular component, and an additive Rayleigh component.

In the case of lognormal fading, the distribution has two parameters, $(\sigma, \hat{x})$,

$$f_A(x) = \frac{\log e}{(2\pi\sigma^2)^{0.5} x} \frac{e^{(\log x - \log \hat{x})^2}}{2\sigma^2} \quad \text{for } x \geq 0$$

where $\log(\overline{x}) = \log(\hat{x}) + (\sigma^2/(2\log e)$ with $\overline{x}$ as the mean and $\hat{x}$ as the median value, and σ as the standard deviation. This distribution is often used to model to describe long-term and large-area fading in long-range tropospheric transmission, transmission over irregular terrains, transmission in built-up areas, and optical transmissions; it is strongly supported by experimental data [48–50].

The Nakagami distribution includes the Rayleigh and half-Gaussian as special cases, and it also approximates the lognormal for large m on the domain $(p_0\epsilon^{-1})^{0.5} \leq x \leq (p_0\epsilon)^{0.5}$ where ε is Euler's constant. The Rice distribution can also be used to approximate other exact or experimentally derived distributions by judicious choice of parameters.

The parameter m indicates the fading figure, where smaller values of m correspond to worse fading. The fading model is classified as follows:

(i) Half-Gaussian fading if $m = 1/2$,
(ii) Q-distribution fading if $1/2 < m < 1$,
(iii) Rayleigh fading if $m = 1$,
(iv) Rician fading if $m > 1$.

The basic relationship between Nakagami and Rician's fading is

$$p_0 = \sigma^2 + \alpha^2 \quad \text{and} \quad m = (1 - r)^{-1}$$

where $r = \alpha^4/p_0$ and $0 < r < 1$.

The random power level of the received signal for the ith user is $P_i = {A_i}^2$ in Nakagami's channel. The probability density function (PDF) of the random variable P_i is

$$f_{P_i}(y) = \frac{1}{\Gamma(m)}\left(\frac{m}{p_0}\right)^m y^{m-1} e^{(-m/p_0)y}$$

for $y \geq 0$ and $m \geq 1/2$.

In the following, the goal is to obtain capture probabilities, C_n, specifically, the probability of capturing a test packet among n other interfering packets, using two different models: one with a joint interference model and the other with an individual interference model.

The joint interference model is commonly used in digital communication systems to address interferences arising from multipath fading in fading channels or from jammers in spread-spectrum communications. In this model, the desired signal is compared to the sum of interfering signals.

On the other hand, the individual interference model is often encountered in FM communications or radio Ham communication environments. In this model, the receiver captures the signal with the strongest power level among others. That is, the desired signal is compared individually with other interfering signals, and it occupies the receiver if it is the strongest.

5.4.2.1 *Joint interference model*

In this model, it is assumed that a test packet can be captured by the receiver in the presence of n interfering packets if the power of the test packet sufficiently exceeds the joint interference power. Let $P_J(n)$ be the random power level of n joint interference packets. The joint interference power is defined by the sum of n independent interferences:

$$P_J(n) = \sum_{i=1}^{n} P_i$$

Then the test packet can be captured by the receiver if and only if

$$P_t \geq \theta P_J(n) \quad \text{for } n \geq 1$$

where θ is called the capture ratio (protection ratio) which depends on the particular modulation scheme and the technical characteristics of the receiver.

Let us find the probability of capturing a particular packet overlapped by n other packets:

$$P_{\mathrm{I}}^{\theta}(n, m) = P[P_t \geq \theta P_J(n)]$$

Let us first find the PDF of $P_J(n)$ by convolving the PDF of $f_{P_i}(y)$ n times. It is shown that

$$f_{P_J}(y) = \frac{1}{\Gamma(nm)} \left(\frac{m}{p_0} \right)^{nm} y^{nm-1} e^{-(m/p_0)y} \quad \text{for } y \geq 0. \tag{5.48}$$

The conditional probability is found to be

$$P\left[P_t \geq \theta P_J(n) \,|\, P_J(n) = p\right] = \frac{1}{\Gamma(m)} \Gamma\left(m, \frac{m\theta}{p_0} p\right) \tag{5.49}$$

where $\Gamma(a, b)$ is the complementary incomplete Gamma function and defined in Appendix A.5.4. Hence, by the principle of total probability, from (A.17) in Appendix A.5.4, the final $P_I^{\theta}(n, m)$ is

$$\begin{aligned} P_{\mathrm{I}}^{\theta}(n, m) = & \frac{1}{nm B(m, nm)} \frac{\theta^m}{(1+\theta)^{nm+m}} \\ & \cdot {}_2F_1\left[1, m(n+1), nm+1; (1+\theta)^{-1}\right]. \end{aligned} \tag{5.50}$$

where $B(a, b)$ is the beta function and ${}_2F_1(a, b, c; z)$ is the Gauss hypergeometric function (for more properties, see Appendix A.5.4). This is a general expression of the capture distribution for the joint interference model.

Thus, instead of using (5.50), more useful expressions for the numerical calculation, using (A.12) in Appendix A.5.4, for $n \geq 1$, are as follows:

$$P_{\mathrm{I}}^{\theta}(n,m) = \frac{1}{nmB(m,nm)} \frac{\theta^{m-1}}{(1+\theta)^{nm+m-1}} \cdot {}_2F_1(1, 1-m, nm+1; -\theta^{-1}) \tag{5.51}$$

and for any integer $n \geq 1$, using (A.8) in Appendix A.5.4,

$$P_{\mathrm{I}}^{\theta}(n,m) = \frac{1}{nmB(m,nm)} \frac{\theta^{m-1}}{(1+\theta)^{nm+m-1}} \cdot \sum_{k=0}^{m-1} (-1)^k \frac{(1-m)_k}{(nm+1)_k} \theta^{-k} \tag{5.52}$$

Now, the capture probability C_n that a packet (any one) is successful from among n interfering packets has the following relationship:

$$C_n \binom{n}{1} P_{\mathrm{I}}^{\theta}(n-1,m) \quad \text{for } n \geq 2$$

where $\binom{n}{k} = \frac{n!}{k(n-k)!}$. So, the capture probability is obtained by

$$C_{\mathrm{I}}^{\theta}(n,m) = \frac{n}{(n-1)mB(m,(n-1)m)} \cdot \frac{\theta^{m-1}}{(1+\theta)^{nm-1}} \cdot {}_2F_1(1, 1-m, (n-1)m+1; -\theta^{-1}) \tag{5.53}$$

(i) Half-Gaussian fading:
For $m = 1/2$, from (5.50),

$$P_{\mathrm{I}}^{\theta}\left(n, \frac{1}{2}\right) = \frac{2}{nB(0.5, 0.5n)} \frac{\theta^{0.5}}{[(1+\theta)^{n+1}]^{0.5}} \cdot {}_2F_1\left(1, \frac{n+1}{2}, \frac{n+3}{2}; \frac{1}{1+\theta}\right)$$

(ii) Rayleigh fading:

$$P_{\mathrm{I}}^{\theta}(n,1) = \frac{1}{(1+\theta)^n} \quad \text{and} \quad C_n = \frac{n}{(1+\theta)^{n-1}}$$

This was previously obtained by Ref. [42] using the direct approach of the Rayleigh fading model.

(iii) Rician fading:
For $r = 1/2$,

$$P_{\mathrm{I}}^{\theta}(n, 2) = \frac{1 + (2n + 1)\theta}{(1 + \theta)^{2n+1}}$$

For $r = 1/3$,

$$P_{\mathrm{I}}^{\theta}(n, 3) = \frac{1 + (3n + 2)\theta + 0.5(3n + 1)(3n + 2)\theta^2}{(1 + \theta)^{3n+2}}$$

5.4.2.2 *Individual interference model*

In this model for interference, the test packet can be captured by the receiver if the power level of the test packet is the strongest one from among n other individual packets. With this definition, the test packet can be captured by the receiver if and only if

$$P_t \geq \theta P_i, \forall i$$

where θ is the protection ratio as defined before. In order to find the capture probability, we need to find

$$P_{\mathrm{II}}^{\theta}(n, m) \equiv P[P_t \geq \theta P_i, \forall i]$$

Since P_i are the random power levels of the independent interferences with the same mean power,

$$P_{\mathrm{II}}^{\theta}(n, m) = \int_0^{\infty} \left[\int_0^{p/\theta} f_{P_i}(x) dx \right]^n f_{P_i}(p)\, dp \tag{5.54}$$

The inner integral is computed by

$$\int_0^{p/\theta} f_{P_i}(x) dx = \frac{1}{\Gamma(m)} \gamma\left(m, \frac{m}{\theta p_0} p\right)$$

where $\gamma(a, b)$ is the incomplete Gamma function defined in Appendix A.5.4. From (5.54), one can find

$$P_{\mathrm{II}}^{\theta}(n, m) = \frac{n}{\Gamma^{n+1}(m)} \int_0^{\infty} \gamma^n \left(m, \frac{y}{\theta}\right) y^{m-1} e^{-y} dy \tag{5.55}$$

This expression is a general formula for the individual interference model. Now, from the definition, the capture probability is

$$C_{\mathrm{II}}^{\theta}(n,m) = \frac{n}{\Gamma^n(m)} \int_0^{\infty} \gamma^{n-1}\left(m, \frac{y}{\theta}\right) y^{m-1} e^{-y} dy \tag{5.56}$$

The asymptotic result of this probability is given in Appendix B.5.4.

(i) **Half-Gaussian fading:**
For $m = 1/2$, from (5.55),

$$P_{\mathrm{II}}^{\theta}\left(n, \frac{1}{2}\right) = \pi^{-0.5(n+1)} \int_0^{\infty} \gamma^{n}\left(\frac{1}{2}, \frac{y}{\theta}\right) y^{-0.5} e^{-y} dy$$

For example, if $n = 1$, then

$$\begin{aligned} P_{\mathrm{II}}^{\theta}\left(1, \frac{1}{2}\right) &= \frac{2}{\pi} \frac{\theta^{0.5}}{1+\theta} {}_2F_1\left(1, 1, \frac{3}{2}; \frac{1}{1+\theta}\right) \\ &= \frac{2}{\pi} \sin^{-1}\left(\frac{1}{(1+\theta)^{0.5}}\right) \end{aligned}$$

(ii) **Rayleigh fading:**
From (5.55),

$$P_{\mathrm{II}}^{\theta}(n,1) = \sum_{i=0}^{n} \binom{n}{i} (-1)^i \frac{\theta}{\theta + i}$$

(iii) **Rician fading:**
For instance, if $r = 1/2$ and $n = 1$, then

$$P_{\mathrm{II}}^{\theta}(1,2) = \frac{3\theta^2}{(1+\theta)^4} {}_2F_1[1, 4, 3; (1+\theta)^{-1}]$$

and if $r = 1/3$ and $n = 1$, then

$$P_{\mathrm{II}}^{\theta}(1,3) = \frac{10\theta^3}{(1+\theta)^6} {}_2F_1\left[1, 6, 4; (1+\theta)^{-1}\right]$$

5.4.2.3 *Special case of individual interference model*

Now, assuming that the interfering events are independent, one can have

$$P_{\mathrm{III}}^{\theta}(n,m) \equiv P\left[\bigcap_{i=1}^{n} P_t \geq \theta P_i\right] = \prod_{i=1}^{n} P[P_t \geq \theta P_i]$$

In a way that is very similar to the reasoning in (5.50 and 5.51), it can be obtained that

$$\begin{aligned} P_{\mathrm{III}}^{\theta}(n,m) &= \frac{1}{\Gamma(m)\Gamma(m)}\left(\frac{m}{p_0}\right)^m \int_0^{\infty} \gamma\left(m, \frac{m}{\theta p_0} p\right) \\ &\quad \cdot p^{m-1} e^{-(m/p_0)p}\, dp \\ &= \frac{1}{mB(m,m)} \frac{\theta^m}{(1+\theta)^{2m}} \cdot {}_2F_1[1, 2m; m+1; (1+\theta)^{-1}] \end{aligned}$$

Hence,

$$P_{\mathrm{III}}^{\theta}(n,m) = \frac{1}{m^n B^n(m,m)} \frac{\theta^{nm}}{(1+\theta)^{2nm}} \cdot {}_2F_1^n[1, 2m; m+1; (1+\theta)^{-1}] \tag{5.57}$$

Note from (A.12) in Appendix A.5.4, (5.57) can be expressed as

$$P_{\mathrm{III}}^{\theta}(n,m) = \frac{1}{m^n B^n(m,m)} \frac{\theta^{n(m-1)}}{(1+\theta)^{n(2m-1)}} \cdot {}_2F_1^n[1, 1-m; m+1; -\theta^{-1}]$$

For integer values of m, the following expression is obtained:

$$\begin{aligned} P_{\mathrm{III}}^{\theta}(n,m) &= \frac{1}{m^n B^n(m,m)} \frac{\theta^{n(m-1)}}{(1+\theta)^{n(2m-1)}} \\ &\quad \cdot \left[\sum_{k=0}^{m-1} (-1)^k \frac{(1-m)_k}{(nm+1)_k} \theta^{-k}\right]^n \end{aligned} \tag{5.58}$$

This is useful for numerical computation.

Note that

$$\begin{aligned} C_{\mathrm{III}}^{\theta}(n,m) &= \frac{n}{m^{n-1} B^{n-1}(m,m)} \frac{\theta^{n(m-1)}}{(1+\theta)^{n(2m-1)}} \\ &\quad \cdot {}_2F_1^{n-1}(1, 1-m; m+1; -\theta^{-1}) \end{aligned} \tag{5.59}$$

This is upper bounded by C_{II}^{θ} and decreases to zero when n is sufficiently large (see Appendix B.5.4).

(i) **Half-Gaussian fading:**
From (5.57)

$$P_{\mathrm{III}}^{\theta}\left(n, \frac{1}{2}\right) = \frac{2^n}{\pi^n}\left[\tan^{-1}\left(\theta^{-0.5}\right)\right]^n$$

(ii) **Rayleigh fading:**
From (5.58),

$$P_{\mathrm{III}}^{\theta}(n, 1) = \frac{1}{(1+\theta)^n}$$

This result is identical to the joint interference case.

(iii) **Rician fading:**
For $r = 1/2$,

$$P_{\mathrm{III}}^{\theta}(n, 2) = \frac{(1+3\theta)^n}{(1+\theta)^{3n}}$$

For $r = 1/3$,

$$P_{\mathrm{III}}^{\theta}(n, 3) = \frac{\left(1+5\theta+10\theta^2\right)^n}{(1+\theta)^{5n}}$$

5.4.3 *Capture probabilities of the general fading model*

In this section, the performance results are presented to compare the proposed models for different fading parameters. The capture probability, C_n, was defined to be the probability that one packet is successfully received despite n overlapping packets on the channel. Hence, using the capture distributions $P_i^{\theta}(n, m)$ obtained in the previous section, we have

$$C_i^{\theta}(n, m) = nP_i^{\theta}(n-1, m) \quad \text{for } i = \mathrm{I}, \mathrm{II}, \mathrm{III} \quad \text{and} \quad n \geq 2 \tag{5.60}$$

This depends explicitly on the threshold θ, the number of transmitted signals n, and the fading parameter m.

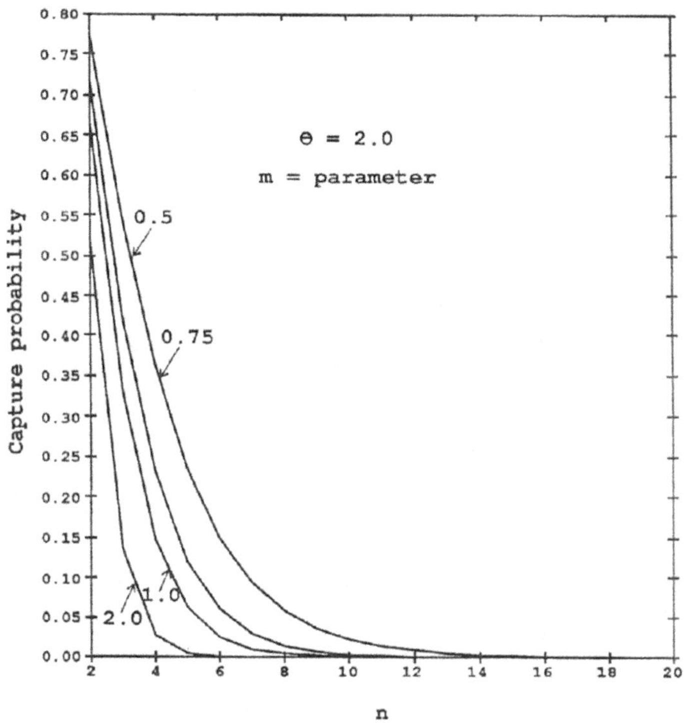

Fig. 5.20. Capture probability for the joint interference model $m = 0.5$ (half-Gaussian), 0.75 (Q-distribution), 1.0 (Rayleigh), and 2.0 (Rician).

Figure 5.20 depicts the capture probabilities $C_{\mathrm{I}}^{\theta}(n, m)$ for the joint interference model. From the numerical computations, it is found that $C_{\mathrm{I}}^{\theta}(n, m) = 0$ when $n \geq 30$, for all m and θ. This implies that the effect of capture is almost null if the number of overlapped packets is about the order of 10. However as the capture ratio increases, the capture probability slowly decreases. It is noted that the smaller the m, the larger the $C_{\mathrm{I}}^{\theta}(n, m)$. The larger fading figure yields the smaller capture probability in general. When $m = 0.5$ (half-Gaussian fading), $C_{\mathrm{I}}^{\theta}(n, m)$ is the largest for all values of θ. In contrast, for $m = 2$ (Rician fading), $C_{\mathrm{I}}^{\theta}(n, m)$ is the smallest for all θ.

Figure 5.21 shows the capture probabilities $C_{\mathrm{II}}^{\theta}(n, m)$ for the individual interference model. Note that when $\theta = 1.43$ is fixed, the half-Gaussian model is the best and the Rayleigh model is the worst. This individual model has a better performance than the joint interference model. Generally, the

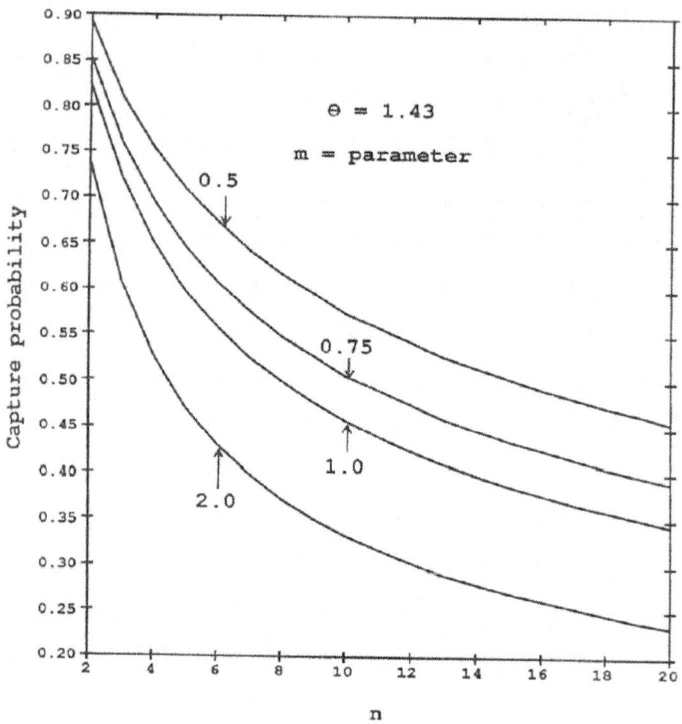

Fig. 5.21. Capture probability for the individual interference model when $m = 0.5$ (half-Gaussian), 0.75 (Q-distribution), 1.0 (Rayleigh), and 2.0 (Rician).

capture effects are very slowly diminishing compared to the joint interference model as n goes to be large.

In Figs. 5.22 and 5.23, three models are compared, where I, II, and III stand for the $C_{\mathrm{I}}^{\theta}(n, m)$ (the joint interference model), $C_{\mathrm{II}}^{\theta}(n, m)$ (the individual interference model), and $C_{\mathrm{III}}^{\theta}(n, m)$ (a special case), respectively, as defined in the previous section. When m is small, $C_{\mathrm{III}}^{\theta}(n, m)$ is the lower bound and when m gets larger, C_{I}^{θ} is the lower bound.

In general, for all θ, m,

$$C_{\mathrm{I}}^{\theta}(n, m) \leq C_{\mathrm{II}}^{\theta}(n, m) \quad \text{and} \quad C_{\mathrm{III}}^{\theta}(n, m) \leq C_{\mathrm{II}}^{\theta}(n, m).$$

These results are proven in Appendix B.5.4.

In the literature [42, 43], the joint interference model is common for the analysis of interference effects on digital communication systems. But the individual interference model is a meaningful model in random access

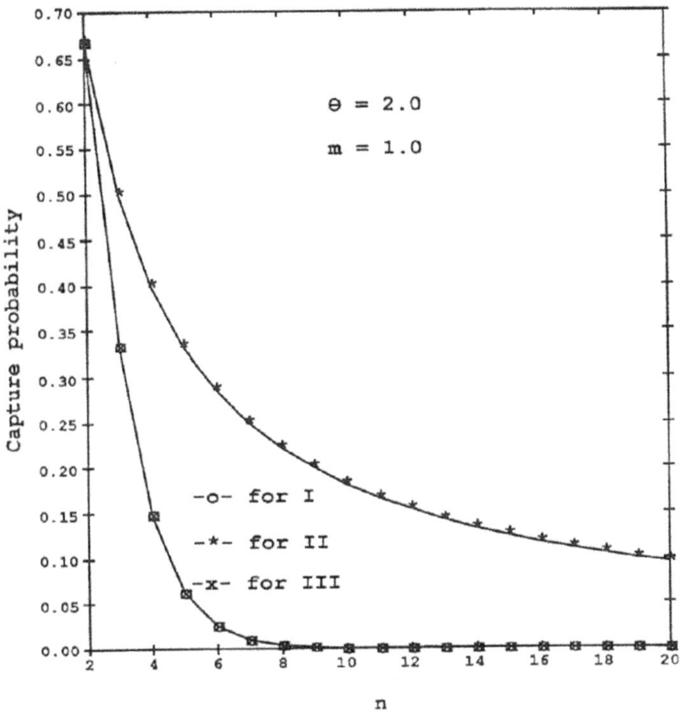

Fig. 5.22. Capture probability for $m = 1.0$ (Rayleigh fading).

systems for capture effects, as seen in FM systems [41], where the strongest signal captures the receiver. It is observed that the joint interference model is strongly dependent on n, while the individual interference model is less dependent on n.

5.4.4 *Throughput of the semi-exponential model*

Several authors [44, 45] have used the semi-exponential model for C_n in order to compute the channel throughput of random access communication networks. The asymptotic results of C_n are given in Appendix B.5.4. They are exponentially decreasing when n increases and thus can be modeled as the semi-exponential model.

In order for C_n to have a functional form of exponential type, curve-fitting methods [51] are employed. Numerical data obtained for C_n are used for fitting a function of exponential type. The fitting is done in a

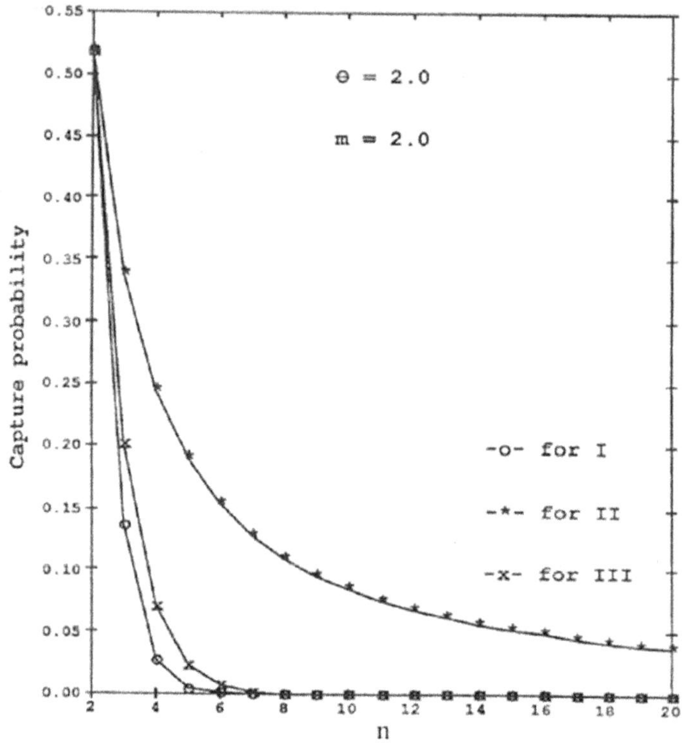

Fig. 5.23. Capture probability for three models, $m = 2.0$ (Rician fading).

series of iterations in which the parameters are systematically adjusted by the Marquardt–Levenberg method until a least square is reached. This technique combines the advantages of the Gauss–Newton procedure and the method of steepest descent. The function to be fitted is of the following form:

$$C_n = ae^{-bn} \quad \text{for } n \geq 2 \tag{5.61}$$

where a and b are decided according to the convergence criterion of a least-square method.

In Figs. 5.24 and 5.25, we compare the capture probabilities in the exact model and the approximated semi-exponential model. Here, the exact model was computed by assuming

$$C_n = 0 \quad \text{whenever } C_n < 10^{-10} \tag{5.62}$$

but the approximated model was the model when using ae^{-bn} for C_n.

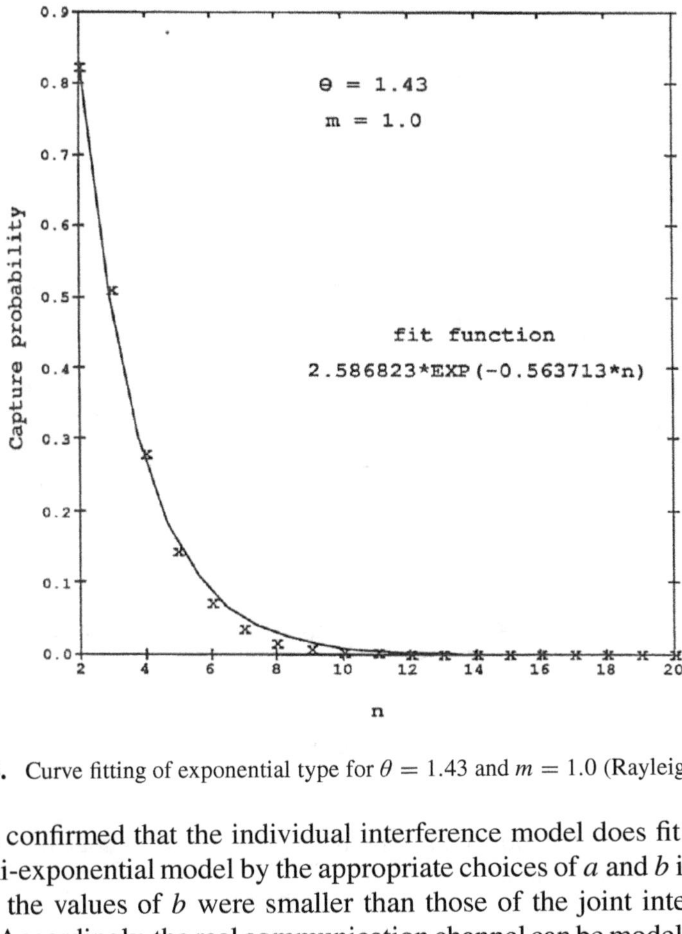

Fig. 5.24. Curve fitting of exponential type for $\theta = 1.43$ and $m = 1.0$ (Rayleigh fading).

It is confirmed that the individual interference model does fit well for the semi-exponential model by the appropriate choices of a and b in (5.61). In fact, the values of b were smaller than those of the joint interference model. Accordingly, the real communication channel can be modeled by the exponential capture model if the coefficients (a, b) are properly identified. The least-square fitting techniques are very useful while estimating the channel capacity for a certain fading channel model.

To compute the channel throughput, the exponential capture model may be used. Using (5.61), the channel throughput of the controlled slotted ALOHA system with the capture channel can be obtained by Ref. [45]:

$$S^{\theta}(G) = e^{-G}[G + a(e^{\overline{b}G} - \overline{b}G - 1)] \tag{5.63}$$

where $\overline{b} = e^{-b}$.

Figure 5.26 depicts the channel throughput versus the channel traffic. The solid lines indicate the exact model and the dotted lines indicate the

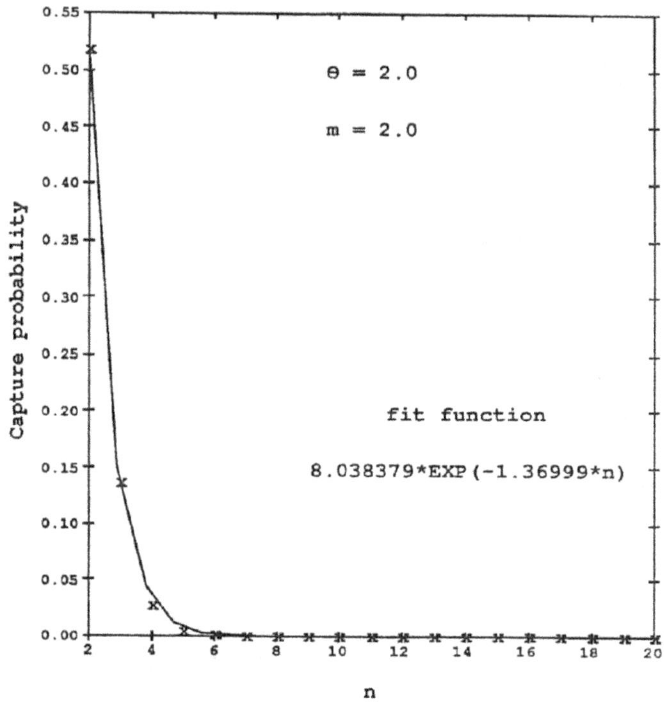

Fig. 5.25. Curve fitting of exponential type for $\theta = 2.0$ and $m = 2.0$ (Rician fading).

approximated one (exponential capture model). They are seen to have very small errors. The channel capacity is

$$S^{\theta}(G^*) = e^{-G^*}[G^* + a(e^{\overline{b}G^*} - \overline{b}G^* - 1)] \tag{5.64}$$

where

$$G^* = 1 + \frac{a}{1 - a\overline{b}} - \frac{a(1 - \overline{b})e^{-\overline{b}G^*}}{1 - a\overline{b}} \tag{5.65}$$

In Table 5.2, the channel capacity and the optimal channel traffic are obtained using (5.64) and (5.65) for the different values of θ and m. When $\theta = 1$ and $m = 0.5$, the maximum channel capacity for all fading models is obtained. For instance, when $\theta = 3.0$ and $m = 2.0$, which is the Rician fading model with about 5 dB protection ratio, the capacity is 0.434,

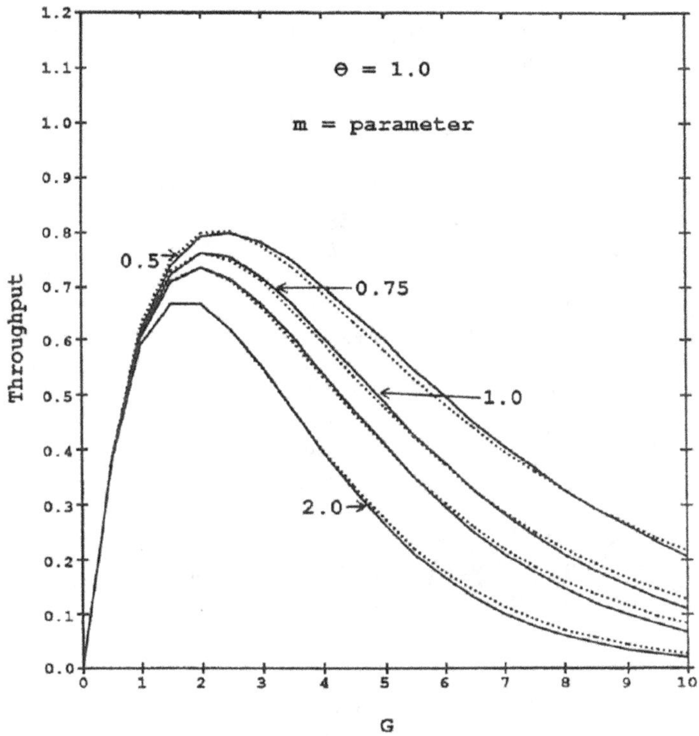

Fig. 5.26. Throughput comparison between the actual model and the exponential model when $\theta = 1.0$ and $m = 0.5$ (half-Gaussian), 0.75 (Q-distribution), 1.0 (Rayleigh), and 2.0 (Rician).

which is an about 18% improvement over the conventional slotted ALOHA systems.

5.4.5 *Summary*

In this section, Nakagami's fading channel is investigated, which is arising in radio communication systems. Note that the Nakagami fading channel can represent a unified treatment of properties for fading media. Computational data are given for a broad range of system parameters that serve figures of merit for the specification of system design. It is shown that the Rayleigh fading results in a higher capture probability than the Rician fading, and the half-Gaussian fading has the largest fading figure and the

Table 5.2. Channel capacity and optimal channel traffic with θ and m.

θ	m	G^*	S^*
1.0	0.5	2.285830	0.8044067
1.0	0.75	2.082197	0.7626277
1.0	1.0	1.958473	0.7340415
1.0	2.0	1.725791	0.6752217
1.43	0.5	1.981671	0.6995068
1.43	0.75	1.798299	0.6547457
1.43	1.0	1.686832	0.6242667
1.43	2.0	1.476698	0.5601977
2.0	0.5	1.763402	0.6271111
2.0	0.75	1.597515	0.5820269
2.0	1.0	1.496489	0.5514442
2.0	2.0	1.307794	0.4881576
3.0	0.5	1.567362	0.5626438
3.0	0.75	1.421153	0.5192551
3.0	1.0	1.332782	0.4904033
3.0	2.0	1.174511	0.4336635

highest capture probability. It is also shown that the capture probabilities for a radio fading channel can be modeled as the semi-exponential model.

Appendix A.5.4

In this appendix, we give the definitions and properties of some special functions [52] used in this section.

The Gamma function is defined by

$$\Gamma(x) = \int_0^\infty e^t t^{x-1} dt \quad \text{for } x > 0 \tag{A.1}$$

Note that

$$\Gamma(x+1) = x\Gamma(x) \quad \text{and} \quad \Gamma(1/2) = \pi^{0.5} \tag{A.2}$$

The incomplete Gamma function is defined by

$$\gamma(a, b) = \int_0^b e^{-t} t^{a-1} dt \tag{A.3}$$

and the complementary incomplete Gamma function is defined by

$$\Gamma(a, b) = \int_b^{\infty} e^{-t} t^{a-1} dt \tag{A.4}$$

Note that

$$\gamma(a, b) + \Gamma(a, b) = \Gamma(a) \tag{A.5}$$

The beta function is defined by

$$B(a, b) = \int_0^1 t^{a-1}(1-t)^{b-1} dt \quad \text{for} \quad a, b > 0 \tag{A.6}$$

Note that

$$B(a, b) = B(a, b) \quad \text{and} \quad B(a, b) = \frac{\Gamma(a)\Gamma(b)}{\Gamma(a+b)} \tag{A.7}$$

The hypergeometric function is defined by

$${}_2F_1(a, b, c; z) = \sum_{k=0}^{\infty} \frac{(a)_k (b)_k !}{(c)_k k!} z^k \quad \text{for } |z| < 1 \tag{A.8}$$

where z is a complex variable, a, b, and c are parameters which can take arbitrary real or complex values if $c \neq 0, -1, -2$, and we define the following quantities:

$$(a)_0 = 1, (a)_k = a(a+1)\cdots(a+k+1), \quad k = 1, 2, \ldots \tag{A.9}$$

This function has many properties and some of them are given as follows: ${}_2F_1(a, b, c; z)$ has the integral representation,

$${}_2F_1(a, b, c; z) = \frac{\Gamma(c)}{\Gamma(b)\Gamma(c-b)} \cdot \int_0^1 t^{b-1}(1-t)^{c-b-1}(1-tz)^{-a} dt \tag{A.10}$$

It has the symmetry property from the definition of (A.8)

$${}_2F_1(a, b, c; z) = {}_2F_1(b, a, c; z) \tag{A.11}$$

From (A.10),

$$\lim_{z \to 1} {}_2F_1(a, b, c; z) = \frac{\Gamma(c)\Gamma(c-a-b)}{\Gamma(c-a)\Gamma(c-b)}.$$

By linear transformation, we have

$$ {}_2F_1(a, b, c; z) = (1-z)^{-a} {}_2F_1\left(a, c-b, c; \frac{z}{z-1}\right) \quad \text{(A.12)} $$

and

$$ {}_2F_1(a, b, c; z) = (1-z)^{c-a-b} \cdot {}_2F_1(c-a, c-b, c; z) \quad \text{(A.13)} $$

Some transcendental functions can be represented by the hypergeometric function as such:

$$ {}_2F_1\left(a, a+\frac{1}{2}, 2a; z\right) = \frac{1}{(1+z)^{0.5}}\left[\frac{1+(1-z)^{0.5}}{2}\right]^{1-2a} \quad \text{(A.14)} $$

$$ {}_2F_1\left(\frac{1}{2}, 1, \frac{3}{2}; -z^2\right) = \frac{\tan^{-1}(z)}{z} \quad \text{(A.15)} $$

and

$$ {}_2F_1\left(\frac{1}{2}, \frac{1}{2}, \frac{3}{2}; -z^2\right) = \frac{\sin^{-1}(z)}{z} \quad \text{(A.16)} $$

Note that

$$ \int_0^\infty x^{a-1} e^{-bx} \Gamma(\alpha, \beta x) dx = \frac{\beta^a \Gamma(a, \alpha)}{\alpha (b+\beta)^{a+\alpha}} \cdot {}_2F_1\left(1, a+\alpha, \alpha+1; \frac{b}{b+\beta}\right) \quad \text{(A.17)} $$

and

$$ \int_0^\infty x^{a-1} e^{-bx} \gamma(\alpha, \beta x) dx = \frac{\beta^a \Gamma(a, \alpha)}{\alpha (b+\beta)^{a+\alpha}} \cdot {}_2F_1\left(1, a+\alpha, \alpha+1; \frac{\beta}{b+\beta}\right) \quad \text{(A.18)} $$

Appendix B.5.4

In this appendix, we give the asymptotic results of capture probability. Our purpose is to show that for $i = \text{I, II, III}$

$$C_i^{\theta}(n, m) \to 0 \text{ as } n \to \infty \text{ for any fixed } \theta > 1 \text{ and } m > 0.$$

Note that the asymptotic expansion of the Gamma function gives

(i) $\Gamma(x) = (2\pi)^{0.5} e^{-x} x^{x-0.5}[1 + O(x^{-1})]$

(ii) $\Gamma(a + x) = (2\pi)^{0.5} e^{-x} x^{x+a-0.5}[1 + O\left(x^{-1}\right)]$

The following two lemmas can be shown by using (i) and (ii).

Lemma 1.

$$B(a, x) = \Gamma(a) x^{-a}[1 + O(x^{-1})]$$

Lemma 2.

$${}_2F_1(1, 1 - m, (n - 1)m + 1; \theta^{-1}) = [1 + O(n^{-1})]$$

for any fixed θ and m.

Proposition 1. *For any fixed $\theta > 1$ and $m > 1/2$*

$$C_{\text{I}}^{\theta}(n, m) = (1 + \theta^{-1}) \frac{m^{m-1}}{\Gamma(m)} [(n - 1)\theta r^n]^m \cdot [1 + 0 * \left(n^{-1}\right)]$$

where $r = 1/(1 + \theta) < 1$

Proof. From (5.53),

$$C_{\text{I}}^{\theta}(n, m) = \frac{n}{(n - 1)m B[m, (n - 1)m]}$$
$$\cdot \frac{\theta}{(1 + \theta)^{nm-1}} {}_2F_1^{m-1}[1, 1 - m, (n - 1)m + 1; -\theta^{-1}]$$

Using Lemma 1,

$$C_{\text{I}}^{\theta}(n, m) = \frac{(1 + \theta^{-1})}{m\Gamma(m)} \left(\frac{n}{n - 1}\right) \left[\frac{m\theta\,(n - 1)}{(1 + \theta)}\right]^m$$
$$\cdot [1 + O\left(n^{-1}\right)]{}_2F_1(1, 1 - m, (n - 1)m + 1; \theta^{-1})$$

Using Lemma 2,

$$C_{\mathrm{I}}^{\theta}(n,m) = \frac{\left(1+\theta^{-1}\right)m^{m-1}}{\Gamma(m)}\left[\frac{(n-1)\theta}{(1+\theta)^n}\right]^m \cdot [1+O(n^{-1})].$$

This is the desired result with $r = 1/(1+\theta)$ □

Proposition 2. $C_{\mathrm{II}}^{\theta}(n,m) \to 0$ *as* $n \to \infty$ *for any fixed* θ *and* m.

Proof. From (5.56), we have

$$C_{\mathrm{II}}^{\theta}(n,m) = \frac{1}{\Gamma(m)}\int_0^{\infty} n\left[\frac{\gamma\left(m,\frac{y}{\theta}\right)}{\Gamma(m)}\right]^{n-1} y^{m-1}e^{-y}dy$$

Note that by definitions

$$\frac{\gamma(m,y)}{\Gamma(m)} < 1 \quad \text{for } 0 < y < \infty \quad \text{and} \quad \frac{\gamma(m,y)}{\Gamma(m)} = \begin{cases} 0 & \text{for } y = 0 \\ 1 & \text{for } y = \infty \end{cases}$$

Letting

$$f_n(y) = n\left[\frac{\gamma\left(m,\frac{y}{\theta}\right)}{\Gamma(m)}\right]^{n-1} y^{m-1}e^{-y}$$

and define

$$f(y) = 0 \quad \text{and} \quad g(y) = 2y^{m-1}e^{-y}$$

We may show that

$$f_n(y) \to f(y) \quad \text{as} \quad n \to \infty \quad \text{for all } y \geq 0$$

and

$$|f_n(y)| \leq g(y) \quad \text{for all } y \geq 0 \quad \text{with}$$
$$\int_0^{\infty} g(y)dy = 2\Gamma(m) < \infty \quad \forall y \geq 0$$

Hence, we are led to the result by Lebesgue's dominated convergence theorem. □

Proposition 3.

(i) $C_{\mathrm{III}}^{\theta}(n,m) \leq C_{\mathrm{II}}^{\theta}(n,m)$.
(ii) $C_{\mathrm{III}}^{\theta}(n,m) \to 0$ *as* $n \to \infty$.

Proof.

(i) This is true by the definitions of P_{II}^{θ} and $P_{\mathrm{III}}^{\theta}$.
(ii) This is just from (i) and Proposition 2. □

5.5 Slotted ALOHA in Wireless Body Area Networks

The IEEE 802.15.6 standard for wireless body area networks (WBANs) specifies a slotted ALOHA-based protocol as a medium access control (MAC), with provisions for different user priorities. In this section, we present an analytical model for WBANs to estimate the saturation throughput of this protocol, considering a heterogeneous network comprising multiple priorities of users. Numerical results are generated to validate the predictions from the model through computation and simulation [56].

5.5.1 *Introduction*

A WBAN is a network of wireless sensors placed around, on, and/or inside the human body. These networks serve various applications, including monitoring vital signs, wearable heart rate-monitoring sensors, and implanted glucose sensors. Beyond medical applications, the WBAN technology can be utilized in fields, such as fitness monitoring, gaming, and wearable computing.

The IEEE 802.15.6 standard, established in 2012, was developed to meet the requirements of WBANs [53]. Each WBAN includes a hub that coordinates up to 64 nodes. The standard defines an MAC layer supporting three different physical layers. The ultra-wideband physical layer (UWB PHY) employs slotted ALOHA as a random access protocol [54]. The UWB PHY specification is specifically designed to provide robust performance for WBANs.

The slotted ALOHA protocol adopted in IEEE 802.15.6 differs from the traditional approach. It is tailored to handle a relatively small number of users by mitigating contention through a unique reduction of the retransmission probability. Additionally, it incorporates specific considerations for user priorities [56].

The CSMA/CA protocol of WBANs has been analyzed in Refs. [55–57]. The WBAN considers the different prioritized users in the network. Analysis for such heterogeneous users considering CSMA/CA of IEEE 802.11 is found in Ref. [58].

The ALOHA protocol does not perform carrier sensing. The slotted ALOHA system with an infinite user population under the conventional collision channel model is unstable [59]. Several strategies have been made to stabilize the channel [60, 61]. However, many of today's networks often limit the number of simultaneous users in the system at any time. In this sense, the system itself is stable in terms of users, such as WBANs.

This section is outlined as follows. Section 5.5.2 reviews the slotted ALOHA protocol in IEEE 802.15.6. In Section 5.5.3, the Markov model is introduced and solved. Section 5.5.4 validates the model, and Section 5.5.5 presents concluding remarks.

5.5.2 *Slotted ALOHA for the IEEE 802.15.6 standard*

The IEEE 802.15.6 standard defines eight user priorities: $UP_i, i = 0, \ldots, 7$. For slotted ALOHA, a parameter, contention probability (CP), is maintained to access the medium. Prioritized access for the traffic of differing users is attained through the CP, where CP $\in [CP_{\min}, CP_{\max}]$. The values of $CP_{\min}$ and $CP_{\max}$ depend on UP_i, as shown in Table 5.3.

A sensor node contends to transmit its packet as follows:

(i) A node generates a random number z from the interval [0,1].
(ii) The CP value is set to $CP_{\max}$ for a new packet.
(iii) A node obtains a contended allocation by the current ALOHA slot if $z \leq CP$.
(iv) If the transmission fails, the CP value is halved for an even number of retransmissions and is fixed to the previous value for an odd number of retransmissions. If halving CP makes the new CP smaller than $CP_{\min}$, the CP value is set to $CP_{\min}$.

The changes in CP value are illustrated in Table 5.4 for UP_0, UP_5, and UP_7. The CP value does not change when it gets settled with $CP_{\min}$ at the

Table 5.3. Contention probability (CP) thresholds for slotted ALOHA access.

User Priority Class (i)	0	1	2	3	4	5	6	7
$CP_{\max}$	$\frac{1}{8}$	$\frac{1}{8}$	$\frac{1}{4}$	$\frac{1}{4}$	$\frac{3}{8}$	$\frac{3}{8}$	$\frac{1}{2}$	1
$CP_{\min}$	$\frac{1}{16}$	$\frac{3}{32}$	$\frac{3}{32}$	$\frac{1}{8}$	$\frac{1}{8}$	$\frac{3}{16}$	$\frac{3}{16}$	$\frac{1}{4}$

Table 5.4. Changes of CP value with the retransmissions of a packet and the value of m.

UP_i	Initial transmission	1st retransmission	2nd retransmission	3rd retransmission	4th retransmission	m
0	$CP = CP_{max} = \frac{1}{8}$	$CP = \frac{1}{8}$	$CP = \frac{1}{16} = CP_{min}$	$CP = CP_{min} = \frac{1}{16}$	$CP = CP_{min} = \frac{1}{16}$	2
5	$CP = CP_{max} = \frac{3}{8}$	$CP = \frac{3}{8}$	$CP = \frac{3}{16} = CP_{min}$	$CP = CP_{min} = \frac{3}{16}$	$CP = CP_{min} = \frac{3}{16}$	2
7	$CP = CP_{max} = 1$	$CP = 1$	$CP = \frac{1}{2}$	$CP = \frac{1}{2}$	$CP = \frac{1}{4} = CP_{min}$	4

mth retransmission. The value of m can be found as

$$m = 2\left\lceil \log_2\left(\frac{CP_{max}}{CP_{min}}\right)\right\rceil \tag{5.66}$$

It may be noted that each ALOHA slot is of equal length.

5.5.3 *Markov model for slotted ALOHA WBAN*

In this subsection, an analytical model is given to assess the saturation throughput of IEEE 802.15.6 slotted ALOHA. In the analysis, a finite number of nodes are assumed, each always having a packet available for transmission. An ideal channel condition is assumed. A node contends to access the medium according to the protocol described in the following section.

The discrete-time Markov chain model depicted in Fig. 5.27 represents the states that a node may occupy during packet transmission attempts. The decreasing probability of transmission stated in the standard and the division of the time into a sequence of well-defined packets of fixed length are keys to the formulation of the model.

In the discrete-time Markov chain, nodes assume states $\{0, 1, 2, \ldots, m\}$ based on varying contention probabilities. For a tagged node in state k, slot-wise event probabilities are denoted as follows: $p(k, i)$ denotes the node remaining idle in the current slot without attempting to transmit, $p(k, s)$ denotes a successful transmission in the current slot, and $p(k, c)$ represents a transmission in the current slot resulting in a collision.

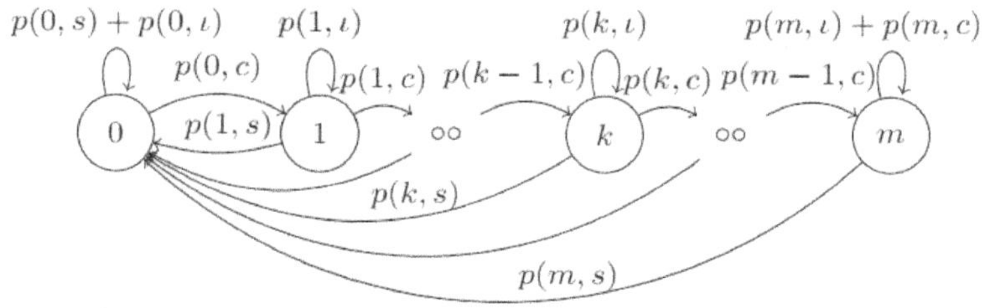

Fig. 5.27. The discrete-time Markov chain for nodes.

The one-step transition probabilities $Pr\{i|j\} = \lim_{t\to\infty} Pr\{S(t+1) = i|S(t) = j\}$ for the slot-wise temporal state $S(t)$ of a node in the chain are

$$\begin{cases} \mathrm{P_r}(0|0) = p(0,s) + p(0,i) \\ \mathrm{P_r}(k+1|k) = p(k,c), & 0 \le k \le m-1 \\ \mathrm{P_r}(0|k) = p(k,s), & 1 \le k \le m \\ \mathrm{P_r}(k|k) = p(k,i), & 1 \le k \le m-1 \\ \mathrm{P_r}(m|m) = p(m,c) + p(m,i) \end{cases} \tag{5.67}$$

The first equation in (5.67) represents either a successful transmission or an idle event at state 0. The second equation accounts for a collision, which transitions the node to the next state unless it is already at the last state, *m*. A successful transmission returns the node to state 0, as indicated in the next equation. The fourth equation signifies that a node remains in the same state if it does not attempt to transmit. Finally, the fifth equation addresses an idle or a collision event when a node reaches the *mth* state.

One tagged node is selected from each class and analyzed according to the proposed model. It is assumed that all N nodes in the network have the same payload size. If there are n_i nodes in class i, then the total number of nodes in the network will be $N = \sum_i n_i$. All nodes within each class have the same values for $CP_{\min(i)}$ and $CP_{\max(i)}$.

Let τ_i be the probability that the tagged node transmits in a generic slot regardless of the fate of that transmission. A packet transmitted by a node belonging to class i will experience a collision only when one or more nodes of any class transmit in the same slot. Therefore, the collision probability for the tagged node can be obtained as

$$\gamma_i = 1 - (1-\tau_i)^{(n_i-1)} \prod_{j=0, j\neq i}^{7} (1-\tau_j)^{n_j} \tag{5.68}$$

The standard specifies that the probability of transmission is halved with an alternative number of attempts for the same packet. If the first transmission attempt with the base probability $CP_{\max(i)}$ is made for a given packet by the tagged node, then the attempt in state k is made with a probability $1 - p_i(k, i)$ such that the following holds:

$$1 - p_i(k, i) = \max\left(\frac{CP_{\max(i)}}{2^{\lfloor k/2 \rfloor}}, CP_{\min(i)}\right) \tag{5.69}$$

The probability of successful transmission and a collision by the tagged node in a slot, respectively, can be obtained as

$$p_i(k, s) = (1 - \tau_i)^{(n_i - 1)} \prod_{j=0, j \neq i}^{7} (1 - \tau_j)^{n_j} \{1 - p_i(k, i)\} \tag{5.70}$$

$$p_i(k, c) = \left\{1 - (1 - \tau_i)^{(n_i - 1)} \prod_{j=0, j \neq i}^{7} (1 - \tau_j)^{n_j}\right\} \{1 - p_i(k, i)\} \tag{5.71}$$

With stationary state probabilities represented by $\pi_{i,k}$, $(0 \leq k \leq m_i)$, the balance equations, as portrayed in Fig. 5.27, are given by

$$\begin{aligned} \pi_{i,0} &= \pi_{i,0} \{p_i(0, s) + p_i(0, i)\} + \sum_{k=1}^{m_i} \pi_{i,k} p_i(k, s) \\ &= \frac{1}{p_i(0, c)} \sum_{k=1}^{m_i} \pi_{i,k} p_i(k, s) \end{aligned} \tag{5.72}$$

$$\begin{aligned} \pi_{i,k} &= \pi_{i,k-1} p_i(k - 1, c) + \pi_{i,k} p_i(k, i) \\ &= \pi_{i,k-1} \frac{p_i(k - 1, c)}{1 - p_i(k, i)}, \ (0 < k < m_i) \end{aligned} \tag{5.73}$$

$$\begin{aligned} \pi_{i,m_i} &= \pi_{i,m_i-1} p_i(m_i - 1, c) + \pi_{i,m_i} \{p_i(m_i, i) + p_i(m_i, c)\} \\ &= \pi_{i,m_i-1} \frac{p_i(m_i - 1, c)}{p_i(m_i, s)} \end{aligned} \tag{5.74}$$

where

$$p_i(m_i, i) + p_i(m_i, c) + p_i(m_i, s) = 1$$

In addition, the normalization relation is

$$\sum_{k=0}^{m_i} \pi_{i,k} = 1 \tag{5.75}$$

We write (5.73) and (5.74) in terms of $\pi_{i,0}$ as

$$\pi_{i,k} = \frac{\pi_{i,0}\gamma_i^{\,k}\{1 - p_i(0,i)\}}{1 - p_i(k,i)}, \; (0 < k < m_i) \tag{5.76}$$

$$\pi_{i,m_i} = \frac{\pi_{i,0}\gamma_i^{\,m_i}\{1 - p_i(0,i)\}}{(1-\gamma_i)\{1 - p_i(m_i,i)\}} \tag{5.77}$$

Putting $\pi_{i,k}$ and π_{i,m_i} from the above equations in (5.75) produces

$$\pi_{i,0} = \frac{1}{\sum_{k=0}^{\frac{m_i}{2}-1}\left(2^k\gamma_i^{\,2k} + 2^k\gamma_i^{\,2k+1}\right) + \frac{2^{\frac{m_i}{2}}\gamma_i^{\,m_i}}{1-\gamma_i}} \tag{5.78}$$

The quantity representing the transmission probability in an arbitrary slot, τ_i can now be expressed as

$$\begin{aligned}\tau_i &= \sum_{k=0}^{m_i} \pi_{i,k}\{1 - p_i(k,i)\} \\ &= \pi_{i,0}\left(1 + \gamma_i + \cdots + \gamma_i^{\,m_i-1} + \frac{\gamma_i^{\,m_i}}{1-\gamma_i}\right) CP_{\max(i)}\end{aligned} \tag{5.79}$$

Further algebraic deduction yields

$$\tau_i = \frac{CP_{\max(i)}(1 - 2\gamma_i^2)}{(1-\gamma_i^2)\left\{1 - \left(\sqrt{2}\gamma_i\right)^{m_i}\right\} + (1 - 2\gamma_i^2)\left(\sqrt{2}\gamma_i\right)^{m_i}} \tag{5.80}$$

The quantity τ_i is expressed in terms of γ_i, $CP_{\max(i)}$, and m_i, which is a function of only unknown γ_i.

Equations (5.68) and (5.80) provide pairs of coupled non-linear equations for eight user priority classes. These can be solved numerically for γ_i and τ_i for seven classes. The probability of successful transmission in a slot is obtained only when a node transmits a packet and there is no collision.

Thus, the normalized per-class throughput is obtained by

$$\begin{aligned}\eta_i &= n_i \tau_i (1 - \gamma_i) \\ &= n_i \tau_i (1 - \tau_i)^{(n_i - 1)} \prod_{j=0, j \neq i}^{7} (1 - \tau_j)^{n_j}\end{aligned} \tag{5.81}$$

5.5.4 *Performance analysis*

The pairs of non-linear coupled equations (5.68) and (5.80) are solved for a fixed point using MATLAB.

To investigate the normalized saturation throughput in a heterogeneous scenario, the model is applied with two representative classes of users. Two pairs for (γ_i, τ_i) in numerical values are obtained from the associated system of non-linear equations for these classes. We select a lower-priority class UP_0 and a higher-priority class UP_5 to illustrate the outcomes. The throughput results are generated under two conditions: (a) when there are equal numbers of nodes for each user class and (b) when there are different numbers of nodes for different user priority classes.

The throughput for the user priority classes considered is compared with equal numbers of nodes in each class. These comparisons are plotted in Fig. 5.28 as a function of the network size, representing the load in

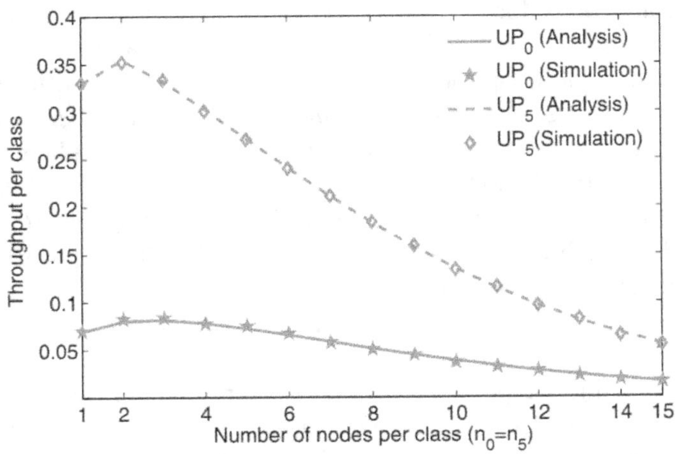

Fig. 5.28. Normalized saturation throughput considering equal numbers of nodes of UP_0 and UP_5.

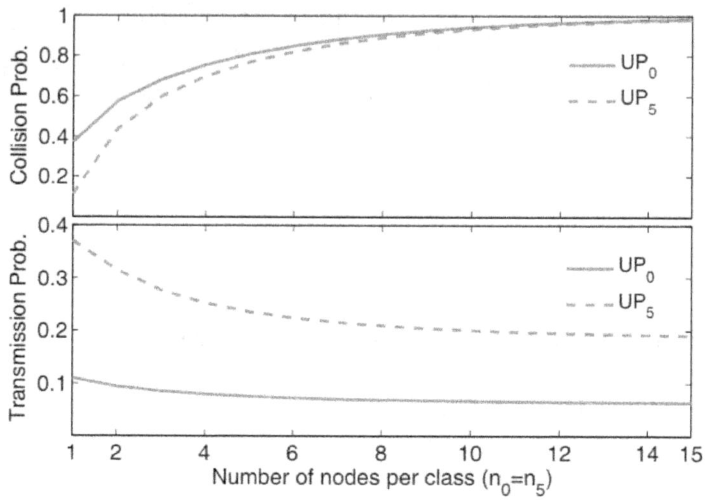

Fig. 5.29. Probability of collision and probability of transmission.

saturation condition. The general trend of the throughput curves shows an increase from very small network sizes ($n_0 = n_5 = 1$) toward the peak. After reaching the peak, the throughput decreases exponentially with higher loads (i.e., larger network sizes).

The higher-priority class UP_5 with higher $CP_{\max}$ and higher $CP_{\min}$ can obtain a contended allocation more frequently than the lower-priority class UP_0. Consequently, the higher-priority class achieves a higher throughput compared to the lower-priority class. The throughput disparities between these two categories are more pronounced when the number of nodes is smaller. The patterns of the throughput curves can be explained by the transmission and collision probabilities shown in Fig. 5.29. As expected, UP_5 has a higher transmission probability (τ_5) as it has a higher $CP_{\max}$ and higher $CP_{\min}$ than those of UP_0. Declination of collision probability for the higher class is apparent, considering (5.68).

The more practical heterogeneous scenario may have different numbers of nodes for different user priority classes. Such cases are presented in Fig. 5.30 showing the per-class throughput for UP_0 and UP_5. The values of n_0 are from 1 to 15, and the values of n_5 are from 1 to 5. The highest throughput for UP_0 is 0.283 at ($n_0 = 13, n_5 = 1$) in this scenario, as shown in Fig. 5.30(a). However, Fig. 5.30(b) shows that the highest throughput for UP_5 is 0.399 at ($n_0 = 1, n_5 = 3$).

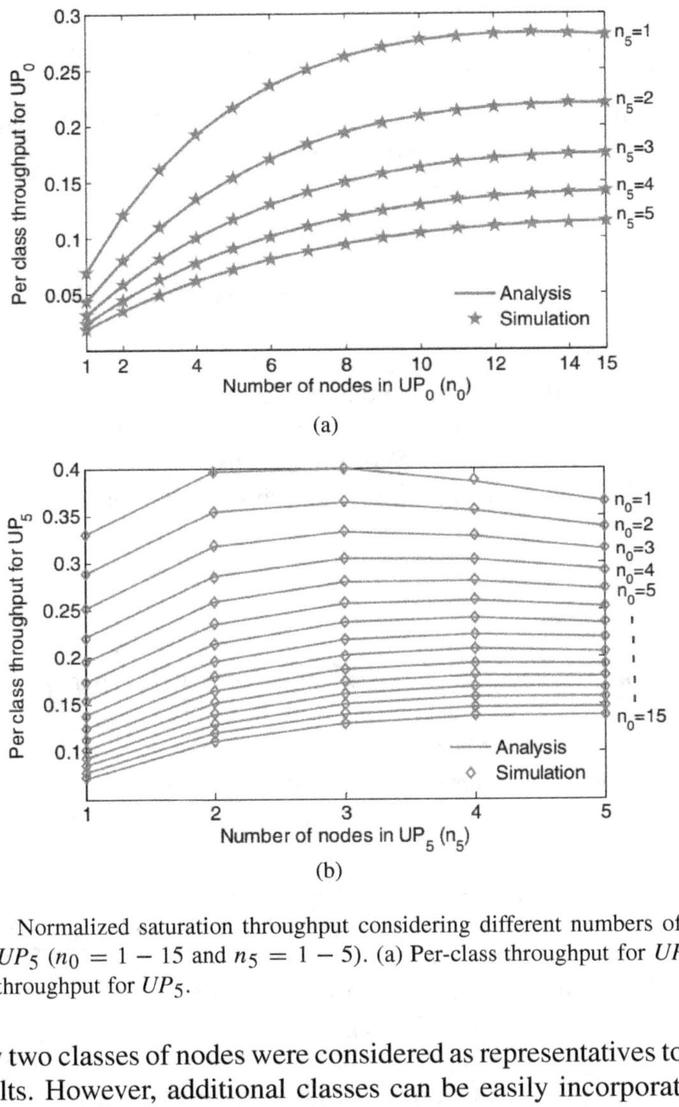

Fig. 5.30. Normalized saturation throughput considering different numbers of nodes of UP_0 and UP_5 ($n_0 = 1 - 15$ and $n_5 = 1 - 5$). (a) Per-class throughput for UP_0 and (b) Per-class throughput for UP_5.

Only two classes of nodes were considered as representatives to explain the results. However, additional classes can be easily incorporated for a heterogeneous network using the proposed analytical model. Furthermore, the model is applicable for homogeneous setups as well.

5.5.5 *Summary*

The analytical model for the IEEE 802.15.6 standard presented in this section can be employed to quantitatively determine the protocol parameters in accordance with the criticality and the general traffic characteristics of the packets generated in certain application scenarios. Further work in this regard can investigate a more realistic scenario of non-saturated traffic, especially differences in delay performance in such settings. The model can also be extended to incorporate different access phases with respective particulars taken into account.

5.6 CSMA/CD in Wireless Personal Area Networks

This section introduces an analytical model for the slotted carrier-sense multiple access with collision avoidance (CSMA/CA) algorithm used in the contention access period (CAP) of the beacon-enabled mode of medium access control (MAC) protocol specified in the IEEE 802.15.4 standard. The analysis is validated through simulation and compared with another related model.

5.6.1 *Introduction*

The IEEE 802.15.4 standard [62] is designed for wireless personal area networks (WPANs) comprising low-rate, low-power devices. Its non-beacon-enabled mode of the MAC protocol utilizes an unslotted CSMA/CA mechanism, which has been extensively studied in the classical literature [63]. In contrast, the beacon-enabled mode structures the channel access timeline with a sequence of superframes separated by periodic beacons. An integral component of each superframe is the CAP, which employs a non-persistent slotted CSMA/CA protocol in a distinct manner. Additionally, the superframe includes provisions for a contention-free period (CFP) comprising guaranteed time slots (GTS) and an inactive period aimed at facilitating power-saving modes. While the GTS allocation mechanism lies beyond the MAC specification, both the optional CFP and the inactive period play significant roles, leading to extensive investigations primarily focused on the behavior of the CAP.

The slotted non-persistent CSMA/CA mechanism employed in the CAP differs significantly from the distributed coordination function (DCF) of IEEE 802.11, despite sharing a common feature in their utilization of binary exponential backoff. Unlike IEEE 802.11 (WLAN), where nodes constantly monitor the channel state, nodes in the CAP do not engage in continuous channel monitoring, either physically or virtually.

This characteristic results in constant slot lengths within the channel state model, regardless of whether the channel is idle or busy, and eliminates the provision for backoff freezing at nodes during busy channel periods. Importantly, each contending node is required to perform physical carrier sensing via clear channel assessment (CCA) to detect two successive idle slots before being permitted to transmit. Failure in carrier sensing during either of these CCAs leads to a deferral of transmission, with the node uniformly selecting a random backoff value from the exponential range of the next backoff stage. The specification constrains the backoff exponent between a minimum (aMinBE = 3) and a maximum (aMaxBE = 5) and limits the number to a maximum (macMaxCSMABackoffs = 5) of backoff stages.

Analytical models offer distinctive insights into protocol operation, prompting numerous analyses of the 802.15.4 MAC protocol with varying levels of accuracy [64]. Several studies utilize Markov chains for node state modeling, albeit often presenting challenges due to their intricate nature [65, 66]. A relatively innovative approach, as shown in Ref. [67], adopts a renewal theoretic framework, offering a novel perspective on protocol analysis.

This work introduces a relatively straightforward analytical model for the CAP, utilizing conditional probability arguments (similar to those in Ref. [68] applied to the 802.11 DCF) to model the node state while employing a discrete-time Markov chain to model the channel state [63]. The assumptions and approximations employed here align with those in Ref. [67]. To validate the model, its predictions, along with those of Ref. [67], are compared with the results obtained from Monte Carlo simulations [63].

5.6.2 *Analytical model*

Consider a single-hop wireless network where n identical nodes contend under saturation conditions to transmit frames. The analysis is approached by formulating key parametric relationships involving σ, the probability of the tagged node sensing the channel in a slot, and ϕ, the probability of idle-sensing failure during such an attempt. The statistical approximation of their independence is central to the model. Consequently, the framework is built upon two sub-models: the node-state model and the channel-state model.

5.6.2.1 *Node-state model*

Let S denote the event that a node is sensing the medium/carrier, with $P(S)$ being the corresponding event probability. Additionally, let $(b = i)$ represent the event that a node is in backoff stage $i \in (0, \ldots, M)$, where $P(b = i)$ is the associated event probability and M is the maximum backoff stage. The principle of conditional probability yields

$$P(b = i, S) = P(b = i \mid S) \cdot P(S) = P(S \mid b = i) \cdot P(b = i) \quad (5.82)$$

That is,

$$P(S) \cdot \frac{P(b = i|S)}{P(S|b = i)} = P(b = i) \quad (5.83)$$

Using the total probability rule,

$$P(S) \sum_{i=0}^{M} \frac{P(b = i|S)}{P(S|b = i)} = \sum_{i=0}^{M} P(b = i) = 1 \quad (5.84)$$

The probability that the tagged node is in the state of sensing is given by

$$\sigma = P(S) = 1 \Big/ \sum_{i=0}^{M} \frac{P(b = i|S)}{P(S|b = i)} \quad (5.85)$$

The conditional probability $P(b = i|S)$ represents the probability that a node, currently sensing the channel, has been in backoff stage i. Let ϕ denote the failure probability in a sensing attempt; this corresponds to the event that the sensing node finds the channel busy either in the first CCA or in the second (in case the channel were idle in the first CCA). Thus, the nodes in the conditioned states follow a simple Markov chain with the following transition probabilities among them:

$$\begin{cases} P(b(k+1) = i|b(k) = i-1) = \phi, & i = 1, \ldots, M \\ P(b(k+1) = 0|b(k) = i) = 1-\phi, & i = 0, \ldots, M-1 \\ P(b(k+1) = 0|b(k) = M) = 1, & i = M \end{cases} \quad (5.86)$$

Given M, the maximum number of backoff stages, equation (5.86) can be readily solved to obtain the steady-state probability of the tagged sensing node being in backoff stage i.

$$P(b = i|S) = \frac{(1-\phi)\phi^i}{1-\phi^{M+1}} \tag{5.87}$$

The probability that the tagged node, currently in backoff stage i, is busy sensing the channel/carrier can be obtained by applying the extension of the elementary renewal theorem [69]. To achieve this, each sensing cycle the node undergoes can be considered as a renewal cycle, comprising three components: (i) the average time the node at stage i waits during backoff (denoted by $E[b_i]$), (ii) the mean number of slots the node spends in carrier sensing (denoted by $\overline{n}_\sigma$, for which the expression is derived later from the channel-state model), and (iii) the expected duration of transmission in this sensing cycle, given by $0 \times \phi + L(1-\phi) = L(1-\phi)$. Hence,

$$P(S|b = i) = \frac{1}{\mathbb{E}[b_i] + \overline{n_\sigma} + (1-\phi)L} \tag{5.88}$$

Here, L denotes the frame length in slots, the adopted backoff mechanism determines $\text{E}[b_i]$, and with the binary exponential backoff, where the backoff value is uniformly randomly chosen upon the tagged node entering backoff stage i, it is expressed as for $i \in (0, 1, \ldots, M)$:

$$\mathbb{E}[b_i] = \begin{cases} (2^{aMinBE+i} - 1)/2 & \text{if } (aMinBE + i) \le aMaxBE \\ (2^{aMinBE} - 1)/2 & \text{otherwise} \end{cases} \tag{5.89}$$

Using (5.85), (5.87) and (5.88), the probability that the tagged node is in the state of sensing is obtained by

$$\sigma = \frac{1-\phi^{M+1}}{(1-\phi)\sum_{i=0}^{M}\phi^i(\mathbb{E}[b_i] + \overline{n_\sigma} + (1-\phi)L)} \tag{5.90}$$

5.6.2.2 *Channel-state model*

To establish the other relationship involving σ and ϕ, a discrete-time Markov chain (see Fig. 5.31) is employed. During operation in the CAP, it is imperative that the channel is sensed to be idle for two successive slots before a node initiates a transmission. To accommodate the state-wise memoryless property inherent in a Markov model, the channel state is modeled in terms of states represented on an adjacent-slot-pair basis.

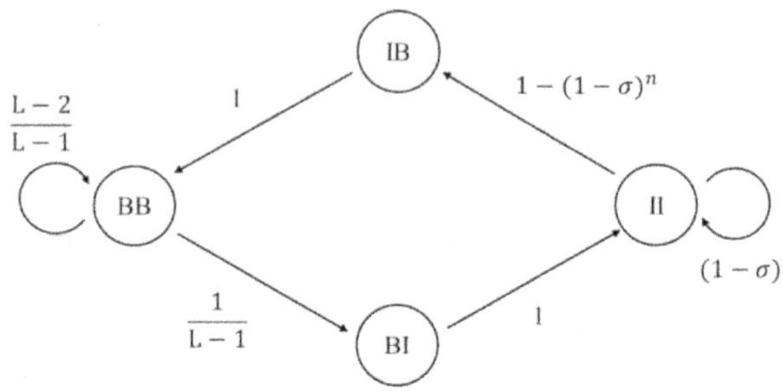

Fig. 5.31. Adjacent-slot-pair-wise channel state model.

The channel transitions from the state of successive idle-slot pair (II) to becoming busy (IB) with a probability of $1-(1-\sigma)^n$ as one or more nodes conduct carrier sensing, while it remains in state II with a probability of $(1-\sigma)^n$. Assuming the frame length in slots to be at least 3 (i.e., $L \geq 3$), the channel retains the busy slot-pair state (BB) with a probability of $(L-2)/(L-1)$, under the assumption of uniformity. Conversely, it takes a transition to state BI with a probability of $1/(L-1)$. According to the state transitions thus modeled, the stationary channel state probabilities are related as follows:

$$\begin{cases} P_{BB} = P_{IB} + P_{BB} \cdot \frac{L-2}{L-1}, & P_{IB} = P_{II}(1-(1-\sigma^n)) \\ P_{II} = P_{BI} + P_{II} \cdot (1-\sigma)^n, & P_{BI} = P_{BB} \cdot \frac{1}{L-1} \\ \text{And,} \;\; P_{BB} + P_{IB} + P_{BI} + P_{II} = 1 \end{cases} \tag{5.91}$$

Solving the above equations yields

$$\begin{cases} P_{BB} = \frac{(L-1)\cdot\tau}{1+\tau(1+L)}, & P_{II} = \frac{1}{1+\tau(1+L)} \\ P_{IB} = P_{BI} = \frac{\tau}{1+\tau(1+L)} \end{cases} \tag{5.92}$$

Since in a sensing attempt, a node utilizes 1 slot with certainty and an additional slot only when the channel state is II or IB, then

$$\overline{n_\sigma} = 1 + (P_{II} + P_{IB}) = \frac{2+\tau(2+L)}{1+\tau(1+L)} \tag{5.93}$$

As successful carrier sensing is only possible during channel state II, the sensing failure probability is now given by

$$\phi = 1 - P_{II} = \frac{\tau(1+L)}{1+\tau(1+L)} \tag{5.94}$$

Using (5.90) and (5.94), one can numerically solve for σ and ϕ.

5.6.3 *Performance analysis*

Given the framework outlined, the MAC throughput and the mean frame service time (or the average access delay) can now be deduced.

5.6.3.1 *Throughput*

Conditioning on the channel states, the normalized network throughput can be expressed with 'Tx' denoting transmission as

$$\Gamma = \mathbb{E}\{\text{per} - \text{slot T}_X\} \times P\{\text{a given T}_X \text{ is successful}\} \tag{5.95}$$

Using equations (5.92) and (5.94), and adjusting for per-slot statistics,

$$\mathbb{E}\{\text{per} - \text{slot T}_X\} = \frac{1}{2}\{2.P_{BB} + 1.P_{BI} + 1.P_{IB} + 0.P_{II}\} = L\tau(1-\phi) \tag{5.96}$$

and

$$\begin{aligned} P\{\text{a given T}_X \text{ is successful}\} &= \frac{P\{\text{exactly one node T}_X\}}{P\{\text{one or more nodes T}_X\}} \\ &= \frac{n\sigma(1-\sigma)^{n-1}}{1-(1-\sigma)^n} = \frac{n\sigma(1-\sigma)^{n-1}}{\tau} \end{aligned} \tag{5.97}$$

Therefore, from (5.95), (5.96), and (5.97), the normalized throughput is as follows:

$$\Gamma = nL\sigma(1-\sigma)^{n-1}(1-\phi) \tag{5.98}$$

5.6.3.2 *Mean frame service time*

The mean frame service time or, equivalently, the average access delay (T), is measured as the mean time elapsed (in slots) from the moment a frame becomes ready for transmission to the point of its successful delivery.

As in Ref. [67], in the event that a node fails to transmit despite reaching the maximum backoff stage, it is considered to restart contention with the same frame from the minimum backoff stage. Consequently, adjustments for frame drop are deemed unnecessary. Given that Γ/L accounts for the

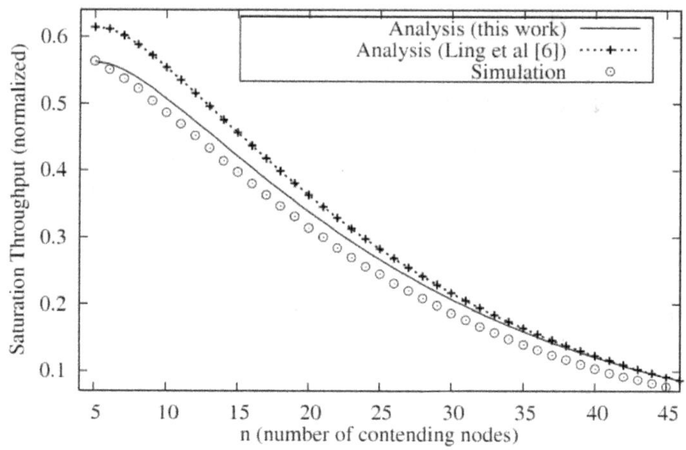

Fig. 5.32. Saturation throughput: $L = 8$, $aMaxBE = 5$, $aMinBE = 3$, $M = 5$.

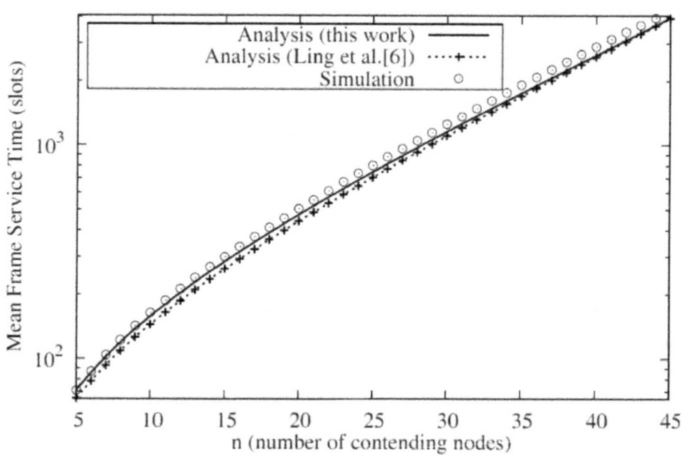

Fig. 5.33. Mean frame service time: $L = 8$, $aMinBE = 3$, $aMaxBE = 5$, $M = 5$.

mean frame service rate (in frames per slot), from the application of Little's formula,

$$T = \frac{n}{\Gamma/L} = \frac{1}{\sigma(1-\sigma)^{n-1}(1-\phi)} \tag{5.99}$$

5.6.4 *Model validation*

The pair of non-linear simultaneous equations (5.90) and (5.94) is solved for a fixed point using MATLAB employing the contraction-mapping method. The analytical results for the throughput (Fig. 5.32) and the mean frame service time (Fig. 5.33) are plotted against the corresponding estimates from discrete-event C++ simulation. The analytical predictions from Ref. [67] are also compared (marked by Ling *et al.* [67]). Our model closely matches the simulation in both measures, whereas Ref. [67] slightly overestimates the throughput for smaller network sizes.

5.6.5 *Summary*

A relatively simple yet accurate model has been proposed for the distinctive slotted CSMA/CA mechanism employed in the CAP of the IEEE 802.15.4 MAC. This model holds the potential for extension to related protocol variants and inclusion of further details, such as acknowledged traffic and unsaturated network conditions.

References

[1] W. Stallings, *Wireless Communications and Networkings*, Prentice Hall, 2002.
[2] N. Abramson, The ALOHA System – Another Alternative for Computer Communications, *Proc. Fall Joint Comput. Conf. A FIPS*, 37, 281–285, 1970.
[3] A. S. *Tanenbaum, Computer Networks*, Prentice-Hall, Inc., Englewood Cliffs, N.J., 1981.
[4] M. Schwartz, *Computer-Communication Network Design and Analysis*, Prentice-Hall, Inc., Englewood Cliffs, NJ., 1977.
[5] L. Kleinrock, *Queueing Systems*, Vol. 1, Theory, John Wiley & Sons, Inc., New York. 1975.
[6] N. Abramson, The throughput of packet broadcasting channels, *IEEE Trans. Commun.*, 25, 117–128, Jan 1977.
[7] N. Abramson, Packet switching with satellites, *A FIPS Conf, Proc*, 42, 695–702, June 1973.
[8] W. Crowther, R. Rettberg, D. Walden, S. Ornstein, and, F. Heart, A system for broadcast communication: Reservation ALOHA, *Proc. Sixth Hawaii Int. Conf. Syst. Sci.*, 371–374, Jan 1973.
[9] R. Peterson, R. Ziemer, and D. Borth, *Introduction to Spread Spectrum Communications*, Prentice Hall, 1995.
[10] R. Dixon, Spread spectrum systems with commercial application, John Wiley, 1994.
[11] N. Abramson, "Wideband random access for the last mile", IEEE Personal Communications, 3(6), 1996.

[12] W. Stallings, "Local Network Performance," IEEE Commun. Mag" vol. 22, No. 2, Feb. 1984, pp. 27–36.

[13] W. Bux, "Local Area Subnetworks: A Performance Comparison," IEEE Trans. Commun, vol. COM29, no. 10, Oct. 1981, pp. 1465–1473.

[14] L. Kleinrock, F. Tobagi, "Packet switching in radio channels: part 1 carrier sense multiple access modes and their throughput-delay characteristics," IEEE Transactions on Communications, COM-23 (12), 1975.

[15] R. G. Gallager, A perspective on multiple access channels, *IEEE Trans. Inf. Theory*, 31(2), 124–142, Mar 1985.

[16] D. Bertsekas and R. Gallager, *Data Networks*, Prentice Hall, 1992.

[17] B. S. Tsybakov, V. A. Mikhailov, Ergodicity of a Slotted ALOHA System, *Probl. Peredachi Inf.*, 15(4), 73–87, 1979; *Problems Inform. Transmission*, 15(4), 301–312, 1979.

[18] R. L. Rivest, Network control by Bayesian broadcast, *IEEE Trans. on Inf. Theory*, 33(3), 323–328, May 1987.

[19] V. A. Mikhailov, *Methods of Random Multiple Access. Candidate Engineering Thesis*. Moscow Institute of Physics and Technology, Moscow, USSR, 1979.

[20] F. P. Kelly, Stochastic models of computer communication systems, *J. Roy. Stat. Soc., Ser.* B47(3), 379–395, 1985.

[21] J.N. Tsitsiklis, Analysis of a multiple access control scheme, *IEEE Trans. Automat. Control*, 32(11), 1017–1020, Nov 1987.

[22] D. Aldous, Ultimate instability of exponential back-off protocol for acknowledgement-based transmission control of random access communication channels, *IEEE Trans. Inf. Theory*, 33(2), 219–233, Nov 1987.

[23] J. Goodman, A. G. Greenberg, N. Madras and P. March, Stability of binary exponential backoff, *J. Association for Computing Machinery*, 35(3), 579–602, Jul 1988.

[24] L. G. Roberts, ALOHA packet system with and without slots and capture, *Comput. Commun. Rev.*, 5, 28–42, Apr 1975.

[25] D. H. Davis and A. Gronemeyer, Performance of slotted ALOHA random access with delay capture and randomized time of arrival, *IEEE Trans. Commun.*, 28(5), 703–710, May 1980.

[26] A. Polydoros and J. Silvester, Slotted random access spread-spectrum networks: An analytic framework, *IEEE J. Select. Areas Commun.*, 5, 989–1002, Jul 1987.

[27] A. Shwartz and M. Sidi, Erasure, capture and noise errors in controlled multiple-access networks, EE Pub. 600, Technion-Israel Inst. Technol., Haifa, Israel, Aug 1986.

[28] A. Shwartz and M. Sidi, Erasure, capture and noise errors in controlled multiple-access networks, in *Proc. 25th Conf. Decision Contr.*, Athens, Greece, Dec 1986.

[29] M. Sidi and I. Cidon, Splitting protocol in presence of capture, *IEEE Trans. Inform. Theory*, 31, 295–301, Mar. 1985.

[30] K. S. Kwak and R. Rao, Transmission control of capture ALOHA systems, in *Proc. IEEE ICC*, June 1988.

[31] S. Ghez, S. Verdu, and S. C. Schwartz, Stability properties of slotted ALOHA with multipacket reception capability, *IEEE Trans. Automat. Contr.*, 33, 640–649, Jul 1988.

[32] L. Cidon, H. Kodesh, and M. Sidi, Erasure, capture and random power level selection in multiple-access systems, *IEEE Trans. Commun.*, 36, 263–271, Mar 1988.

[33] K. S. Kwak, Capture Statistics in Multiple Access Communication Systems, *IEICE Trans. Fundamentals*, E75-A(6), 701–711, June 1992.

[34] J. N. Tsitsiklis, Analysis of a multi-access control scheme, *IEEE Trans. Automat. Contr.* 32, 1017–1020, Nov 1987.

[35] N. Abramson, The throughput of packet broadcasting channels, *IEEE Trans. Commun.*, 25, 117–128, Jan 1976.

[36] J. J. Metzer, On improving utilization in ALOHA networks, *IEEE Trans. Commun.*, 24, 447–448, Apr 1976.

[37] D. Bertsekas and R. Gallager, *Data Networks*. Prentice Hall, 1987.

[38] M. Kaplan, A Sufficient Condition for Nonergodicity of a Markov Chain, *IEEE Trans. Inform, Theory*, 25(4), 470–471, Jul 1979.

[39] B. Hajek and T. van Loon, Decentralized dynamic control of a multiaccess broadcast channel, *IEEE Trans. Automat. Contr.*, 27, 559–569, June 1982.

[40] V. A. Mikhailov, Geometric analysis of the stability of Markov chains in R: and its application to throughput evaluation of the adaptive random access algorithm, *Probl Peredachi Inf*, 24(I), pp. 61–73, Jan-Mar 1988 (translated in Problems of Information Transmission, Jul 1988).

[41] E. J. Baghdady, Theory of Stronger-Signal Capture in FM Reception, *Proc., IRE*, 4, 728–738, Apr 1958.

[42] J. C. Arnbak and W. van Blitterswijk, Capacity of slotted ALOHA in Rayleigh fading channels, *IEEE J. Sel. Areas Commun.*, 5, 261–269, Feb 1987.

[43] J. A. Robert and T. J. Healy, Packet radio performance over slow Rayleigh fading channels, *IEEE Trans. Commun.*, 28, 279–286, 1980.

[44] D. H. Davis and A. Gronemeyer, Performance of slotted ALOHA random access with delay capture and randomized time of arrival, *IEEE Trans. Commun.*, 28(5), 703–710, May 1980.

[45] K. S. Kwak and R. R. Rao, Controlled ALOHA with geometric capture probabilities, *IEEE Trans. Automatic Control*, 35(11), 1270–1274, 1990.

[46] M. L. Nakagami, The m-distribution: A general formula of intensity distribution of rapid fading, *Statistical Method of Radio Wave Propagation*, Pergamon Press (I960).

[47] L. Rayleigh, On the resultant of a large number of vibration of the same pitch and of arbitrary phase, *Phil. Mag.*, 10, 77–78, Aug 1880.

[48] H. B. James and P. I. Wells, Some tropospheric scatter propagation measurements near the Radio-Horizon, *Proc., IRE*, 43, 1336–1340, Oct 1955.

[49] U. Charash, Reception through Nakagami fading multipath channels with random delays, *IEEE Trans. Commun.*, 27, 657–670, Apr 1979.

[50] H. Suzuki, A statistical model for urban radio propagation, *IEEE Trans. Commun.*, 25, 673–680, Jul 1977.

[51] E. A. Robinson, *Least Squares Regression Analysis in Terms of Linear Algebra*, Goose Pond Press, 1981.

[52] Y. L. Luke, *The Special Function and Their Applications*, Vol. I & II, Academic Press, 1969.

[53] "Part 15.6: wireless body area networks, IEEE standard for local and metropolitan area networks," IEEE Std 802.15.6, 2012.

[54] S. Ullah, H. Higgins, B. Braem, B. Latre, C. Blondia, I. Moerman, S. Saleem, Z. Rahman, and K. S. Kwak, A comprehensive survey of wireless body area networks, *J. Medical Systems*, 36(3), 1065–1094, 2012.

[55] S. Rashwand, J. MišIć, and H. Khazaei, Performance analysis of IEEE 802.15.6 under saturation condition and error-prone channel, in *Proc. 2011 IEEE Wireless Communications and Networking Conference*, 1167–1172.

[56] M.S. Chowdhury, K. Ashrafuzzaman, and K.S. Kwak, Saturation throughput Analysis of IEEE 802.15.6 Slotted Aloha in Heterogeneous Conditions, *IEEE Wireless Comm. Letters*, 3(3), 257–260, 2014.

[57] B. H. Jung, R. U. Akbar, and D. K. Sung, Throughput, energy consumption, and energy efficiency of IEEE 802.15.6 body area network (BAN) MAC protocol, in *Proc. 2012 IEEE Personal Indoor and Mobile Radio Communications*, 584–589.

[58] D. Malone, K. Duffy, and D. Leith, Modeling the 802.11 distributed coordination function in non-saturated heterogeneous conditions, *IEEE/ACM Trans. Netw.*, 15(1), 159–172, 2007.

[59] D. Bertsekas and R. Gallager, *Data Networks*. Prentice Hall, 1987.

[60] Y. Yu, X. Cai, and G. B. Giannakis, On the instability of slotted ALOHA with capture, *IEEE Trans. Wireless Commun.*, 5(2), 257–261, 2006.

[61] M. Ferguson, On the control, stability, and waiting time in a slotted ALOHA random-access system, *IEEE Trans. Commun.*, 23(11), 1306–1311, 1975.

[62] Part 15.4: Wireless Medium Access Control (MAC) and Physical Layer (PHY) Specifications for Low-Rate Wireless Personal Area Networks (LR-WPANs), IEEE 802.15.4 Std., 2003.

[63] K. Ashrafuzzaman and K.S. Kwak, On the performance analysis of the contention access period of IEEE 802.15.4 MAC, *IEEE Comm. Letters*, 15(9), 986–988, Sept. 2011.

[64] J. He, Z. Tang, H. Chen, and Q. Zhang, An accurate and scalable analytical model for IEEE 802.15.4 slotted CSMA/CA networks, *IEEE Trans. Wireless Commun.*, 8(1), 440–448, 2009.

[65] P. Park, P. Di Marco, P. Soldati, C. Fischione, and K. Johansson, A generalized Markov chain model for effective analysis of slotted IEEE 802.15.4, in *Proc. 6th IEEE MASS*, 130–139, 2009.

[66] S. Pollin, M. Ergen, S. Ergen, B. Bougard, L. Der Perre, I. Moerman, A. Bahai, P. Varaiya, and F. Catthoor, Performance analysis of slotted carrier sense IEEE 802.15.4 medium access layer, *IEEE Trans. Wireless Commun.*, 7(9), 3359–3371, 2008.

[67] X. Ling, Y. Cheng, J. Mark, and X. Shen, A renewal theory based analytical model for the contention access period of IEEE 802.15.4 MAC, *IEEE Trans. Wireless Commun.*, 7(6), 2340–2349, 2008.

[68] G. Bianchi and I. Tinnirello, Remarks on IEEE 802.11 DCF performance analysis, *IEEE Commun. Lett.*, 9(8), 765–767, 2005.

[69] R. Gallager, *Discrete Stochastic Processes*, Springer Science & Business Media, Dec 6, 2012.

Chapter 6

Non-Orthogonal Multiple Access Systems

6.1 Introduction

Each new generation of cellular networks brings forth advancements in standards, techniques, and features, distinguishing it from its predecessors. The forthcoming cellular systems, including the fifth generation (5G) and beyond the next generation (NG), are expected to enable advanced services, such as multimedia applications, Internet of Things (IoT)-based applications, and vehicle-to-everything (V2X) connectivity. These innovative use cases are fueling the exponential growth of mobile traffic, thereby posing radio spectrum scarcity as a primary challenge that 5G and beyond must confront.

Multiple access, a fundamental building block in wireless communication systems, profoundly influences the utilization of available spectrum, system throughput, and latency. In the cellular radio context, multiple access refers to a technique where multiple users share a common radio resource to establish communication links with a base station (BS). Time-division multiple access (TDMA), frequency-division multiple access (FDMA), and code-division multiple access (CDMA) have been widely used in the previous generations of cellular networks, categorized as orthogonal multiple access (OMA). In OMA, users' access is ideally orthogonal, ensuring minimal interference, and radio resources in time, frequency, or code domains are assigned to multiple users [1].

In TDMA, each user is assigned a dedicated time slot for signal transmission. Users' signals are differentiated at the receiving end by identifying the corresponding time slots. Similarly, FDMA divides the available bandwidth (BW) into non-overlapping frequency subchannels, allocating each subchannel to a distinct user. At the receiving end, users' signals are differentiated by identifying the respective frequency bands. In contrast, CDMA assigns a unique orthogonal spreading sequence to each user, and users transmit their signals sharing the same time–frequency resources. Users' signals are separated at the receiving end using decorrelation based on the assigned code, treating other users' signals as noise.

OMA-based systems theoretically avoid inter-user interferences due to orthogonal resource allocation. However, the limited number of orthogonal resources constrains OMA systems in serving a large number of users, as mandated by 5G/6G. Non-orthogonal multiple access (NOMA), on the other hand, permits inter-user interference in resource allocation, enabling multiple users to share the same resource block. To mitigate interference, schemes like successive interference cancellation (SIC) are applied, demonstrating NOMA's potential to handle numerous connections while offering superior sum capacity and user fairness. NOMA-based cellular networks are projected to support diverse data-hungry applications.

The concept of NOMA in the 5G cellular context was first proposed, showing its superior performance. The appealing advantages of NOMA ignited substantial research interest. Hence, understanding NOMA and its applications in cellular communication systems is crucial. NOMA achieves superior spectral efficiencies by superposition coding (SC) at the transmitter with SIC at the receiver [2, 3]. On top of that, the evolution of wireless networks to NG poses new challenges for energy efficiency (EE) since the entire network will be ultra-dense. With an extreme increase in the number of infrastructure nodes, the total energy consumption may simply surpass an acceptable level. Although substantial energy is basically consumed by the hardware, NOMA has an inherent ability to adapt the transmission strategy according to the traffic and users' channel state information (CSI), achieving a good operating point where both spectrum efficiency (SE) and EE become optimum.

Over the past few years, NOMA has attracted a great deal of attention from researchers trying to meet new requirements in 5G and beyond. As a consequence, many research efforts in this field already exist, covering various topics, such as performance analysis methods, fairness analysis, EE, and user pairing. Researchers are attempting to further enhance the performance of other existing wireless technologies, such as cooperative communications, multiple-input multiple-output (MIMO), light communications, and relay networks by using NOMA. However, NOMA is still in its infancy. At this stage, a comprehensive knowledge of the up-to-date research status of NOMA is extremely useful to researchers who want to do more in this area. In this regard, this chapter examines the fundamentals of NOMA and the various issues that need to be addressed to transform multi-access techniques through NOMA innovation.

6.2 Fundamentals of NOMA

Different NOMA solutions fall into two primary approaches, broadly classified as power-domain NOMA and code-domain NOMA. Figure 6.1 illustrates a simple classification of the existing NOMA techniques.

Power-domain NOMA achieves multiplexing in the power domain, while *code-domain NOMA* accomplishes multiplexing in the code domain. In code-domain NOMA, unlike the basic CDMA systems, users share the entire available resources, including time and frequency.

Code-domain NOMA utilizes user-specific spreading sequences, which can be sparse sequences or non-orthogonal cross-correlation sequences with low correlation coefficients. This category can be further divided into distinct classes, including the following:

- **Low-density spreading-CDMA (LDS-CDMA):** Utilizing sparse sequences, LDS-CDMA mitigates the interference impact on each chip of basic CDMA systems [4, 5].
- **Low-density spreading-based-OFDM (LDS-OFDM):** This is a combination of LDS-CDMA and orthogonal frequency-division multiplexing (OFDM), where information symbols are first spread across low-density spreading sequences and then transmitted on a set of subcarriers [6, 7].

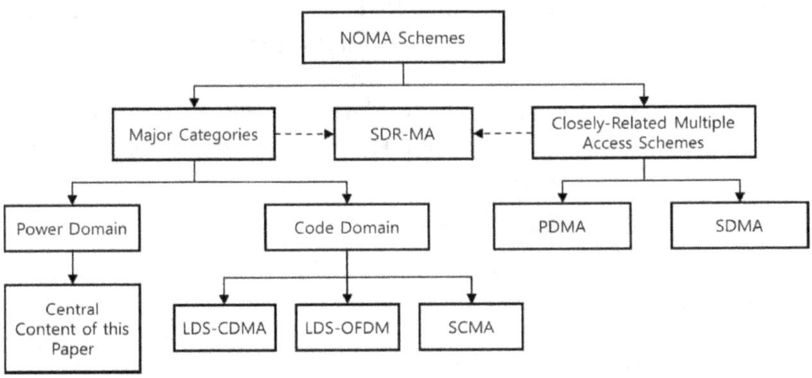

Fig. 6.1. A simple classification of NOMA techniques.

- **Sparse code multiple access (SCMA):** Unlike LDS-CDMA, SCMA allows direct mapping of information bits to different sparse codewords, combining both bit mapping and bit spreading. It provides a low-complexity reception technique and offers improved performance compared to LDS-CDMA [8, 9].

This classification provides insights into the diverse approaches within code-domain NOMA, emphasizing their unique features and advantages.

There are several other multiple access techniques closely related to NOMA, such as pattern division multiple access (PDMA) [10] and spatial division multiple access (SDMA) [11–13]. PDMA operates across various domains, aiming to maximize diversity and minimize overlaps among users. At the transmitter, PDMA designs non-orthogonal patterns, and multiplexing occurs in the code domain, spatial domain, or a combination of both.

In the case of SDMA, its principle draws inspiration from basic CDMA systems. Unlike using user-specific spreading sequences, SDMA distinguishes between users by employing user-specific channel impulse responses (CIRs). This approach is particularly valuable in scenarios where the number of uplink users significantly exceeds the number of corresponding receiving antennas at the BS. However, accurate CIR estimation becomes challenging when dealing with a large number of users.

The concept of software-defined radio for multiple access (SDR-MA) enables the coexistence of various NOMA schemes [15]. This technique allows for a flexible configuration of multiple access schemes, supporting heterogeneous services and applications in 5G [14]. It's essential to note that while this list provides insights into different forms of NOMA, it is not exhaustive. The primary focus of this chapter remains on power-domain NOMA [58].

The subsequent section provides a brief overview of superposition coding (SC) and SIC, as these fundamental techniques are pivotal for grasping the essence of the NOMA paradigm, particularly in the context of power-domain NOMA.

6.2.1 *PD-NOMA*

6.2.1.1 *A downlink scenario in PD-NOMA*

Figure 6.2 illustrates a simple NOMA system with a single BS and two users, each equipped with a single antenna. Let x_1 and x_2 be the signals

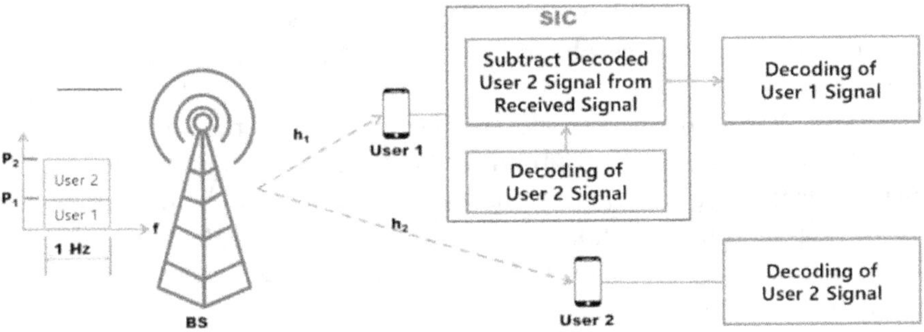

Fig. 6.2. Two-user power-domain downlink NOMA.

transmitted from the BS to users 1 and 2, respectively. The BS transmits a superposition-coded signal as

$$s = \sqrt{P_1 x_1} + \sqrt{P_2 x_2} \tag{6.1}$$

where $P_i, i = 1, 2$ is the transmit power for user i, and the message signal $x_i, i = 1, 2$ has unit power (i.e., $E\{|x_i|^2\} = 1$). The total transmit power of users 1 and 2 is denoted as P. In practice, P is predefined and divided into P_1 for user 1 and P_2 for user 2 based on the adopted power allocation (PA) scheme. The received signal at user i is given by

$$y_i = h_i s + n_i \tag{6.2}$$

where h_i is the channel gain between the BS and user i, and n_i represents Gaussian noise plus interferences with power spectral density $N_{f,i}$. For a multi-cell scenario, inter-cell interference is also included in n_i.

To distinguish between users' signals, SIC is employed at the receivers. The optimal decoding order of SIC is based on the decreasing strengths of the users' channels, determined by $|h_i|^2/N_{f,i}$. This order enables each user to significantly reduce interferences from the signals of other users whose decoding orders follow that user. Consequently, user 1, known as the strong user due to its maximum channel strength $|h_1|^2/N_{f,1}$, can effectively cancel interference from user 2, referred to as the weak user with the least channel strength $|h_2|^2/N_{f,2}$. It's important to note that the BS periodically performs SIC ordering based on the CSI feedback received from users. Users receive updated information on the SIC ordering from the BS.

The PA is adjusted based on the channel strength, with a user having a weaker channel (i.e., the weak user) being allocated higher power compared to a user with a stronger channel (strong user). This strategy aims to enhance the signal-to-interference-plus-noise ratio (SINR). In a two-user NOMA scenario where $|h_1|^2/\mathrm{N}_{f,1} > |h_2|^2/\mathrm{N}_{f,2}$ (and thus $P_1 < P_2$), only user 1 performs SIC. It initially decodes x_2, the signal of user 2, and subtracts it from the received signal y_1. Subsequently, it decodes its own signal. User 2, on the other hand, treats x_1, the signal of user 1, as noise and directly decodes its own signal from y_2 without SIC. Assuming a perfect SIC, the achievable data rate of the NOMA user i, R_i^{NOMA} for a transmission bandwidth of 1 Hz can thus be expressed as

$$R_1^{\text{NOMA}} = \log_2\left(1 + \frac{P_1|h_1|^2}{N_{f,1}}\right) \tag{6.3}$$

$$R_2^{\text{NOMA}} = \log_2\left(1 + \frac{P_2|h_2|^2}{P_1|h_2|^2 + N_{f,2}}\right) \tag{6.4}$$

The achievable sum capacity is denoted as $R^{\text{NOMA}} = R_1^{\text{NOMA}} + R_2^{\text{NOMA}}$, where R_1^{NOMA} and R_2^{NOMA} are the individual data rates for users 1 and 2 in NOMA, respectively. Equations (6.3) and (6.4) indicate that the BS can regulate the data rate of each user by adjusting the PA coefficients α_1 and α_2, where $\alpha_1 = P_1/P$ and $\alpha_2 = P_2/P$. To facilitate a comparative analysis of the data rate performances between NOMA and OMA, we examine the two-user FDMA scheme illustrated in Fig. 6.3. In this configuration, the 1 Hz transmission bandwidth is partitioned for the two users: user 1 utilizes W Hz, while user 2 uses the remaining $(1-W)$ Hz of the bandwidth. The PA ratio (α_1:$\alpha_2 = P_1$:P_2) remains consistent with that of the NOMA scheme. Consequently, the achievable data rate of the OMA user i, R_i^{OMA}, can be

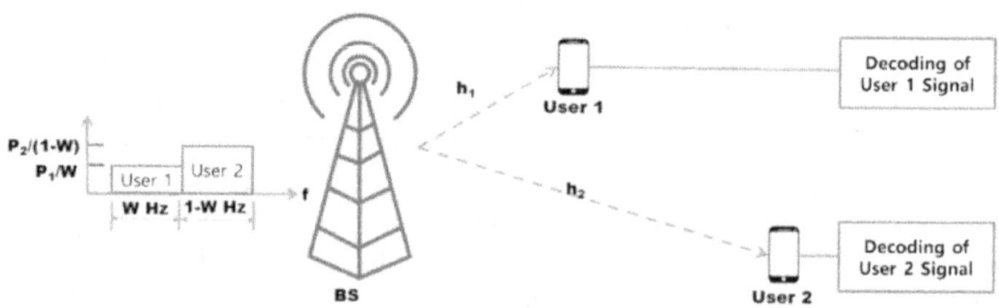

Fig. 6.3. Two-user downlink FDMA.

expressed as

$$R_1^{\mathrm{OMA}} = W\log_2\left(1 + \frac{P_1|h_1|^2}{WN_{f,1}}\right) \tag{6.5}$$

$$R_2^{\mathrm{OMA}} = (1-W)\log_2\left(1 + \frac{P_2|h_2|^2}{(1-W)N_{f,2}}\right) \tag{6.6}$$

The achievable sum capacity is $R^{\mathrm{OMA}} = R_1^{\mathrm{OMA}} + R_2^{\mathrm{OMA}}$. The OMA user suffers from interference from the signal of the other user, unlike NOMA, as indicated by equation (6.4). In Fig. 6.4, we present the data rate regions of both downlink NOMA and OMA by plotting the data rate of one user against the other at different PA ratios, assuming $h_1 = 10$ and $h_2 = 1$. As observed, the rate region of NOMA is significantly wider compared to OMA. For instance, if we anticipate the data rate of user 2 to be 1 bit/Hz, the achievable rate of user 1 in NOMA is approximately 9 bits/Hz, surpassing that of OMA with an equal bandwidth sharing. This discrepancy arises because the data rate of NOMA user 1 is limited by bandwidth rather than power. Despite allocating lower transmission power

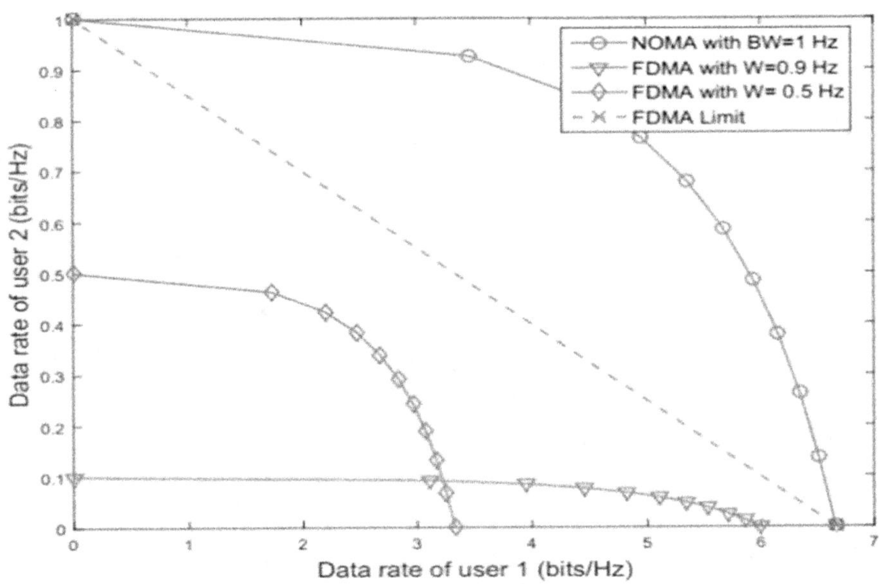

Fig. 6.4. Rate region in two-user downlink NOMA and OMA.

to user 1 compared to user 2, its high channel gain (i.e., $|h_1|^2/N_{f,1}$) and the use of SIC enable the user to exploit the full bandwidth efficiently. While it's possible to increase the rate of OMA user 2 by assigning a significant fraction of bandwidth, it results in a substantial reduction in the data rate of OMA user 1.

Furthermore, as depicted in Fig. 6.5, NOMA consistently outperforms OMA in terms of the sum capacity, irrespective of the bandwidth allocation ratio. It is noteworthy that we considered an asymmetric channel condition (different channel gains for users 1 and 2) for the above comparisons in data rate regions. In a symmetric channel environment, where both users experience the same interference and noise, OMA and NOMA provide identical data rate regions. Multiplexing in the power domain does not become advantageous in this scenario [16]. Consequently, compared to OMA, NOMA in cellular downlink exhibits higher sum capacity and achieves a better trade-off between system efficiency and user fairness, especially when channel conditions differ among users. It has been demonstrated that other forms of OMA techniques, including TDMA and orthogonal frequency-division

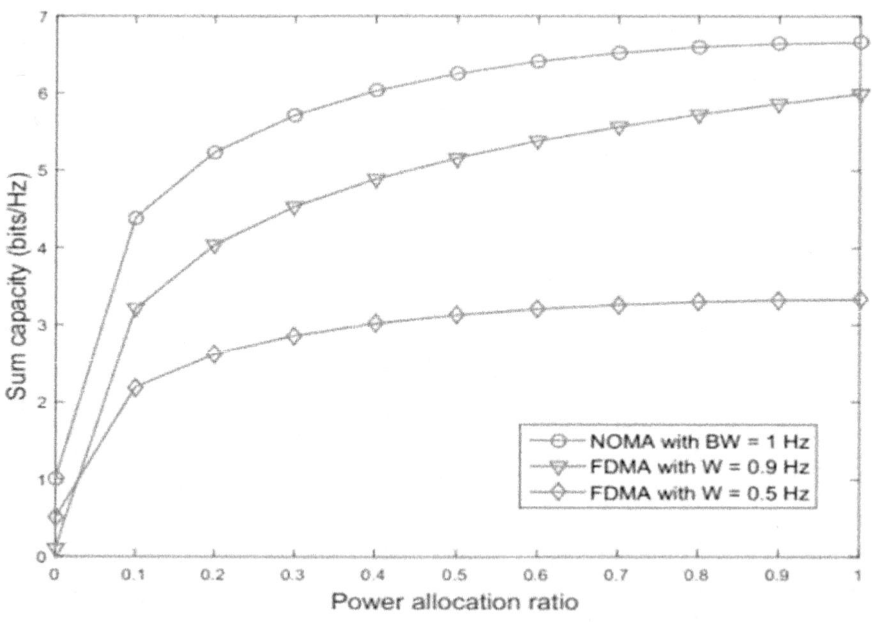

Fig. 6.5. Sum capacity in two-user downlink NOMA and OMA.

multiple access (OFDMA), are also significantly surpassed by NOMA in terms of throughput under different settings [17].

Continuing the generalization of the K-user scenario for downlink NOMA, where

$$|h_1|^2/N_{f,1} > |h_2|^2/N_{f,1} \ldots . > |h_k|^2/N_{f,k}$$

with a 1 Hz bandwidth, the data rate of user 1 remains the same as (6.3). For any other user k, the data rate can be expressed as

$$R_k^{\text{NOMA}} = \log_2\left(1 + \frac{P_k|h_k|^2}{\sum_{j=1}^{k-1} P_j|h_k|^2 + N_{f,k}}\right) \tag{6.7}$$

While theoretically, there is no strict limitation on the number of NOMA users, practical considerations often restrict its application to a small number of users, typically two or three, within a cluster. This limitation is imposed by practical challenges such as the degradation in bit error rate due to error propagation, primarily originating from SIC. Additionally, as the number of users in a cluster increases, end-user devices demand more computing power and higher energy consumption, posing feasibility challenges in practical implementations, particularly for resource-constrained devices.

It's noteworthy that power-domain NOMA in downlink has been standardized in the 3rd-Generation Partnership Project (3GPP) LTE Release 13, recognized as multiuser superposition transmission (MUST) [18]. This standardization underscores the significance of NOMA as a viable and established technique in contemporary wireless communication systems.

6.2.1.2 *Uplink scenario in PD-NOMA*

Figure 6.6 illustrates an uplink NOMA scheme, where users 1 and 2 simultaneously transmit their signals x_1 and x_2, respectively, to the BS. The received signal at the BS is given by

$$y = \sum_{i=1}^{2} \sqrt{P_i} h_i x_i + n \tag{6.8}$$

where P_i is the transmit power for user i, with $E\{|x_i|^2\} = 1$, and n represents Gaussian noise plus interferences with power spectral density N_f.

Generally, the BS broadcasts a downlink reference signal, based on which each user performs channel estimation. Consequently, users can

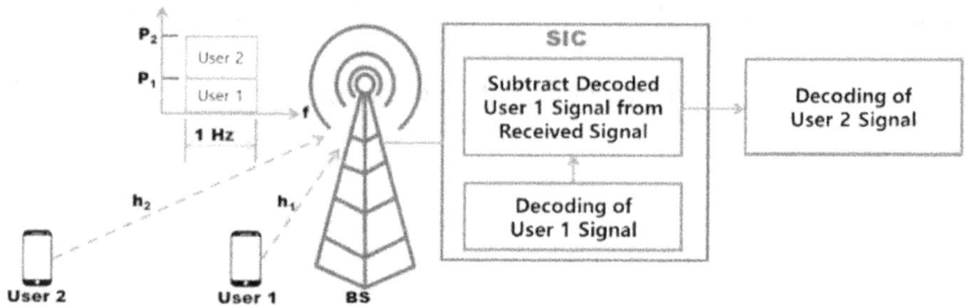

Fig. 6.6. Two-user power-domain uplink NOMA.

adjust their transmit power to either P_1 or P_2 depending on their channel gains. User 1 is designated as the strong user, experiencing a higher channel gain compared to user 2, the weak user. Upon receiving the superposed signal, the BS employs SIC to separate the users' signals. Specifically, the BS first decodes the signal of user 1 by treating the signal of user 2 as noise [19] and then subtracts the decoded signal of user 1 ($\widehat{x_1}$) from the received signal (y). Subsequently, from the remaining signal, the BS decodes the signal of user 2 ($\widehat{x_2}$). In uplink NOMA, user 1 experiences interference from user 2, whereas user 2 receives no interference from user 1 since user 1's signal is removed before decoding user 2's signal. In contrast, in downlink NOMA, user 2 experiences interference from user 1, whereas user 1 does not suffer interference from user 2 since user 2's signal is removed before decoding user 1's signal. If SIC is perfect, the achievable data rate of the NOMA user i, $\mathrm{R}_i^{\mathrm{NOMA}}$, for a transmission bandwidth of 1 Hz is expressed as

$$R_1^{\mathrm{NOMA}} = \log_2\left(1 + \frac{P_1|h_1|^2}{P_2|h_2|^2 + N_f}\right) \tag{6.9}$$

$$R_2^{\mathrm{NOMA}} = \log_2\left(1 + \frac{P_2|h_2|^2}{N_f}\right) \tag{6.10}$$

The achievable sum capacity in uplink NOMA is denoted as $R^{\mathrm{NOMA}} = R_1^{\mathrm{NOMA}} + R_2^{\mathrm{NOMA}}$. It's noteworthy that the same equations (6.5) and (6.6) can be applied to calculate the achievable data rates of uplink OMA users 1 and 2 by replacing $N_{f,i}$ with N_f. Similar to downlink NOMA, the superiority of uplink NOMA in terms of achievable data rate and sum capacity can be demonstrated by comparing equations 6.9 and (6.10) with their OMA counterparts [16]. This comparison reinforces the efficiency and capacity benefits of NOMA in both uplink and downlink scenarios.

In the context of a K-user uplink NOMA scenario with $|h_1|^2/N_f > |h_2|^2/N_f \ldots > |h_K|^2/N_f$ (thus $P_1 < P_2 < \ldots < P_K$) and a 1 Hz bandwidth, the data rates of user $k(k = 1, 2, \ldots, K-1, K)$ can be expressed as

$$R_k^{\text{NOMA}} = \log_2\left(1 + \frac{P_k|h_k|^2}{\sum_{j=k+1}^{K} P_j|h_j|^2 + N_f}\right) \tag{6.11}$$

$$R_K^{\text{NOMA}} = \log_2\left(1 + \frac{P_K|h_K|^2}{N_f}\right) \tag{6.12}$$

Unlike downlink NOMA, uplink NOMA can accommodate a relatively larger number of users because the BS can usually be equipped with the required computing power and energy. Additionally, the BS can implement more computing-intensive decoding schemes to mitigate the impact of interferences. Consequently, uplink NOMA is preferred over downlink NOMA for scenarios involving massive machine-type communications [20]. This preference is driven by the enhanced capacity of the BS to handle multiple users and apply advanced decoding techniques.

6.2.2 *CD-NOMA*

In contrast to power-domain NOMA, code-domain NOMA achieves multiplexing in the code domain. The concept of code-domain NOMA draws inspiration from classic CDMA systems, where multiple users share the same time–frequency resources while employing unique user-specific spreading sequences. However, when compared to the traditional CDMA, the distinctive characteristic of code-domain NOMA lies in the utilization of sparse sequences (alternatively referred to as low-density sequences) or non-orthogonal low cross-correlation sequences for spreading. The most basic form of code-domain NOMA is the LDS-CDMA [21].

To comprehend the functioning of LDS-CDMA, let's begin with a brief overview of the operational principle of a classical CDMA system. In a CDMA system, a user-specific high-rate pseudo-noise spreading sequence is initially generated. This sequence is then segmented into a number of chips. Subsequently, the low-rate symbol of the user is spread across each chip before transmission. Simultaneously, other users in the system also transmit their symbols using their unique spreading sequences. On the receiving end, the signals received during all the chips need to be combined to separate the different signals by despreading with the same spreading sequences used at the transmitter.

In this context, the elements of the spreading sequences are generally non-zero, indicating that the spreading sequences are not sparse. Consequently, the signals received from all active users overlap at each chip, and each user experiences inter-user interferences from all other users. This type of CDMA system is termed non-orthogonal, where the spreading sequences are non-orthogonal. Conversely, orthogonal CDMA systems employ orthogonal spreading sequences, making it straightforward to eliminate interferences, requiring only a simple correlation receiver to detect a user's signal. However, an orthogonal CDMA system can only support as many users as the number of chips.

In contrast, non-orthogonal CDMA systems have more codes than the number of chips in a sequence. Still, due to the non-orthogonality of the codes, they suffer from inter-user interferences. Therefore, the receiving side in a non-orthogonal CDMA system demands more complex multi-user detection (MUD) techniques. The concept of LDS-CDMA arises as a solution, employing sparse spreading sequences where the number of non-zero elements in the spreading sequence is much lower than the number of chips. This departure from the conventional spreading sequences allows LDS-CDMA to support more users than its orthogonal counterpart while reducing interference. Since all symbols are modulated onto sparse spreading sequences before transmission, each user spreads its data over a small number of chips, resulting in fewer superposed signals and, consequently, reduced effective interferences at each chip compared to the number of active users. Therefore, the use of low-density spreading sequences enables LDS-CDMA to offer improved system performance when compared to the conventional CDMA.

Another family of code-domain NOMA is LDS-OFDM, which can be thought of as an amalgamation of LDS-CDMA and OFDM [22]; the original data streams are first multiplied with the low-density spreading sequences and then transmitted over different subcarriers. The operation of LDS-CDMA thus becomes analogous to multi-carrier CDMA (MC-CDMA), keeping all the benefits of OFDM-based MC transmissions in terms of its intersymbol interference avoidance, together with MUD-mediated LDS-CDMA operating at a lower complexity. SCMA [23] is a recent code-domain NOMA technique based on LDS-CDMA. In contrast to LDS-CDMA, the procedure of bit to quadrature amplitude modulation symbol mapping and spreading are combined together, and input data streams are directly mapped into a multidimensional codeword of an SCMA codebook set. Compared to LDS-CDMA, SCMA provides a low-complexity reception technique and offers better performances. The reader is referred to Refs. [23, 24] for a comprehensive discussion of the code-domain NOMA techniques, which highlights both transmitter schemes and multi-user detectors at the receiver, emphasizing their advantages and drawbacks.

6.2.3 *Key technologies in NOMA*

6.2.3.1 *Superposition coding (SC)*

Superposition coding (SC) was initially proposed in Ref. [25] as a technique to simultaneously communicate information to multiple receivers from a single source. In essence, it enables the transmission of information to several users concurrently.

To implement SC practically, the transmitter needs to encode information relevant to each user. For instance, in a two-user scenario, the transmitter requires two point-to-point encoders that map inputs to complex-valued sequences for the two-user signal. A schematic diagram illustrating how SC is performed is provided in a two-dimensional vector space as in Fig. 6.7, where the quadrature phase-shift keying (QPSK) constellation of user 1 with higher transmitting power is superposed on that of user 2 with lower transmitting power. It's important to note that SC is a recognized non-orthogonal scheme that achieves capacity on a scalar Gaussian broadcast channel (BC).

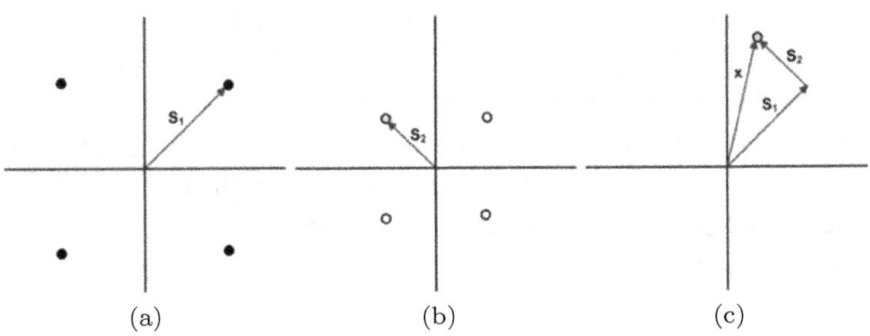

Fig. 6.7. An example of SC encoding: (a) signal constellation of user 1, (b) signal constellation of user 2, (c) constellation of the superposed signal.

An effective strategy for SC is introduced in Ref. [26] and is a design technique using off-the-shelf single-user coding and decoding blocks. In the superposition encoding phase, two point-to-point encoders, f_1:$\{0, 1\}^{2^{R_1}} \rightarrow C^T$ and f_2:$\{0, 1\}^{2^{R_2}} \rightarrow C^T$, first map the respective input bits to two output bit sequences, $S_1(n)$ and $S_2(n)$, each with a block length of T. Here, R_1 and R_2 denote the transmission rates of user 1 and user 2, respectively. C is the code library. Then, a summation device provides an output sequence as

$$X(n) = \sqrt{P\beta_1}\, S_1(n) + \sqrt{P\beta_2}\, S_2(n) \tag{6.13}$$

where β_i denotes a fraction of the total power P assigned to user i, subject to the constraint on $\beta_1 + \beta_2 = 1$.

6.2.3.2 *Successive interference cancellation (SIC)*

The SIC technique, initially proposed by Cover [25], is employed to decode superposed information at each receiver. SIC leverages specifications regarding the differences in signal strength among the signals of interest. The fundamental principle of SIC involves the successive decoding of user signals. Upon decoding one user's signal, it is subtracted from the combined signal before proceeding to decode the next user's signal. This iterative process allows for the decoding of one user's signal while treating the other user's signal as interference. However, before applying SIC, users are ordered based on their signal strengths. This ensures that the receiver first decodes the stronger signal, subtracts it from the combined signal, and subsequently isolates the weaker signal from the residual interference. It is noteworthy that during decoding, each user treats the other interfering users as noise in signal reception.

The technique for decoding the superposed signal (Fig. 6.7) at the receiving side is depicted in Fig. 6.8. Initially, the constellation point of user 1 is decoded from the received signal. Then, the decoding of the constellation point of user 2 is performed with respect to the decoded constellation point of user 1. In brief, the particular process involved in decoding the superposed messages can be mathematically expressed as follows [26]: From encoders,

$$f_1:\{0, 1\}^{2^{R_1}} \rightarrow C^T \text{ and } f_2:\{0, 1\}^{2^{R_2}} \rightarrow C^T$$

(1) At user 1, a single-user decoder g_1:$C^T \rightarrow \{0, 1\}^{2^{R_2}}$ decodes the message $S_1(n)$ by treating $S_2(n)$ as noise.

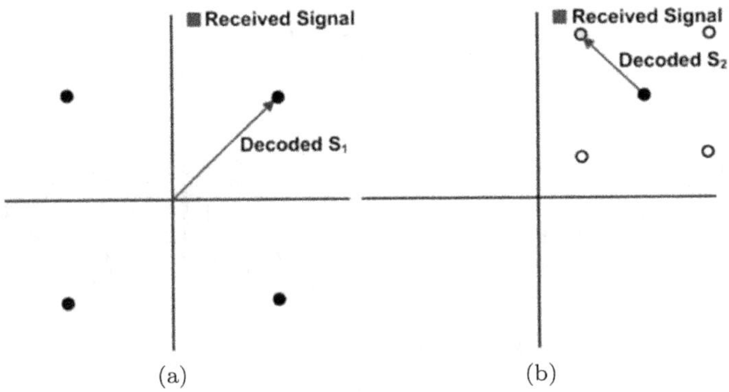

Fig. 6.8. An example of SC decoding: (a) decoding the signal of user 1 and (b) decoding the signal of user 2.

(2) User 2 performs the following steps to successively recover its message from its received signal $Y_2(n)$:

(a) decoding user 1's message $S_1(n)$ by using the single-user decoder $g_1{:}C^T \rightarrow \{0, 1\}^{2^{R_1}}$;

(b) subtracting $\sqrt{P\beta_1}h_2S_1(n)$ from the received signal $Y_2(n)$,

$$Y_2'(n) = Y_2(n) - \sqrt{P\beta_1}h_2S_1(n) \tag{6.14}$$

where h_2 is the complex channel gain at user 2;

(c) decoding user 2's messages $S_2(n)$ by applying another single-user decoder $g_2 : C^T \rightarrow \{0, 1\}^{2^{R_2}}$ on $Y_2'(n)$.

6.2.3.3 *User pairing concept with SC and SIC*

(i) SC procedure: The two key operations that make NOMA possible are superposition coding which must be done at the transmitter side and SIC at the receiver side. In this subsection, we will see about superposition coding.

Consider two users, U_1 and U_2, intending to communicate concurrently over the same frequency band. Let $\underline{m}_1$ and $\underline{m}_2$ denote the data messages of U_1 and U_2, respectively. For simplicity, let us assume that each user has just 4 bits of data to send although this assumption is far from practical scenarios but is sufficient to understand the basic working of NOMA.

Let $\underline{m}_1 = [1010]$ and $\underline{m}_2 = [0110]$. Graphically, the data messages for $\underline{m}_1$ and $\underline{m}_2$ in pulse forms are depicted in Fig. 6.9.

Assuming pulse modulation via binary phase shift keying (BPSK), where 0's map to -1's and 1's map to $+1$'s, the modulated signals for $\underline{m}_1$ and

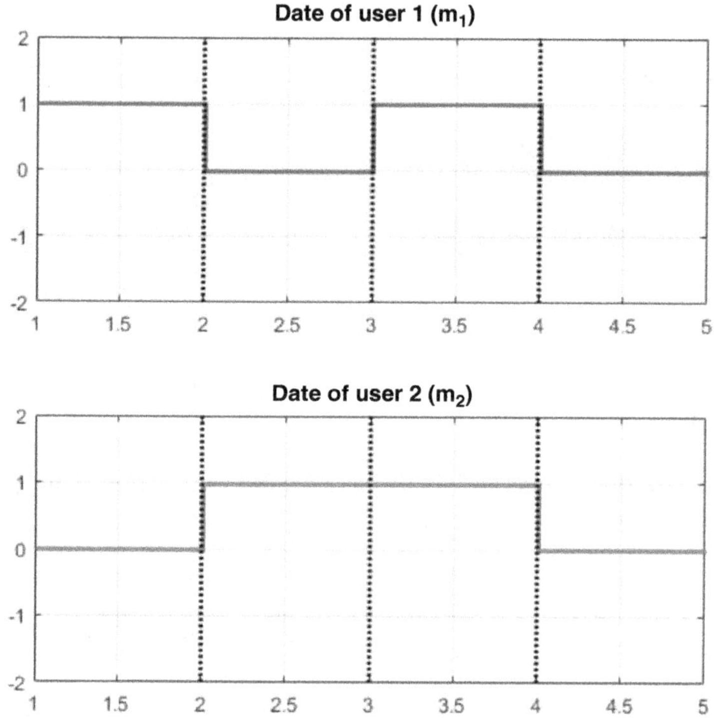

Fig. 6.9. Message data for $\underline{m}_1$ and $\underline{m}_2$.

$\underline{m}_2$ are mapped to $x_1 = [+1, -1, +1, -1]$ and $x_2 = [-1, +1, +1, -1]$, respectively. These modulated pulse signals are illustrated in Fig. 6.10.

Both x_1 and x_2 exhibit a peak amplitude of ± 1, implying unit power (1 W), as power is the square of amplitude. Before transmission, PA to users is determined based on CSI or user proximity. Employing a fixed allocation strategy with weighting factors a_1 and a_2 such that $a_1 + a_2 = 1$, with $a_1 = 0.75$ and $a_2 = 0.25$ assigned to U_1 and U_2, respectively.

NOMA requires superposition coding at the transmitter side. Superposition coding implies power-domain multiplexing. Consequently, the weighted pulse signals as shown in Fig. 6.11 are

$$\sqrt{a_1}x_1 = [0.866 \ -0.866 \ 0.866 \ -0.866]$$

$$\sqrt{a_2}x_2 = [-0.5 \ 0.5 \ 0.5 \ -0.5]$$

where $\sqrt{a_i}$ is the amplitude of the signal x_i.

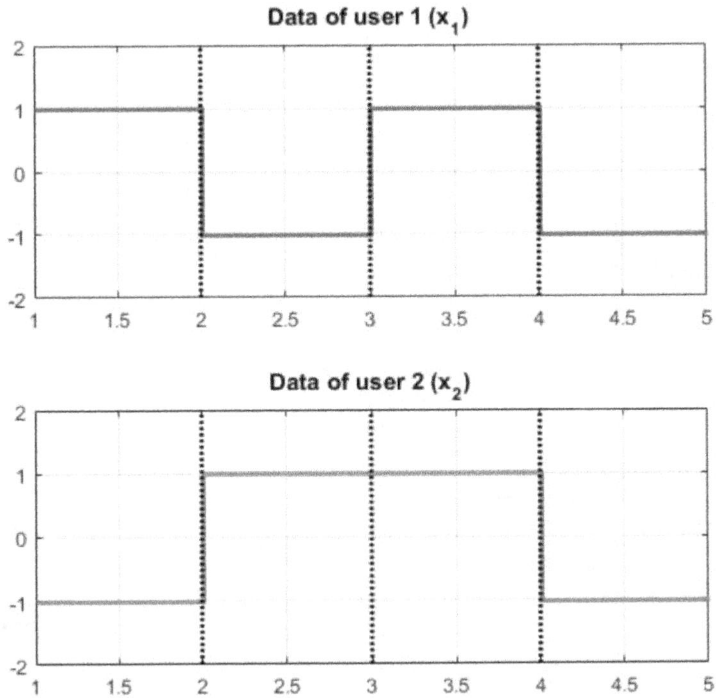

Fig. 6.10. Pulse modulated signals x_1 and x_2 of $\underline{m}_1$ and $\underline{m}_2$.

The transmitted pulse signal after superposition coding, denoted as y, the linear combination of two signals, is expressed as follows:

$$y = \sqrt{a_1}x_1 + \sqrt{a_2}x_2 = [0.366 - 0.366\ 1.366 - 1.366]$$

A graphical representation of y, the superposition coded signal, is shown in Fig. 6.12.

(ii) SIC procedure: At the receiver side, SIC is employed to decode the superposition coded signal. SIC operates as an iterative algorithm, decoding data in a sequence based on decreasing power levels. The user with the highest power is decoded first, followed by users with progressively lower power levels until all data is successfully decoded. The SIC process involves the following steps:

- **Step 1:** Directly decode y to obtain the signal with the higher power weight. For example, if $a_1 > a_2$, the direct decoding of y results in x_1.

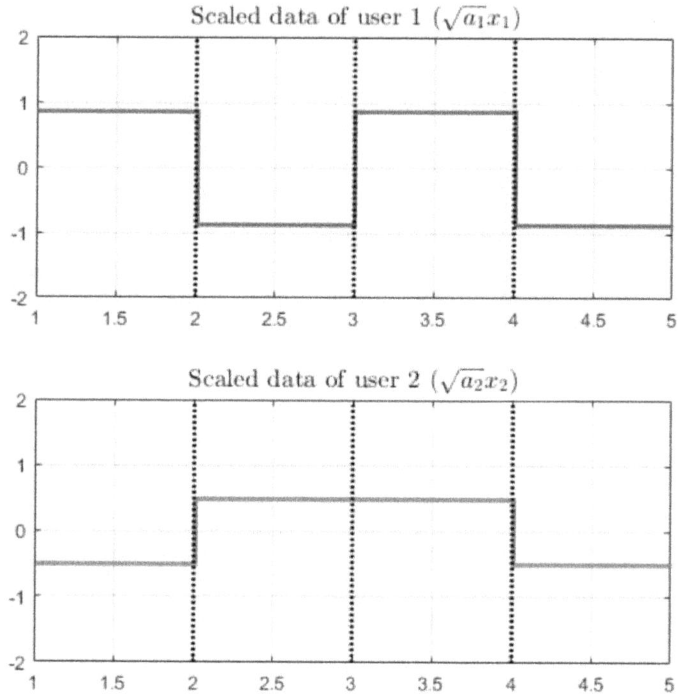

Fig. 6.11. Weighted pulse signals x_1 and x_2.

- **Step 2:** Multiply the signal decoded in Step 1 by its corresponding weight and subtract it from y. For instance, if x_1 is decoded in the previous step, subtracting $\sqrt{a_1}x_1$ from y yields $\hat{x}_2 = y - \sqrt{a_1}x_1$.
- **Step 3:** Decode the signal obtained in Step 2 to retrieve the other signal originally multiplexed with the lower power weight. For example, $\hat{x}_2$ obtained from the previous step would result in x_2.

Consider the previous example, assuming a perfectly ideal SIC scenario. Given PA weights $a_1 = 0.75$ and $a_2 = 0.25$, U_1 is allocated more power weight than U_2.

The decoding process for BPSK is that for each symbol the received pulse amplitude is just compared to the optimal threshold; if the amplitude exceeds zero, we decode it as 1, and 0 otherwise. For BPSK, the optimal threshold is zero for our case. We assume that the received signal is

$$y = x + n$$

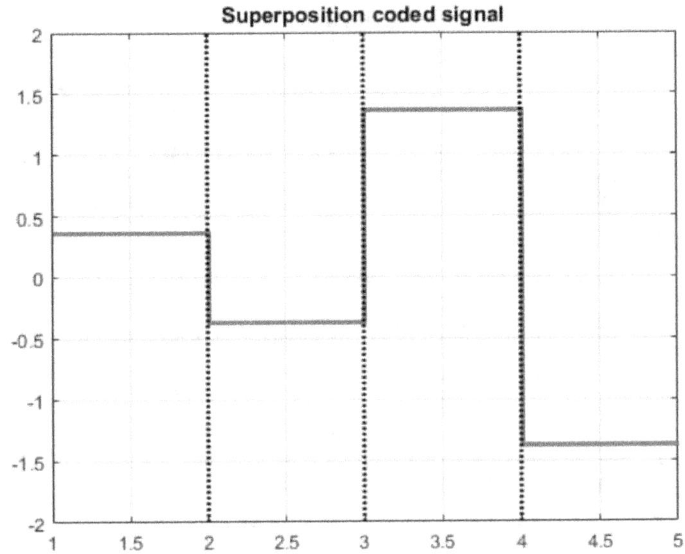

Fig. 6.12. Transmitted pulse signal after superposition coding.

where n is the white noise plus other interference components. If we have a perfect SIC, with $n = 0$, then

$$y = \sqrt{a_1}x_1 + \sqrt{a_2}x_2 = [0.366 - 0.366\ 1.366 - 1.366]$$

Figure 6.13 shows a plot of the received signal y, with the threshold denoted as a solid black horizontal line.

The decoding process proceeds symbol by symbol, employing a threshold of zero. The observations for each symbol are as follows:

- The first and third symbols lie above the threshold, leading to the decision that the first and third transmitted bits are ones.
- The second and fourth symbols lie below the threshold, resulting in the decision that the second and fourth transmitted bits are zeros.

Hence, the decoded sequence in order is 1010, corresponding to x_1. It is noteworthy that Step 1 utilizes the SIC process with known CSI (h_1, h_2). BPSK demodulation is directly performed on y in Step 1. The presence of the other component x_2 in y can be safely neglected, treating it as interference.

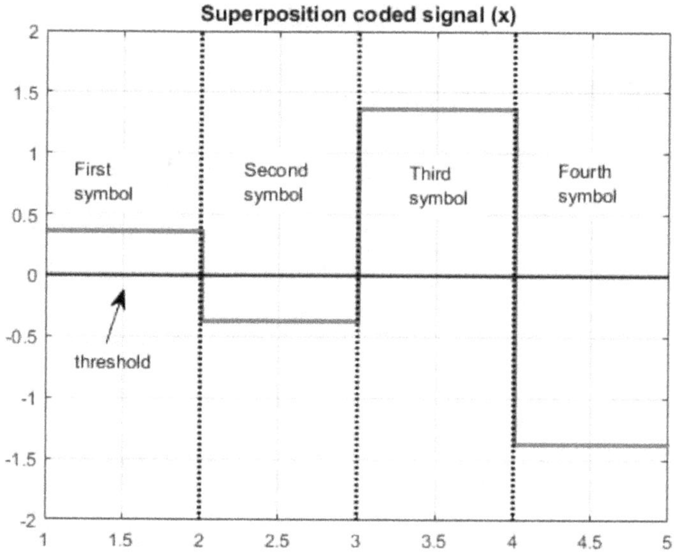

Fig. 6.13. Transmitted pulse signal after superposition coding.

In Step 2, the x_1 component is multiplied by its corresponding power weight and subtracted from y. A careful consideration is required as x_1 has been obtained as 1010. However, it's important to note that x_1 is present in y in its BPSK-modulated form. Specifically, x_1 in BPSK modulation is represented as $1 -1\ 1 -1$, not as 1010. Therefore, the subtraction involves deducting the BPSK-modulated version of x_1 from y, resulting in $\hat{x}_2 = y - \sqrt{a_1}x_1$.

The graphical representation of this subtraction is depicted in Fig. 6.14.

In Step 3, after y is subtracted by $\sqrt{a_1}x_1$, $\sqrt{a_2}x_2$ is left and shown in Fig. 6.14. Adhering to the BPSK rule, the first and fourth symbols would be demodulated as zeros and the other symbols would be demodulated as ones. Thus, the demodulated signal is 0110 in order, which is the same as x_2. The combination of superposition coding (SC) and SIC has effectively and successfully recovered the two separate data streams in the power domain.

(iii) User-pairing procedure: While NOMA enables the simultaneous connection of multiple users to a shared resource (such as frequency or time), there exists a limitation on the number of users that can efficiently utilize this resource. Beyond a certain threshold, increasing the number of users per carrier may lead to a decrease in the network's overall throughput.

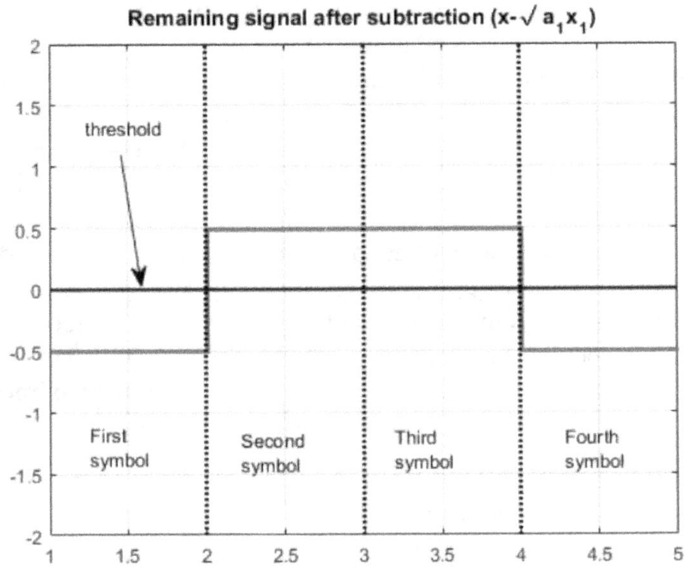

Fig. 6.14. Signal sequences of $y - \sqrt{a_1}x_1$.

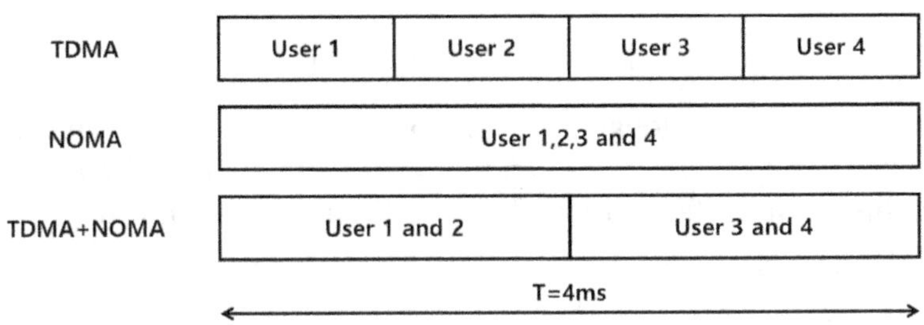

Fig. 6.15. A hybrid NOMA architecture.

One strategic solution to this challenge is the implementation of hybrid NOMA. Hybrid NOMA integrates NOMA with an OMA technique. Consider the example of NOMA combined with TDMA, as depicted in Fig. 6.15.

Assuming a frame duration of 4 ms and the goal of supporting four users within this frame, TDMA would divide the 4 ms frame into four 1 ms slots, assigning one slot to each user. In contrast, pure NOMA would allocate the entire 4 ms frame to the four users. However, this approach elevates the complexity of SIC and the processing delay.

Hybrid NOMA provides a balanced solution by dividing the 4 ms frame into two 2 ms slots, accommodating two NOMA users in each sub-frame. This optimization manages the trade-off between user capacity and system efficiency, enhancing the overall performance of the network.

The pivotal question arises: How should NOMA users be paired within each available orthogonal resource? The decision on user pairing holds significance and impacts the overall system performance. There arise several pairing strategies, and three potential configurations are possible: $\{(1, 2), (3, 4)\}$ or $\{(1, 3), (2, 4)\}$ or $\{(1, 4), (2, 3)\}$. The optimal user-pairing strategy depends on factors, such as channel conditions, interference patterns, and the specific goals of the network. An in-depth analysis is necessary to determine the most suitable pairing scheme for maximizing the benefits of hybrid NOMA.

(iv) Simple two-user pairing strategies: As shown in Fig. 6.16, let's consider a downlink communication scenario with four users. Let d_1, d_2, d_3, and d_4 denote the distances of U_1, U_2, U_3, and U_4, respectively, from the BS. U_1 is the nearest user and U_4 is the farthest user. Therefore, when the pairing is done based on the distances, their channel conditions may be ordered as

$$|h_1|^2 > |h_2|^2 > |h_3|^2 > |h_4|^2$$

We have two orthogonal resource blocks (time/frequency/subcarriers), and we have to assign two users per block. There are two simple ways to realize this: (a) near–far pairing (N–F) and (b) near–near and far–far pairing (N–N and F–F).

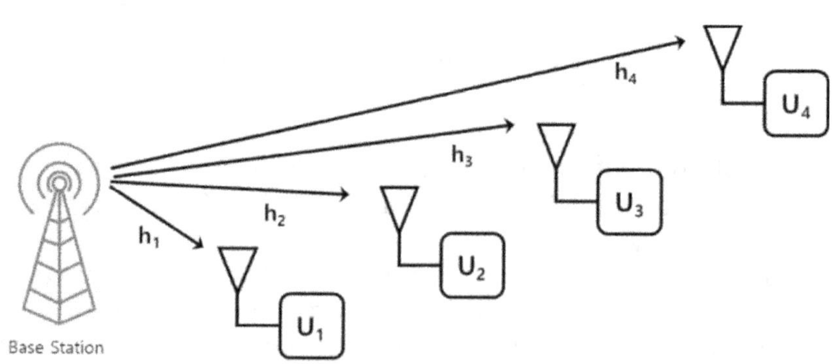

Fig. 6.16. Downlink communication scenario with four users.

(a) Near–far pairing (N–F): In the N–F pairing strategy, the nearest user to the BS is paired with the farthest user, followed by pairing the next nearest user with the next farthest user, and so on. In our example, since user U_1 is the nearest to the BS and user U_4 is the farthest, user U_1 will be paired with user U_4 in one resource block, and the next nearest user will be paired with the next farthest user in the subsequent block.

For the first pair of users, where U_1 is the near user and U_4 is the far user, the PA weights should be chosen as $a_1 < a_4$, where a_1 and a_4 are the weights for users U_1 and U_4, respectively. This setup ensures that user U_1 performs SIC, while user U_4 performs direct decoding. Similarly, for the second pair of users, where U_2 is the near user and U_3 is the far user, $a_2 < a_3$ should be chosen, where a_2 and a_3 are the weights for users U_2 and U_3, respectively. Here, user U_2 performs SIC, while user U_3 performs direct decoding.

The achievable rates for the users in the first pair are

$$R_{1,nf} = \frac{1}{2}\log_2\left(\frac{P\alpha_1|h_1|^2}{\sigma^2}\right)(\text{afterSIC})$$

$$R_{4,nf} = \frac{1}{2}\log_2\left(1 + \frac{P\alpha_4|h_4|^2}{P\alpha_1|h_4|^2 + \sigma^2}\right)$$

Similarly, for the second pair,

$$R_{2,nf} = \frac{1}{2}\log_2\left(\frac{P\alpha_2|h_2|^2}{\sigma^2}\right)(\text{afterSIC})$$

$$R_{3,nf} = \frac{1}{2}\log_2\left(1 + \frac{P\alpha_3|h_3|^2}{P\alpha_2|h_3|^2 + \sigma^2}\right)$$

The sum rate of the N–F scheme will be

$$R_{nf} = R_{1,nf} + R_{2,nf} + R_{3,nf} + R_{4,nf}$$

(b) Near–near, far–far pairing (N–N and F–F): In the near–near and far–far (N–N and F–F) pairing strategy, users are grouped based on their proximity to the BS. The nearest user is paired with the next nearest user, and the farthest user is paired with the next farthest user. Let's explore the application of N–N and F–F pairing in a scenario with two orthogonal resource blocks.

Now, in the first pair of users, U_1 is nearest to the BS when compared to U_2. Therefore, we have to choose $a_1 < a_2$. U_1 should perform SIC, while U_2 will perform direct decoding. Similarly, U_3 is closer to the BS than U_4. So, we have to choose $a_3 < a_4$. U_3 should perform SIC, while U_4 will perform direct decoding.

The achievable rates for the users in the first pair are

$$R_{1,nn} = \frac{1}{2}\log_2\left(\frac{P\alpha_1|h_1|^2}{\sigma^2}\right) \text{(afterSIC)}$$

$$R_{2,nn} = \frac{1}{2}\log_2\left(1 + \frac{P\alpha_2|h_2|^2}{P\alpha_1|h_2|^2 + \sigma^2}\right)$$

Similarly, for the second pair,

$$R_{3,ff} = \frac{1}{2}\log_2\left(\frac{P\alpha_3|h_3|^2}{\sigma^2}\right) \text{(afterSIC)}$$

$$R_{4,ff} = \frac{1}{2}\log_2\left(1 + \frac{P\alpha_4|h_4|^2}{P\alpha_3|h_4|^2 + \sigma^2}\right)$$

The sum rate of the N–N and F–F schemes will be

$$R_{nn} = R_{1,nn} + R_{2,nn} + R_{3,ff} + R_{4,ff}$$

6.2.4 *Characteristics of NOMA*

In comparison to OMA, NOMA offers several key advantages, which has been summarized in the following section [58].

6.2.4.1 *Advantages*

- **High spectral efficiency (SE):** NOMA excels in SE by serving multiple users with the same resource block. This results in improved system throughput.
- **Massive connectivity:** NOMA aligns with the requirements of massive IoT, referring to the connectivity of a massive scale of devices, objects, and machines, even in remote locations. 5G and beyond are expected to support such massive connectivity, and NOMA proves beneficial by efficiently utilizing resources to serve a large number of users/devices, not strictly limited by the available orthogonal resources.
- **Improved user fairness:** NOMA's PA capability enables a system to strike a balance between user fairness and throughput. Through

appropriate PA, NOMA ensures that even cell-edge users can enjoy higher data rates without compromising the overall system throughput. This feature contributes to a fair distribution of resources among users.

- **Low latency:** In the context of latency, OMA relies on access-grant requests, where an uplink user initiates the process by sending a scheduling request to the BS. Subsequently, the BS responds with a clear signal in the downlink channel. This grant-based access procedure introduces latency, even with additional signaling overhead, which is suboptimal for NG connectivity. Particularly in scenarios with massive connectivity, the limitations of orthogonal pilots make the access-grant process more intricate. In contrast, NOMA introduces the concept of grant-free multiple access in uplink schemes. This involves blindly detecting active users and decoding their data streams. Grant-free NOMA offers a significant reduction in latency compared to OMA. Moreover, the concurrent service of multiple users in NOMA contributes to the overall reduction in latency. Therefore, NOMA's grant-free approach aligns with the imperative need for low-latency communication in advanced connectivity scenarios.

6.2.4.2 *Disadvantages*

Although SC-SIC achieves the capacity region of the scalar Gaussian BC, its receiver complexity escalates with the growing number of users, exacerbating SIC error propagation. In a K-user single-input single-output (SISO) BC, the user with the strongest channel necessitates K-1 SIC layers to decode and eliminate interference from the K-1 messages of all other co-scheduled users before accessing its intended message.

One strategy to mitigate this complexity is to limit the number of SIC layers per user by grouping users into smaller clusters, employing SC-SIC within each cluster, and scheduling user groups via OMA. However, this approach may lead to performance degradation and increased latency. While compelling a user to decode another user's entire message is effective in degraded SISO BCs, it proves inefficient in multi-antenna networks for most scenarios due to the suboptimal utilization of multiple antennas and SIC receivers. According to Ref. [88], NOMA in multi-antenna networks faces limitations, such as each user contends with inter-user interference from all other users.

6.3 Resource Allocations

In wireless communication systems, the key resources include time, frequency, space, code, and power. NOMA systems leverage these resources by accommodating multiple users in each time, frequency, or code resource block, constituting a NOMA cluster. Consequently, the effective grouping of users into NOMA clusters and the allocation of power within these clusters hold much importance. The subsequent discussion will discuss user clustering and PA in various NOMA systems.

6.3.1 *User pairing and clustering*

User clustering in NOMA systems is inherently a challenging problem, characterized by its combinatorial nature [27]. Particularly in the downlink, the assignment of NOMA users to different orthogonal resource blocks, such as subcarriers or subchannels, has been identified as a non-deterministic polynomial-time (NP)-hard problem [27–29]. While addressing this problem in its entirety may be intractable, studies have shown that the performance gain of NOMA, particularly with fixed power allocation (F-PA), over OMA increases as the difference between the channel gains of the users widens [30]. This positive correlation has been leveraged to propose greedy user clustering methods.

For instance, in Ref. [30], a pairing strategy is introduced, aligning the user with the highest channel gain with the one possessing the lowest channel gain, and proceeding in a similar fashion for the other users. However, this method may result in an uneven distribution of performance gains across different clusters. To mitigate this, Ref. [31] suggests an initial division of users into two groups based on their channel gains. Subsequently, the user with the highest channel gain in one group is paired with its counterpart in the other group, and so forth. It has been proven that this clustering method can yield an optimal solution when combined with suitable PA [32]. Nonetheless, Ref. [32] is limited to scenarios with frequency-flat channels.

The user clustering methods based on channel gain, while possessing low complexity, may exhibit unstable performance due to their heuristic nature. Striking a balance between complexity and effectiveness calls for the adoption of systematic frameworks. In the following, we outline some promising approaches in this regard:

(i) Monotonic optimization: Monotonic optimization emerges as a valuable strategy in addressing the non-convex nature of resource allocation problems within NOMA systems. In scenarios where intra-cluster interference complicates the optimization process, exploiting monotonicity becomes crucial. The principle of monotonic optimization involves leveraging the monotonic nature of certain objective functions or constraints to navigate non-convex problem spaces. In the context of NOMA, researchers have applied monotonic optimization techniques to devise optimal solutions for joint power and subcarrier allocation problems, as demonstrated in the work presented in Ref. [33]. By harnessing the monotonicity property, the optimization algorithm can make strategic adjustments to the power and subcarrier allocations, aiming to improve the overall system performance.

(ii) Combinatorial relaxation: Combinatorial relaxation offers a powerful strategy for addressing the inherent complexity of user clustering problems in NOMA systems. At the heart of this approach lies the transformation of binary decision variables, associated with the assignment of users to clusters, into continuous variables. By relaxing these binary constraints, the original NP-hard problem can often be reformulated as a convex optimization problem, which is computationally tractable and amenable to efficient solution methods. In this process, techniques such as rounding are employed to recover the binary nature of the variables after optimization. However, it's worth noting that while this relaxation technique facilitates the solution of the problem, it may introduce a performance gap compared to the original NP-hard formulation. This performance difference underscores the trade-off inherent in leveraging combinatorial relaxation: While it offers computational advantages, there may be compromises in the optimality of the solution obtained.

(iii) Game theory and matching theory: Game theory has emerged as a powerful tool for addressing user clustering challenges in NOMA systems, offering a framework to model the interactions among users and resource blocks. In recent research, users are conceptualized as players, while resource blocks are treated as the action profile, forming the basis for various game-theoretic formulations. In one approach, a coalition game is formulated, as demonstrated in Ref. [34], where users make decisions to maximize their own utility without consideration for the overall system performance. Building on this, an enhanced coalition game is proposed in Ref. [35], leveraging particle swarm optimization to adjust the utility function of each user toward a global optimum. This modification aims to align individual user strategies with the collective goal of improving system-wide performance. Beyond coalition games, a Stackelberg game framework is introduced in Refs. [35, 36] for millimeter-wave (mmWave) NOMA systems, where user clustering acts as the leader and PA as the follower. This hierarchical approach enables strategic decision-making, with the leader influencing the behavior of the follower to achieve the desired system outcomes. Despite their potential, game theory-based approaches face challenges in distributed implementation and ensuring equilibrium among participants. These shortcomings underscore the need for further research to refine and optimize these methodologies for real-world NOMA systems.

Matching theory, a robust mathematical tool for addressing combinatorial user allocation challenges, emerges as a promising solution to overcome the complexities associated with NOMA systems [37]. Notably, conventional matching algorithms like the Hungarian algorithm face limitations in handling NOMA's intra-cluster interference intricacies [38]. To navigate this, researchers turn to the many-to-many two-sided matching theory as an effective approach for resource allocation in NOMA systems. Subsequently, iterative swap-based matching algorithms are introduced, offering a practical means to optimize user allocations [29, 38, 39].

The computational demands associated with the aforementioned approaches may pose challenges for practical implementation. To mitigate this, an effective strategy involves excluding user groups unsuitable for NOMA multiplexing, preventing unnecessary comparisons of candidate user pairs [40]. Furthermore, existing user clustering algorithms often overlook the possibility of insufficient strong users for pairing, leaving weak users unpaired. To address this scenario, a hybrid approach is feasible, allowing leftover users to be accessed via OMA. However, this approach comes at the expense of depriving these users of the benefits offered by NOMA. An alternative solution involves the concept of virtual user clustering [41], where a frequency band is shared by two weak users and a strong user: Half of the bandwidth is allocated to the strong user and one weak user, while the remaining half is used by the strong user and the other weak user. This innovative approach enables NOMA implementation while efficiently managing scenarios with limited strong users.

6.3.2 *Power allocation*

PA is a critical aspect of NOMA, as it involves the multiplexing of users in the power domain [42, 43]. The effectiveness of PA directly influences system performance, impacting factors such as interference management, rate distribution, and user admission. Inappropriate PA strategies can result in an unfair distribution of rates among users and even system outages due to successive SIC failures. When devising PA strategies, considerations must include user channel conditions, the availability of CSI, QoS requirements, total power constraints, and overall system objectives. Several widely recognized PA performance metrics include the number of admitted users, sum rate, EE, user fairness, outage probability, and total power consumption. Therefore, the primary goal of PA in NOMA is to achieve either a higher number of admitted users, increased sum rate, and EE, or a balanced fairness while minimizing power consumption. In the following, we present PA in both single-carrier (SC) and multi-carrier (MC) systems:

(i) PA in SC-NOMA: In downlink single-carrier NOMA (SC-NOMA) systems, the optimal PA strategy for maximizing the sum rate typically assigns all power to the user with the strongest channel [59, 72, 73]. However, this approach leads to significant unfairness among users and reduces the number of users admitted to the system. To strike a balance between system throughput and user fairness, it is more common to allocate more power to the weaker user in NOMA setups. This allows the stronger user to effectively mitigate interference from the weaker user through SIC, while its interference to the weaker user remains relatively minimal.

Fixed power allocation (F-PA) represents the simplest PA algorithm, distributing power to users based on a fixed ratio determined by their positions in the channel ordering. However, since F-PA does not consider users' specific channel gains, it may fail to meet various QoS requirements. Fractional transmit power control addresses this by allocating power to users inversely proportional to their channel gains, powered with a decaying factor. Nevertheless, applying the same decaying factor to all users is still suboptimal, and determining the appropriate decaying factor to balance system throughput and user fairness remains an ongoing challenge.

The availability of CSI significantly influences PA in NOMA systems. When perfect CSI is available, researchers have demonstrated in Ref. [28] that the problem of maximizing the weighted sum rate is convex, enabling the derivation of optimal PA through convex optimization techniques. Similarly, for the max–min fairness problem, which is proven to be quasi-convex, optimal PA can be obtained using the bisection method [44, 45]. For energy-efficient PA, the problem can be formulated as a fractional problem, and techniques like Dinkelbach's algorithm have been applied to find solutions [46–48]. However, when only statistical CSI is available, the problem of minimizing the maximum outage probability under a given SIC order becomes non-convex. Nevertheless, optimal PA solutions have been derived in Ref. [44], and based on these, optimal SIC decoding orders have been derived in subsequent research, such as Ref. [49].

The aforementioned approaches may not necessarily ensure a higher throughput for NOMA over OMA. To address this, a cognitive radio (CR)-inspired PA strategy can be employed, where the weak user is treated as a primary user in a CR network [50, 51]. However, this comes at the cost of the strong user, as it is served only after ensuring the QoS of the weak user. To overcome this limitation, dynamic PA has been introduced in Ref. [52], aiming to allocate power in a way that NOMA achieves consistently higher individual user rates than OMA.

(ii) PA in MC-NOMA: In multi-user systems, NOMA is often combined with MC systems, referred to as MC-NOMA. This integration aims to reduce the decoding complexity [53]. In MC-NOMA, a user has the capability to occupy multiple sub-carriers, and reciprocally, a sub-carrier can be assigned to multiple users. MC-NOMA is particularly well suited for 5G and beyond, given the challenge of finding continuous wide bandwidth in these advanced network generations. When compared to OFDMA, MC-NOMA not only enhances the system's SE but also accommodates a larger number of users. The performance of MC-NOMA is influenced by both PA and sub-carrier assignment. In the downlink scenario, the weighted sum rate maximization problem in MC-NOMA has been proven to be NP-hard [58–60]. Conversely, in the uplink, the sum rate problem is demonstrated to be convex [54]. Moreover, an optimal and low-complexity iterative water-filling algorithm is proposed for uplink scenarios. This discrepancy between the downlink and uplink scenarios arises from the fact that in uplink, the BS decodes all user signals, while in downlink, each user decodes its own signals separately.

It's noteworthy that the schemes mentioned above assume perfect CSI, which may be impractical for MC-NOMA systems overloaded with a large number of users. To address this, resource allocation under statistical CSI becomes crucial. In the absence of perfect CSI, an explicit SIC decoding order needs to be derived initially, as the BS cannot directly decide the SIC decoding order [55]. Based on this, PA and sub-PA can be carried out, similar to the scenario with perfect CSI. To further enhance the SE of MC-NOMA systems, the deployment of full-duplex (FD) BS can be considered [56]. This deployment leads to a substantial increase in throughput when compared to full-duplex multi-carrier orthogonal multiple access (FD MC-OMA) and half-duplex (HD) MC-NOMA systems [57].

6.4 Performance Issues

To demonstrate the performance improvements, this section presents numerical results for different NOMA configurations. The analysis focuses on two key performance metrics: outage probability and achievable rate. Additionally, insights into the SE and EE of NOMA are provided [58].

6.4.1 *Outage probability*

In a randomly deployed cellular network, the performance of NOMA in terms of path loss can be assessed in two scenarios. Firstly, when each user has a predetermined data rate based on their assigned QoS, the outage probability serves as a crucial metric, indicating NOMA's ability to fulfill users' QoS requirements. Secondly, when users' rates are dynamically allocated according to channel conditions, the achievable ergodic sum rate becomes a key measure of NOMA performance.

With the appropriate selection of users' data rates and allocated power, NOMA demonstrates superior outage performance compared to other OMA techniques. Additionally, the study highlights NOMA's capability to achieve higher ergodic sum rates. In the case of high signal-to-noise ratio (SNR), the user outage probability in a typical disc-shaped cell with radius R_D is investigated [108].

Figure 6.17 illustrates the outage performance of a NOMA scheme compared to an OMA scheme in a cellular network with randomly deployed users ($K = 2, L = 10, \alpha_p = 2$, and $R_D = 3\,\text{m}$), where K is the number of users in the cell, L is a complexity accuracy parameter, and α_p is the path loss. Users are uniformly distributed, and the targeted data rates are set at 0.1 bits per channel use (BPCU) for the weak user and 0.5 BPCU for the strong user (user 2). For benchmarking, the conventional orthogonal scheme with a targeted rate of 0.6 BPCU (the sum of two users' data rates) is considered. It's essential to note that the results are based on the normalized SNR model.

In the figure, NOMA demonstrates superior performance compared to the comparable scheme, with the diversity order of users being influenced by their channel conditions. The PA ratio between the strong and weak users

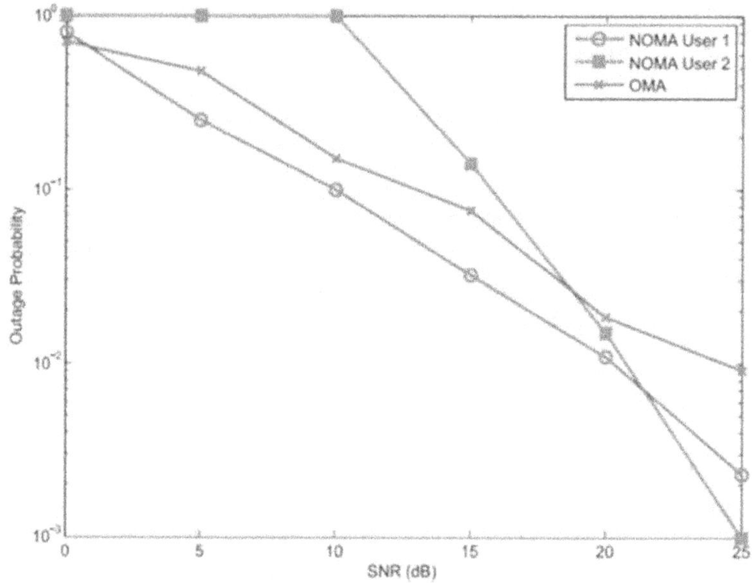

Fig. 6.17. Outage of NOMA with random users in a cell.

is set at 1:4 in this case. The outage probability is primarily valid under high SNR conditions. To assess the relative outage performance, focus on the high SNR regions where both users surpass the OMA scheme. At low SNR levels, the assigned power to the strong user is relatively lower, leading to poorer outage performance. However, as SNR increases, power-domain multiplexing becomes dominant, resulting in superior performance with an enhanced diversity order.

Figure 6.18 illustrates the performance of a typical MIMO NOMA system, focusing on outage probability. The system consists of two clusters, each containing two users equipped with three antennas. The PA ratio is set to 1:4, with target data rates of 3 BPCU for the strong user and 1.3 BPCU for the weak user.

The figure highlights that the MIMO NOMA system exhibits superior performance compared to MIMO with OMA, especially at higher SNR values. By examining the slopes of the performance curves, it becomes apparent that MIMO NOMA achieves the same diversity gain as MIMO OMA. However, MIMO NOMA outperforms MIMO OMA in terms of SE by allowing both users within the same cluster to utilize the same bandwidth.

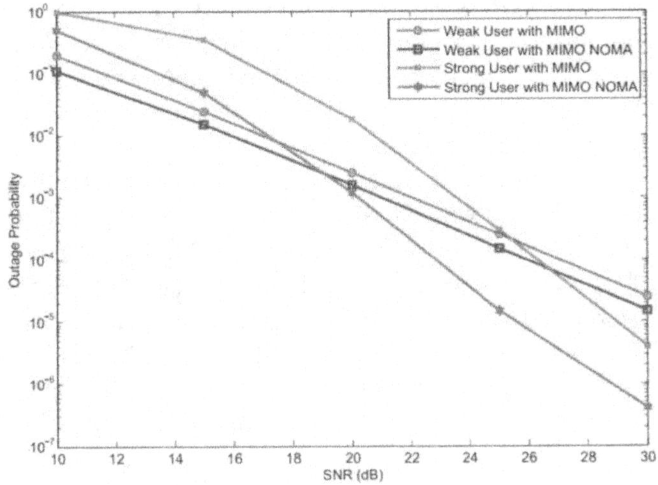

Fig. 6.18. Outage of a MIMO NOMA.

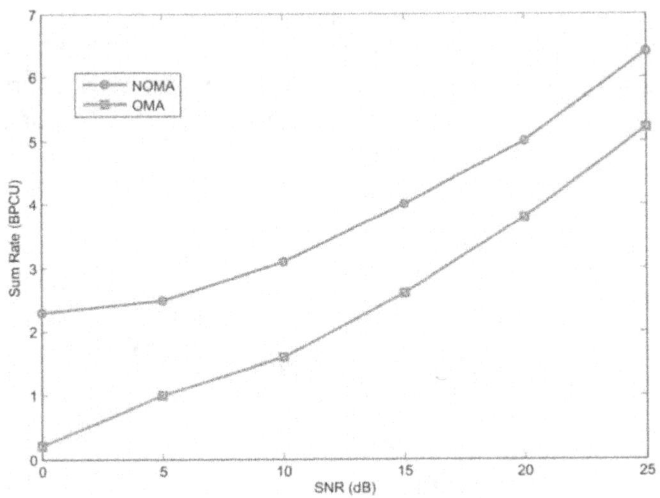

Fig. 6.19. Sum rate performance of NOMA with random users in a cell.

6.4.2 *Achievable rate*

Figure 6.19 depicts the sum rate achieved by a two-user NOMA system with the same parameters as in Fig. 6.17, showcasing the sum of the data rates of NOMA users across various SNR values. The targeted data rates for randomly deployed users are opportunely determined based on their channel conditions. User 2 consistently maintains a higher data rate than user 1 due to its consistently superior channel gain. The figure illustrates that NOMA consistently achieves a higher sum rate compared to the OMA scheme across the entire SNR range.

Figure 6.20 illustrates the rate performance of a cooperative relaying system utilizing NOMA with F-PA (see Section 6.5.1). The figure demonstrates that the cooperative relaying system with NOMA surpasses the performance of a conventional relaying system, especially at high SNR values, where the average power of the channel coefficient from the source to the relay (SR) significantly exceeds that from the relay to the destination (RD). Therefore, the relays should be nominated from the cell-center users in a NOMA-based relay network in order to get the benefits of NOMA.

6.4.3 *Spectral efficiency and energy efficiency*

In the network NOMA scenario (4 users with 2 BSs) described in Fig. 6.23, the power gains of a 4 × 2 downlink channel are [−77, −117; −97, −107; −77, −117; −97, −107] dB. The SE of this network NOMA scheme is depicted in Fig. 6.21. To evaluate the performance of network NOMA in terms of total SE, we compare it with the performance of the entire

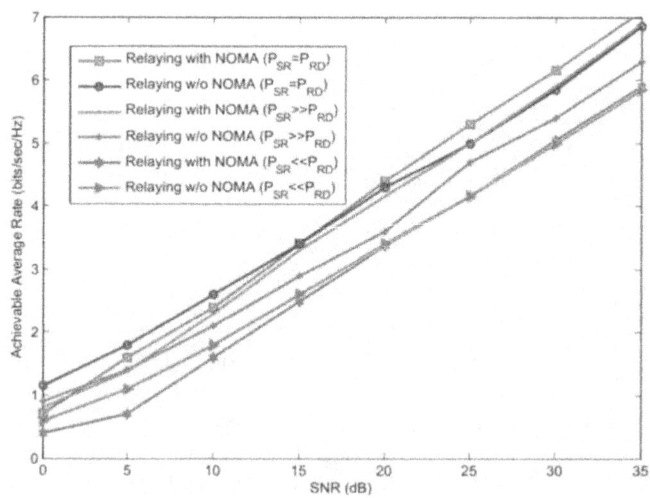

Fig. 6.20. Achievable rate of NOMA with relay.

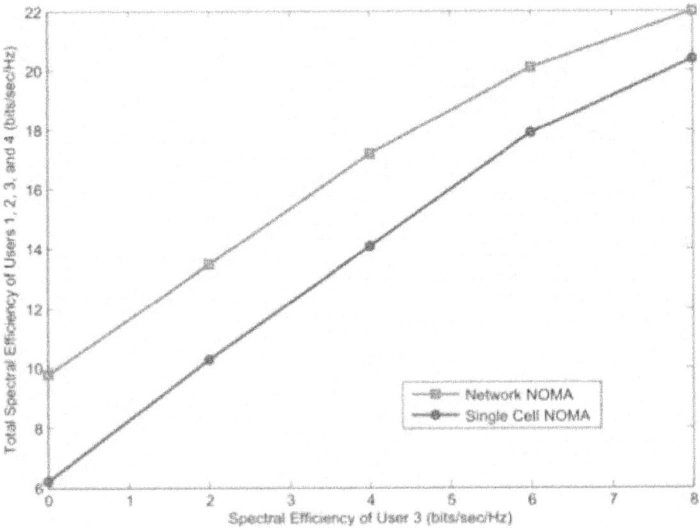

Fig. 6.21. Spectral efficiency of network NOMA.

network based on single-cell NOMA, where BSs do not cooperate with each other. The figure illustrates that network NOMA outperforms conventional NOMA. This improvement is attributed to the mitigation of severe inter-cell interference, especially between cell-edge users, which leads to performance degradation in conventional single-cell NOMA due to the lack of inter-BS cooperation. In network NOMA, the precoder operates such that BSs transmit jointly to cell-edge users, and cell-center users first detect and subtract the signals of cell-edge users to mitigate mutual interference.

In Fig. 6.22, the trade-off between EE and SE in a single-user MIMO NOMA scheme is compared with a single-user MIMO TDMA scheme as an OMA approach. This comparison aims to elucidate the impact on EE performance when changing SE and vice versa. The analysis considers statistical CSI in both cases. The figure illustrates that NOMA outperforms OMA, showcasing a superior EE–SE trade-off. The SE and EE corresponding to the peak operating point offered by the NOMA scheme are higher than those achieved by the OMA scheme. This improvement can be attributed to EE optimization, which maximizes EE under the constraints of the total transmit power and the minimum rate of the weak user, with the assistance of NOMA. Notably, these performance differences become more significant in high SNR regions.

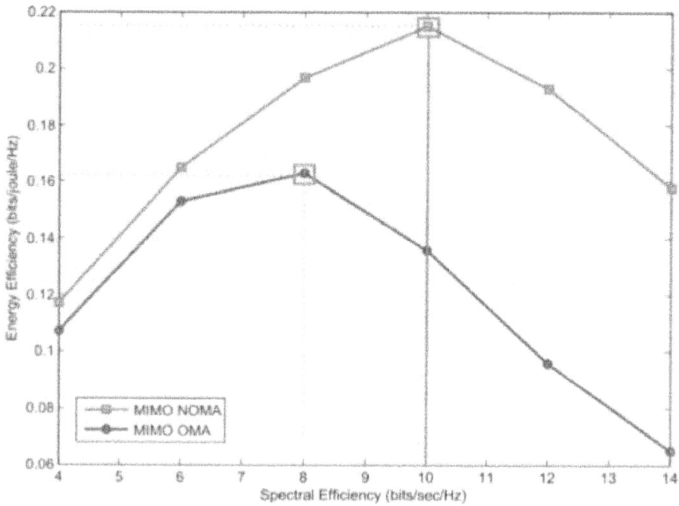

Fig. 6.22. A trade-off between energy efficiency and spectral efficiency.

6.5 Enabling Technologies with NOMA

6.5.1 *Cooperative NOMA (C-NOMA)*

Certainly, cooperative NOMA (C-NOMA) has attracted significant attention due to its promising advantages. The foundation of a C-NOMA scheme, particularly in the context of a downlink system, was initially introduced in Ref. [50]. This approach leverages the prior information inherent in NOMA systems. Let's present the basic form of a C-NOMA, followed by an exploration of various C-NOMA variants.

(i) Basic C-NOMA: In the C-NOMA system outlined in Ref. [50], comprising a downlink scenario with one BS and K users, the operation unfolds in two distinct phases:

- **Direct transmission phase:** Initially, during this phase, the BS disseminates a superposed message to all K users by combining K messages according to the NOMA principle. Subsequently, each user decodes the transmitted messages using the SIC, with the exception of the weakest user, denoted as user K.
- **Cooperative phase:** Following the direct transmission phase, the cooperative phase ensues, involving $(K-1)$ time slots. During the first time slot,

the strongest user (i.e., user 1) first broadcasts the superposed messages to the other $(K - 1)$ weak users by superposing the $(K - 1)$ messages. Then, user 2 successively decodes the messages of the weak users (user 3 to user K) by combining the observations from the first phase and the first time slot of the second phase via diversity combining (e.g., maximum ratio combining) and eventually decodes its own message using the NOMA principle. Similarly, during the second time slot, the second strongest user (i.e., user 2) broadcasts the superposed messages to the other $(K - 2)$ weak users by superposing the $(K - 2)$ messages, and user 3 successively decodes the messages of the weak users (user 4 to user K) by combining the observations from the first phase and the first and second time slots of the second phase via diversity combining and eventually decodes its own message using the NOMA principle.

This iterative process is replicated across the remaining time slots, with varying PA coefficients at each phase and time slot, determined by the local channel conditions. This dynamic allocation contributes to larger SINRs for users, leveraging the combination of signals from both phases. The cooperative phase's integration in C-NOMA significantly enhances reception reliability. Consequently, C-NOMA surpasses non-cooperative NOMA (NOMA without the cooperative phase) regarding both outage probability and outage capacity, securing diversity gain for all users. Despite these advantages, engaging all users in C-NOMA poses challenges due to significant computational complexity and increased communication latency associated with additional time slots. To address these challenges, user grouping is employed to mitigate the impact of these shortcomings. Users within a cell are divided into multiple groups, allowing C-NOMA to be implemented within each group independently. This approach balances the benefits of C-NOMA while managing practical complexities associated with involving the entire user population.

(ii) Cooperative relaying scheme-based NOMA (CRS-NOMA): The C-NOMA concept described above bears resemblance to relay-assisted NOMA, where the strongest user in the cooperative phase serves as a relay for other users. Cooperative relaying systems, capitalizing on spatial diversity, have gained popularity in communication paradigms. In a half-duplex relay system, the destination receives and combines two independent copies of the transmitted signal: one directly from the source and the other from the relay, enhancing the overall SINR and improving signal reception quality.

However, the conventional cooperative relaying suffers from a SE loss due to redundant transmissions.

To address this, a relay scheme utilizing NOMA, termed cooperative relaying scheme with NOMA (CRS-NOMA), has been proposed to enhance SE. In CRS-NOMA, the source broadcasts a superposition-coded signal in the first time slot. The relay then performs the SIC, decoding its own symbol and the symbol destined for the receiver. In the subsequent time slot, the relay forwards the decoded signal to the destination at full transmit power. CRS-NOMA exhibits superior performance over conventional cooperative relaying in high SNR regions, leading to improved sum capacity. While initially designed for Rayleigh fading channels, the performance of CRS-NOMA over Rician fading channels has been explored in related studies. A variant of the CRS-NOMA scheme has been introduced, wherein the destination employs diversity combining and additional SIC to jointly decode the transmitted signal, outperforming the basic CRS-NOMA in scenarios where the source-to-relay channel strength is stronger than the relay-to-destination channel strength. It's noteworthy that C-NOMA schemes have been primarily investigated in downlink settings, demonstrating their advantages. However, the benefits of C-NOMA have also been extended to uplink scenarios, emphasizing its versatility and applicability.

(iii) Uplink and downlink C-NOMA: The concept of cooperative relaying using NOMA has been explored in Ref. [60], where multiple sources collaborate by sharing a common relay to transmit their messages to their respective destinations. In this scheme, the relay initially employs an uplink NOMA approach to receive symbols from the sources and subsequently utilizes a downlink NOMA strategy to broadcast a superposition-coded signal to the destinations. By leveraging both uplink and downlink NOMA principles, Ref. [60] aims to achieve higher ergodic capacity in a cooperative setting.

Expanding on this idea, the SE of joint uplink and downlink NOMA can be further enhanced by introducing the transmission of a new symbol from the source to the destination during the cooperative phase [61]. In this modified scheme, the source initially broadcasts a superposition-coded signal to the destination and relays in downlink NOMA mode. Concurrently, in the subsequent phase, while the relay forwards the symbol decoded via SIC to the destination, the source simultaneously transmits a new symbol to the destination following the uplink NOMA principle. This coordinated utilization of uplink and downlink NOMA mechanisms contributes to improving

SE and enhancing the overall system performance in cooperative relaying scenarios.

Indeed, the utilization of dedicated relays in C-NOMA systems presents a versatile approach to enhance transmission reliability, particularly for NOMA users experiencing challenging channel conditions [62]. Furthermore, dedicated relays can prove beneficial for facilitating reliable communications, especially for NOMA users equipped with multiple antennas [63].

In scenarios involving multiple relays within a cooperative network, the process of relay selection becomes pivotal for ensuring robust communications. While conventional cooperative communication often relies on the max–min criterion for relay selection [64], C-NOMA introduces alternative strategies that offer notable performance improvements. For instance, the two-stage relay selection scheme proposed in Ref. [65] demonstrates significant enhancements in terms of outage probability compared to the max-min strategy. In this scheme, the first stage is dedicated to ensuring the targeted data rate for the users of interest, while the second stage opportunistically maximizes the data rate for other users, resulting in more efficient resource utilization and improved system performance.

6.5.2 *Coordinated NOMA*

In cellular systems, users at the cell edge typically encounter lower data rates compared to those near the BS. Coordinated multipoint (CoMP) transmission and reception techniques, where multiple BSs collaborate to support cell-edge users, are commonly employed to enhance transmission rates for such users. However, when allocating the same channel to multiple cell-edge users, the SE of the system can degrade as the number of these users increases. To address this challenge, a coordinated space–time coding (CSC)-based NOMA scheme [66] has been developed. This scheme employs space–time coding for downlink transmissions to a group of cell-edge users and users near the BS simultaneously, utilizing a common access channel [26]. In essence, BSs transmit Alamouti (space–time) [67]-coded signals to cell-edge users while concurrently transmitting signals to users in proximity to the BS. The CSC-NOMA scheme, incorporating Alamouti coding, offers a favorable transmission rate to cell-edge users (U_c) without compromising the rates for users near BS 1 (U_1) and BS 2 (U_2), thereby enhancing the overall SE. The sum rate, denoted as $R_{c1} + R_{c2} + R_c$, is determined by the rates to U_1, U_2, and U_c, respectively. The mathematical expression for the sum rate is as follows:

$$R_{c1} = E\left[\log_2\left(1 + \frac{|h_{1,1}|^2 P_1}{E[|h_{1,2}|^2]P_2 + \sigma_n{}^2}\right)\right] \tag{6.15}$$

$$R_{c2} = E\left[\log_2\left(1 + \frac{|h_{2,2}|^2 P_2}{E[|h_{2,1}|^2]P_1 + \sigma_n{}^2}\right)\right] \tag{6.16}$$

$$R_c = \min\{Z_1, Z_2, Z_c\} \tag{6.17}$$

where $Z_1 = E[\log_2(1 + \text{SINR}_1)]$, $Z_2 = E[\log_2(1 + \text{SINR}_2)]$, and $Z_c = E[\log_2(1 + \text{SINR}_c)]$. Here, SINR_i represents the signal-to-interference-plus-noise ratio for user U_i in decoding the signal of user U_c. The channel coefficient from BS j to user U_i is denoted as $h_{i,j}$. Importantly, the utilization of Alamouti coding in CoMP communications eliminates the need for instantaneous CSI exchange, providing a significant advantage over coherent transmission schemes that incur excessive backhaul overhead, especially for high-mobility cell-edge users.

6.5.3 *Network NOMA*

Consider a simplified scenario with two cells in a cellular system, as illustrated in Fig. 6.23, where users U_3 and U_4 are served by BS1, while users U_1 and U_2 are served by BS2. Adopting a two-user NOMA scheme, U_3 is paired with U_4, and U_1 is paired with U_2. Given that a cell-edge user in NOMA does not perform SIC before decoding its signal, U_4 and U_1 inherently experience interference from U_3 and U_2, respectively.

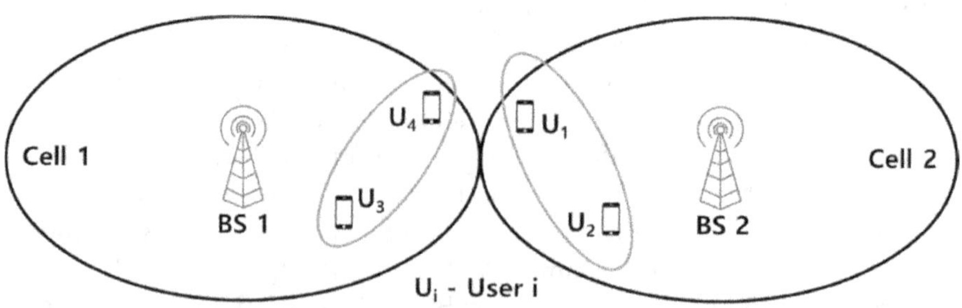

Fig. 6.23. NOMA in a two-cell scenario (network NOMA).

To alleviate this interference impact, BS1 and BS2 allocate more power to U_4 and U_1, respectively. However, this results in inter-cell interference during downlink transmissions and mutual interference between U_4 and U_1 during uplink transmissions. Addressing this challenge associated with employing NOMA in a multi-cell scenario, the straightforward application of single-cell NOMA solutions is not suitable. Instead, single-cell NOMA needs to evolve into network NOMA.

Similar to various inter-cell interference mitigation techniques utilized in OMA, solutions are required in NOMA to mitigate interferences in a multi-cell scenario. One potential solution is to employ joint precoding of users' signals across neighboring cells. However, designing an optimal precoder is challenging, as each BS needs to possess all users' data and CSI. Correlation-based precoder design necessitates dynamic user selection for each NOMA pair. Moreover, multi-user precoding applicable to single-cell NOMA may not be realistic in network NOMA, as a beam generated via geographically separated BSs does not support more than one spatially separated user for intra-beam NOMA.

To address these complexities, a low-complexity precoding scheme for network NOMA was proposed in Ref. [68], leveraging the fact that large-scale fading would differ significantly between links of different cells. Here, the joint precoder is applied only to cell-edge users (e.g., U_4 and U_1, as shown in Fig. 6.23), and the resulting SINR of each user U_i of power P_i is found as

$$\mathrm{SINR}_1 = \frac{[\{(H_{41}(H_{41})^H)^{-1}\}_{1,1}]^{-1} P_1}{|h_{11}|^2 P_0 + |h_{12}|^2 P_2 + N_0 B} \tag{6.18}$$

$$\mathrm{SINR}_2 = \frac{|h_{22}|^2 P_2}{|h_{21}|^2 (P_0 + |w_{0,0}|^2 P_1 + |w_{0,1}|^2 P_3) + N_0 B} \tag{6.19}$$

$$\mathrm{SINR}_3 = \frac{|h_{30}|^2 P_3}{|h_{32}|^2 (P_2 + |w_{1,0}|^2 P_1 + |w_{1,1}|^2 P_3) + N_0 B} \tag{6.20}$$

$$\mathrm{SINR}_4 = \frac{[\{(H_{41}(H_{41})^H)^{-1}\}_{0,0}]^{-1} P_4}{|h_{41}|^2 P_0 + |h_{42}|^2 P_2 + N_0 B} \tag{6.21}$$

where $H_{41} = [h_4, h_1]^T$ with the channel vector of the ith user, $h_i = [h_{i1}, h_{i2}]$, and $h_{ij}\,(i\{1, 2, 3, 4\}, j \in \{1, 2\})$ is the channel response between the jth BS and ith user. The zero-forcing precoder, W, is the normalized pseudo-inverse of H_{41}, i.e.,

$$W = (H_{41})^H [H_{41}(H_{41})^H]^{-1}$$

B is the system bandwidth, and N_0 is the noise spectral density.

In Ref. [69], two coordinated beamforming strategies were introduced to manage inter-cell interference in two-cell MIMO-NOMA networks. Building upon this work, Ref. [70] extends these strategies to accommodate an arbitrary number of cells. Furthermore, the study characterizes the maximum number of users supported by the proposed scheme in multi-cell MIMO networks.

6.5.4 *Multiple-Input Multiple-Output NOMA*

The integration of MIMO technology with NOMA holds substantial significance, as MIMO introduces additional spatial degrees of freedom that can contribute to enhanced system performance. The research and exploration of MIMO-NOMA have garnered considerable attention from both academic [71–74] and industrial [75, 76] perspectives. This intersection of technologies aims to leverage the advantages offered by both MIMO and NOMA to achieve improved efficiency and capabilities in wireless communication systems.

6.5.4.1 *Downlink MIMO-NOMA*

In comparison to SISO-NOMA, the primary challenge in MIMO-NOMA arises from the non-degraded nature of the MIMO channel. Unlike SISO-NOMA, users in MIMO-NOMA cannot be straightforwardly ordered based on their channel strengths in general scenarios. Consequently, achieving capacity in MIMO-NOMA is generally challenging. The utilization of complex dirty paper coding (DPC) stands out as the sole strategy capable of achieving the capacity region of the multiple-input single-output (MISO) Gaussian BC when perfect CSI is available at the transmitter. However, it's important to note that, under specific conditions, MIMO-NOMA can still attain performance comparable to DPC. In Ref. [71], the concept of quasi-degradation is introduced for a two-user MISO-NOMA system with a minimum rate constraint. The study demonstrates that NOMA can achieve performance equivalent to DPC when the channels exhibit quasi-degradation under certain conditions. This highlights the potential for achieving optimal performance in MIMO-NOMA systems with careful consideration of channel characteristics and constraints.

A simplistic approach to address this challenge is to order users based solely on large-scale path loss. Nevertheless, this approach may lead to performance degradation, as it does not leverage small-scale channel information. Maximizing the benefits of spatial degrees of freedom requires the development of an appropriate beamforming/precoding design specific to MIMO-NOMA systems. In this context, a well-conceived beamforming design should consider both the power domain and the angular domain to enhance the system's SE.

Two prevalent MIMO-NOMA designs have gained popularity to address these challenges:

(1) **Cluster-based MIMO-NOMA design:** This design approach, as explored in studies [72–74, 76, 77], organizes users into clusters. The cluster-based MIMO-NOMA design is aimed at optimizing performance and efficiency within user clusters, thereby mitigating the complexities associated with direct user ordering.
(2) **Beamforming-based MIMO-NOMA design:** Another approach, discussed in studies [71, 78–80], focuses on beamforming. This design leverages advanced beamforming techniques to optimize user performance. The consideration of both power and angular domains in the beamforming design is crucial for maximizing SE in MIMO-NOMA systems.

These design strategies reflect the ongoing efforts to overcome the challenges posed by the unique characteristics of MIMO-NOMA, emphasizing the importance of tailored solutions for user ordering and beamforming:

(i) Cluster-based MIMO-NOMA: The cluster-based MIMO-NOMA design operates as follows: Users are initially grouped into clusters based on their channels, with users within the same cluster sharing a common beamforming vector. It is important to note that determining the optimal user clustering often involves exhaustively considering all possible user clustering combinations, a computationally intensive process. A practical heuristic for this task is to group users with high channel correlation and gain difference into the same cluster. This approach leverages the positive impact of high channel correlation on the beamforming design and the favorable conditions for implementing SIC due to significant gain differences [76].

Once users are clustered, the primary challenge lies in designing appropriate beamforming vectors for each cluster. The simplest approach is to use random beamforming, which has proven effective in reducing CSI feedback requirements [51, 77]. In Ref. [77], the authors propose a combination of random beamforming and intra-beam SIC for downlink NOMA transmission. However, due to the presence of inter-cluster interference, different clusters remain coupled. To address this issue, the authors in Ref. [51] suggest using random beamforming at the BS and zero-forcing detection at the user side. This allows the removal of inter-cluster interference at each user, decomposing the MIMO-NOMA system into a set of independent SISO-NOMA arrangements.

(ii) Beamforming-based MIMO-NOMA: Beamforming-based MIMO-NOMA, depicted in Fig. 6.24, operates without grouping users into clusters. Instead, each user (*UE k*) is assigned its own beamforming vector. However, unlike MIMO-OMA, SIC is still executed on the user side to eliminate interference. To ensure successful SIC, the explicit constraint is imposed that the achievable rate for each user cannot exceed the minimum rate among all users that need to decode it. Notably, as demonstrated in Ref. [73], this

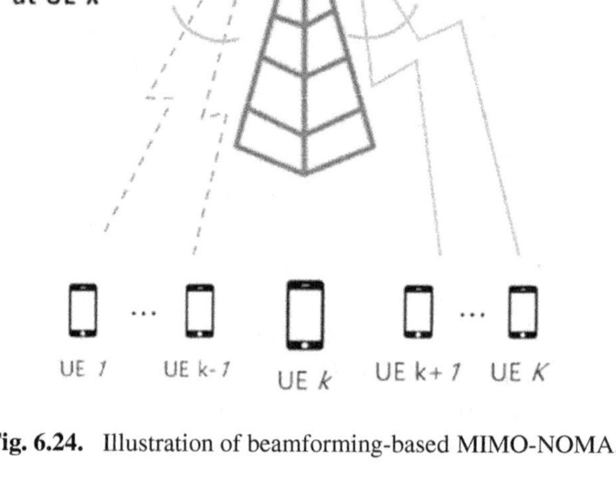

Fig. 6.24. Illustration of beamforming-based MIMO-NOMA.

constraint poses less of an issue in cluster-based MIMO-NOMA, where users in the same cluster share a common beamforming vector.

In Ref. [78], researchers tackle the problem of maximizing the ergodic sum rate in MIMO-NOMA systems characterized by Rayleigh fading and statistical CSI at the transmitter. They put forth both optimal and low-complexity suboptimal PA schemes, subject to total transmit power constraints and minimum rate constraints for the weak user. Numerical evaluations demonstrate the substantial performance gains of these proposed NOMA schemes over the traditional OMA. Additionally, Ref. [79] introduces a layered transmission scheme based on QR factorization. Under instantaneous CSI, an approach to maximizing the sum rate of MIMO-NOMA with layered transmissions is proposed, highlighting the concavity of the sum rate concerning the allocation of powers to multiple layers of users. For statistical CSI, a closed-form expression for the average sum rate is derived, and PA is performed using alternating optimization.

In Ref. [71], optimal precoding strategies are explored for a QoS-constrained optimization problem in a MISO-NOMA network. The study demonstrates that if the BCs exhibit quasi-degradation, the proposed optimization algorithm when combined with superposition coding and SIC, can achieve the system capacity. It's noteworthy that Refs. [71, 78, 79] specifically address scenarios with two users. Expanding the scope, Ref. [80] considers the general scenario with multiple users and utilizes the min-max algorithm for sum-rate maximization. To summarize, while cluster-based MIMO-NOMA is more tractable, beamforming-based MIMO-NOMA has the potential to yield superior system performance.

6.5.4.2 *Uplink MIMO-NOMA*

In an uplink system featuring a multiple-antenna BS serving K single-antenna users, with the number of antennas at the BS denoted by M, the received signal at the BS is expressed as

$$y = \sum_{k=1}^{K} \sqrt{P_k} h_k x_k + n \tag{6.22}$$

where $h_k \in C^M$ is the channel vector between user k and the BS. Additionally, x_k and P_k represent the signal and transmit power for the kth user, respectively, while n represents the additive white Gaussian noise vector.

At the BS, a receiver employing minimum mean square error (MMSE)-based linear filtering followed by SIC can be utilized for signal detection [16]. It has been demonstrated in Ref. [16] that the sum rate of the K users is unaffected by the user ordering and is equal to the system capacity. However, the individual rate is influenced by the user ordering, with users decoded later in the SIC process experiencing increased throughput. To improve system fairness, users with superior channel conditions are prioritized for earlier decoding.

While the MMSE-SIC receiver theoretically achieves the system capacity, practical implementations may encounter challenges related to error propagation. To address this, alternative detection methods, such as zero-forcing, can be employed. In a mmWave massive MIMO system, zero-forcing is utilized among strong users as demonstrated in Ref. [97]. Additionally, an enhanced scheme is proposed to effectively eliminate inter-cluster interference through SIC. It's noteworthy that this enhanced scheme represents a compromise between MMSE-SIC and zero-forcing techniques.

6.5.5 *NOMA with HARQ*

The hybrid automatic repeat reQuest (HARQ) protocol plays a crucial role in link adaptation, aiming to ensure reliable communications through the utilization of retransmission diversity and channel coding gain. In the context of NOMA, where multiple packets may be transmitted concurrently, the potential for collisions exists. In such scenarios, a retransmission request is sent to the users involved in the collision. Despite the collision, each packet can carry a certain amount of information. Leveraging HARQ, all retransmitted signals can be intelligently combined to enhance the SE of NOMA [81].

A comprehensive investigation by Li *et al.* [82] delves into the HARQ design specifically tailored for NOMA with single-user-multiple-input multiple-output (SU-MIMO). The study revealed that designing HARQ for NOMA presents more challenges compared to OFDMA systems. A robust HARQ algorithm needs to address two key challenges in NOMA retransmission: determining when the HARQ combination should occur and how to manage power assignment for retransmission.

Based on the analysis of channel responses to initial transmissions and retransmissions, an optimal HARQ combination timing was identified. Specifically, it was found that the opportune moment for the distant user to perform an opportunistic HARQ combination is when the SINR of the initial transmission is reasonably larger than the retransmission SINR. Regarding retransmission power assignment, one effective approach involves allocating power to maximize the geometric mean user throughput across all users involved in the retransmission process.

In summary, an advanced HARQ design holds the potential to yield improvements in NOMA gain, benefiting both overall cell throughput and cell-edge throughput.

6.5.6 *NOMA in visible light communications*

Visible light communication (VLC) systems face a challenge due to the narrow modulation bandwidth of light sources, limiting their ability to achieve competitive data rates. To address this limitation, optical wireless communications explore various signal-processing techniques, including multicarrier and multi-antenna systems, to enhance data rates. With NOMA emerging as a potential candidate for NG wireless communications, its feasibility in VLC systems has become an area of interest.

Applying NOMA in VLC systems shows promise in enhancing achievable throughput, especially in high-rate VLC scenarios. Studies indicate that NOMA serves as a promising MA scheme for downlink communications in VLC networks. However, existing works primarily focus on scenarios with multiple light-emitting diodes (LEDs) in indoor environments, lacking consideration for real high-speed OFDM-based communications [58].

In a related research direction, investigations into optical OFDM transmission compared the achievable capacities of NOMA and OFDMA in VLC systems. The study also analyzed the impact of cancellation errors in SIC receivers for VLC systems. The achievable data rates for users in a two-user (U_1 and U_2) NOMA system with DC-biased optical OFDM (DCO-OFDM), assuming a frequency-flat fading channel, can be expressed as follows:

$$R_1 = \frac{B}{2}\log_2\left(1 + \frac{{\gamma_0}^2 {L_0}^2 P\beta_{1,n} {G_1}^2}{\epsilon_0 {\gamma_0}^2 {L_0}^2 P^2 (1-\beta_{1,n}) {G_1}^2 + N_0 B}\right) \tag{6.23}$$

$$R_2 = \frac{B}{2}\log_2\left(1 + \frac{{\gamma_0}^2 {L_0}^2 P\beta_{2,n} {G_2}^2}{{\gamma_0}^2 {L_0}^2 P^2 \beta_{2,n} {G_2}^2 + N_0 B}\right) \tag{6.24}$$

where the optical parameters γ_0, L_0, and ϵ_0 are the receiver responsivity, the number of LED chips, and the signal cancellation error, respectively.

G_1 and G_2 are channel gain for U_1 and U_2, respectively, and the symbol $\beta_{i,n}$ denotes the power coefficient for the nth subcarrier for U_i.

6.5.7 *NOMA with other solutions*

(i) NOMA with Raptor Codes: NOMA combined with Raptor Codes presents an innovative approach to enhance communication reliability and efficiency. Raptor Codes for a given integer k and a real number ϵ, initially proposed by Shokrollahi [83], are characterized by their ability to encode a message of k symbols into a potentially limitless sequence of encoding symbols. The distinctive feature of Raptor Codes lies in the fact that any subset of $k(1+\epsilon)$ encoding symbols allows for the recovery of the original message with a high probability.

In the context of cooperative communications, Raptor Codes have demonstrated effectiveness. The integration of Raptor Codes with NOMA has been explored in-depth in a study [107]. In this scenario, an interfering channel equipped with Raptor Codes is introduced alongside an existing primary non-orthogonal wireless channel. The key finding of this investigation was that the coded interference introduced by Raptor Codes does not adversely impact the performance of the main channel. Simultaneously, the interfering signal itself, despite being coded, can be successfully decoded with a high probability.

(ii) NOMA with network coding: Combining NOMA with random linear network coding (RLNC) introduces a robust encoding scheme that facilitates efficient data retransmission. RLNC, characterized by its randomness, eliminates the need for the source to be aware of packets lost by the intended receiver, enhancing the adaptability and resilience of the communication system.

In the context of multicast and broadcast services, various RLNC techniques have been proposed to optimize transmission efficiency. The integration of RLNC with NOMA, specifically in the multicast context, has been explored by Ref. [85]. While conventional NOMA focuses on power-domain multiplexing for multiple receivers in unicast services, NOMA-RLNC extends this concept to leverage power-domain multiplexing for multiple reception groups in multicast services.

The study indicates that NOMA-RLNC brings about improvements in the packet success probability, particularly for multicast services. By superposing multiple coded packets before transmission, NOMA-RLNC enhances the performance of multicast communication scenarios, showing its potential to elevate the reliability and efficiency of such services.

(iii) Coexistence of NOMA and OMA: The simultaneous existence of NOMA and OMA is evaluated with a focus on capacity enhancement, a primary objective of future mobile communications. NOMA emerges as a promising candidate for shaping the future of radio access. However, this does not imply the complete replacement of OMA schemes. In scenarios with a small number of users and where the near–far effect is not a critical consideration, OMA may still be preferred over NOMA, particularly in small-cell deployments.

The expectation is that both OMA and NOMA will coexist to address the diverse requirements of different services and applications within the evolving NG framework. Recognizing the varied needs of different scenarios, the coexistence of these multiple access technologies becomes crucial for the comprehensive fulfillment of 5G/6G network capabilities. In the exploration of NOMA schemes for 5G, as discussed by Dai *et al.* [15], the analysis delves into the principles, key features, and receiver complexities. The concept of software-defined multiple access is underscored as a flexible approach capable of meeting the diverse demands of various services and applications in the dynamic 5G ecosystem.

(iv) NOMA with SWIPT: Beyond the goal of improving SE, a key objective in 5G networks is to maximize EE. Researchers have shown that simultaneous wireless information and power transfer (SWIPT) holds promise as a solution to address EE challenges. An innovative approach involves integrating SWIPT with NOMA, creating a cooperative communication protocol. In this setup, cell-center users play the role of energy-harvesting relays, assisting cell-edge users. Notably, the use of SWIPT maintains the diversity orders for both nearby and distant users, preserving performance compared to traditional NOMA [86].

6.6 Challenges in NOMA

Numerous researchers have devoted efforts to the design and implementation of NOMA techniques, addressing various technological challenges inherent in these methods. Extensive literature illustrates NOMA's compatibility with cooperative communications, relaying, and MIMO systems, demonstrating notable performance enhancements [84]. Within this context, several NOMA solutions, aimed at mitigating challenges associated with multi-cell networks, wireless link adaptation, EE–SE trade-offs, and user pairing, are highlighted:

- **Dynamic user pairing:** As discussed earlier, co-channel interference poses a significant challenge in NOMA systems, given that multiple users concurrently share the same time, frequency, and spreading code resources. To manage this interference, it is impractical to require all users within the system to engage in NOMA simultaneously. Instead, a pragmatic approach involves dividing users into distinct groups, applying NOMA within each group, and allocating orthogonal bandwidth resources to different groups.

  The prevalent static pairing scenario, where the mth and nth users (with a constant and relatively high $|m - n|$) are paired for NOMA implementation, is commonly considered. However, realizing the maximum benefits of NOMA necessitates the exploration of dynamic user pairing and grouping schemes. Addressing this challenge, analytical insights derived by Ding *et al.* [87], particularly for the case with two selected users, can serve as criteria for designing distributed methods.

  The user pairing and power allocation (UPPA) scheme [87, 88], known to enhance performance gains, currently restricts scheduling to only two users. A promising extension would involve devising a PA scheme for multi-user NOMA systems where the scheduler accommodates more than two users. Furthermore, applying UPPA schemes and their variations [89] to practical systems with suitable modulation and coding schemes requires additional research.

  In short, the development of dynamic user pairing strategies and their integration into practical systems stands as a crucial area of exploration to fully exploit the potential benefits offered by NOMA.
- **Impact of transmission distortion:** The transmission of source information, such as voice and video, over communication channels is inherently subject to losses. The transmitted data undergoes distortion during propagation to the receiver. Addressing this inevitable loss, substantial theoretical attention has been dedicated to assessing source fidelity over fading channels. Various frameworks involving source coding and channel coding have been devised to minimize end-to-end distortion. However, the pursuit of source coding diversity and channel coding diversity introduces a nuanced interplay among preferred distortion levels, cost considerations, and system complexity.

A notable study by Choudhury and Gibson [90] compared source distortion concerning two definitions of channel capacity: ergodic capacity and outage capacity. Both information capacity and distortion are contingent upon outage probability. Notably, the outage probability that maximizes the outage rate may not necessarily yield the minimum expected distortion [90, 91]. An intriguing avenue for exploration involves an investigation to optimize the outage probability, allowing a NOMA scheme to achieve the maximum outage rate with acceptable distortion.

An examination of the interplay between outage probability, outage rate, and distortion in NOMA schemes holds promise for refining transmission strategies and achieving a harmonious balance between reliability and fidelity in communication channels.

- **Impact of interference:** While interference analysis is a generic term in wireless communications, we focus on C-NOMA [60], which employs Bluetooth-like short-range communications during the cooperative phase. However, integrating Bluetooth radio into cellular communications introduces significant challenges due to extreme interference from the existing wireless personal area network (WPAN) operations. Bluetooth interference manifests in reduced coverage and throughput, intermittent or complete loss of connectivity, and difficulties in pairing during the user discovery phase.

  In practice, interference in the deployed environment, payload size, and distance between cooperative users all influence channel allocation deployment. Moreover, the formulation of a self-organizing scatternet to manage Bluetooth nodes must be reimagined to align with NOMA requirements, as NOMA users are paired based on CSI. A robust scatternet should facilitate valid routes between nodes with a high probability, even amidst users' mobility that may result in the complete loss of certain wireless links.

  Furthermore, the dynamic nature of user mobility introduces dynamic interference, necessitating a thorough performance analysis of C-NOMA schemes in such dynamic interfering environments. This exploration promises to yield valuable insights into optimizing performance and managing interference in practical NOMA deployments.
- **Resource allocation:** Efficient resource allocation is imperative to meeting the diverse traffic requirements of 5G systems [91, 92], which demand high data rates, ultra-low latency, and reliability despite limited resources. Wireless resource management encompasses a series of processes aimed

at determining the timing and quantity of resources allocated to each user [93], contingent upon the type of resources available.

According to Shannon's information capacity theorem, bandwidth (BW) stands as a crucial wireless resource. Effective BW management involves partitioning the total BW into multiple chunks, with each chunk allocated to specific users or groups, as observed in NOMA deployments. Moreover, the dynamic nature of user traffic necessitates sophisticated algorithms for user-pairing and optimum PA in NOMA to maximize performance while minimizing resource consumption.

Resource allocation in NOMA presents fertile ground for exploration from a mathematical optimization theory perspective. For instance, Lei *et al.* [94] proposed a framework for joint power and channel allocation in NOMA-based 5G networks, considering user power control and SIC implementation to address the power and channel allocation problem. Additionally, the exploration of PA in NOMA-based cognitive radio networks remains an area ripe for further research [95].

In summary, leveraging mathematical optimization techniques and innovative algorithms holds promise for enhancing resource allocation efficiency in NOMA systems, thereby optimizing network performance and resource utilization in 5G and beyond.

- **SIC receiver variations:** In the context of NOMA, the conventional approach involves adopting an SIC receiver at cell-center users, effectively eliminating interference from cell-edge users. However, advancements in NOMA performance are feasible by incorporating sophisticated receivers, particularly at cell-edge users [96]. In this context, an advanced receiver refers to one capable of ideal SIC processing, as the gains in NOMA performance with non-ideal SIC are deemed unsatisfactory.

  Symbol-level SIC (SLSIC) denotes a type of SIC where the cell-edge user's signal is demodulated, and a hard decision is made without channel decoding. Another approach, codeword-level SIC (CWSIC), involves both demodulating and decoding the signal of a cell-edge user. CWSIC exhibits a higher probability of successful cell-edge user signal recovery compared to SLSIC, given the inclusion of channel decoding in signal detection. However, this comes at the expense of increased computational complexity [98].

  NOMA employing an MMSE-based linear filtering followed by SIC (MMSE-SIC) can benefit from an appropriate power control mechanism. This approach helps mitigate inter-cell interference, with fairness assured through proportional fairness (PF)-based multiuser scheduling [99].

- **NOMA with multiple antenna and antenna selection:** The performance of MIMO NOMA systems improves with a larger rank of the MIMO channel matrix, as it leads to a greater number of decorrelated channels. Consequently, evaluating MIMO NOMA hinges on the rank of the channel matrix. However, existing research in MIMO NOMA predominantly focuses on full-rank channel matrices to analyze system performance. For instance, analytical results presented in Ref. [100] establish upper bounds on outage probability and capacity under this assumption. Hence, it is imperative to investigate the design of MIMO NOMA systems with rank-deficient channel matrices.

  Addressing this need, rank adjustment techniques for NOMA can be employed to enhance MIMO NOMA performance [101]. Furthermore, a joint optimization of rank and transmission power assignment holds promise for further performance improvement. Exploring the integration of existing MIMO NOMA techniques with other promising wireless technologies, such as orthogonal frequency- and code-division multiplexing (OFCDM), presents an intriguing avenue for research.

  However, it's noteworthy that many MIMO NOMA algorithms, such as the beamforming algorithm proposed in Ref. [102], exhibit high computational complexity. Therefore, there is an urgent need for research endeavors aimed at complexity reduction in MIMO NOMA systems. Advancing the MIMO NOMA technology entails exploring novel approaches for handling rank-deficient channel matrices, optimizing resource allocation strategies, and mitigating computational complexity, thereby unlocking the full potential of MIMO NOMA in future wireless communication systems.

  Employing multiple antennas at the transmitter side necessitates an equivalent number of parallel radio frequency (RF) chains at the front end, thereby introducing complexities, increased power consumption, and elevated costs. In response to these challenges, transmit antenna selection (TAS) emerges as a preferred approach to streamline transmitter structures. TAS techniques strategically choose the optimal antenna from a set of multiple antennas, utilizing a single RF chain for transmission. This pragmatic solution reduces system complexity, power consumption, cost, and physical size, albeit with an acceptable performance trade-off.

  In light of these considerations, there is a compelling need for research to explore innovative TAS-NOMA schemes for downlink communications. This scenario involves a BS equipped with multiple antennas communicating with multiple users, each equipped with a single antenna.

The primary objective of such schemes should be to enhance the sum rate, considering the opportunistic allocation of targeted user rates based on prevailing channel conditions.

- **Heterogeneous networks:** A Heterogeneous Network (HetNet) represents a wireless network comprising nodes with varying transmission powers and coverage sizes. HetNets hold substantial promise for the NG of wireless networks, delivering enhanced capacity, coverage, and reduced energy consumption. The infrastructure, characterized by a high-density deployment of low-power nodes, can significantly amplify EE compared to the low-density deployment of a limited number of high-power nodes.

  Various research efforts have delved into HetNets, exploring aspects, such as node cooperation, optimal load balancing, and enhanced inter-cell interference coordination [103]. A recent study [104] introduced a system framework for a cooperative HetNet in the context of 5G, aiming to optimize both SE and EE.

  Given the shared objectives of NOMA and HetNets, leveraging NOMA within a specific HetNet configuration can yield extended benefits. Moreover, the non-uniform spatial distribution of mobile users can influence the performance of NOMA [120]. Therefore, conducting investigations into outage performance, ergodic capacity, and user fairness in NOMA schemes with spatially distributed users represents a worthwhile avenue for research.

  In conclusion, the intersection of NOMA and HetNets presents an opportunity for synergistic improvements in wireless network performance, capacity, and EE, warranting further exploration and study.

- **Outage probability analysis:** Outage analysis stands as a foundational aspect in comprehending the performance of any wireless system, as the achievable capacity hinges on the outage behavior of users. Numerous researchers have delved into investigating the outage probability of the basic NOMA scheme. For instance, comparative studies have revealed that NOMA exhibits lower outage compared to OMA for users randomly deployed within a cell, accounting for path loss effects [38].

  Furthermore, investigations have shown that NOMA with beamforming (NOMA-BF) enhances the sum capacity in comparison to conventional multi-user beamforming systems [105]. However, the introduction of beamforming in NOMA alters the outage probability of users, necessitating further analysis of outage performance in NOMA-BF systems.

Similarly, network NOMA [68] primarily addresses inter-cell interference mitigation through a precoder design, yet explicit analysis is required to comprehend the outage behaviors of cell-edge users within this framework. Additionally, other NOMA variants, such as NOMA with VLC [106] and NOMA with coding [84, 85], merit further investigation to analyze their respective outage performance characteristics.

In summary, continued exploration and analysis of outage probability across various NOMA configurations and enhancements are vital for advancing our understanding of NOMA's performance in diverse wireless communication scenarios.

- **Practical channel model:** As the demand for consumer data continues to surge, NG wireless networks must not only employ efficient radio access techniques but also ensure spectrum availability. Presently, it is evident that 5G/6G networks will leverage spectrum allocations in the untapped mmWave frequency bands. Additionally, the backbone infrastructure is poised to transition from traditional copper and fiber connections to mmWave wireless links, facilitating rapid deployment and establishing mesh-like connectivity.

  The mmWave frequencies, spanning between 30 and 300 GHz, represent a new frontier for cellular networks, offering vast bandwidth resources. Understanding the challenges inherent in mmWave cellular communications, particularly channel behavior, is paramount. It serves as a foundational prerequisite for the development of 5G mobile systems and backhaul techniques [109].

  Existing studies on NOMA typically assume wireless links between transmitters and receivers to exhibit Rayleigh fading channels with Additive White Gaussian Noise (AWGN). However, a more realistic analysis can be attained by considering the measured path loss and delay spread values [110], which accurately reflect the characteristics of radio channels in the mmWave band.

  In conclusion, adopting practical channel models, specifically tailored to mmWave cellular communications, is essential for accurately assessing the performance and feasibility of NOMA and other advanced wireless technologies.
- **Uniform fairness:** NOMA, incorporating fairness constraints, demonstrates superior performance compared to OMA methods, particularly in enhancing the performance of users in challenging channel conditions [111]. Additionally, controlling the rate allocated to concurrent

transmissions determines the fairness of a NOMA system, and various scheduling approaches can be employed to achieve this [31].

In mmWave cellular networks, it is observed that beyond a distance of 175 m, a majority of locations face signal outages [109]. Given the environment-dependent nature of outages, the actual impact may be more pronounced in areas with numerous local obstacles. A noteworthy endeavor would be the development of a NOMA scheme that ensures uniform outage experiences for users, especially those positioned beyond 150 m up to the cell boundary in mmWave cellular scenarios. This pursuit aligns with the aim of achieving uniform fairness in the face of varying environmental challenges.

- **Carrier aggregation:** To enhance bandwidth and consequently increase data rates for users, LTE-Advanced employs carrier aggregation (CA) as a fundamental concept [112]. In CA, users are allocated aggregated resources comprising two or more component carriers (CCs), with each CC representing an individual carrier. The arrangement of aggregation can be either contiguous, where CCs are adjacent to each other, or non-contiguous, with a gap between them.

  Integrating CA with NOMA presents an opportunity to harness the advantages of both technologies. However, the user pairing in CA-enabled NOMA differs from that of basic NOMA. In a traditional two-user NOMA scheme, a NOMA user is typically paired with another NOMA user, and each carrier is usually assigned to two users based on channel conditions.

  Exploring the appropriateness of different CA types for NOMA solutions remains an open research question. Therefore, investigating various CA types in the context of NOMA represents an intriguing research direction.
- **Other challenges:** Several challenges must be addressed before NOMA can seamlessly integrate into the future networks. In a downlink scenario, for instance, the transmitter allocates power to users based on their CSI. Establishing a robust mechanism for CSI feedback, accompanied by an effective channel estimation scheme and a proper reference signal design, is crucial for achieving resilient performance.

  The proven benefits of NOMA rely on the ideal condition of perfect acquisition of CSI data at the transmitter side. A potential solution involves the use of a limited feedback channel to acquire CSI, albeit at the expense of additional bandwidth required to convey various channel quality indicators [113]. Thus, proper user selection and PA schemes become imperative when the transmitter operates under imperfect CSI and/or with limited feedback.

In the realm of multicarrier communications, the peak-to-average-power ratio (PAPR) can lead to the power amplifier (PA) operating within a non-linear region, causing significant signal distortion in the PA output. Understanding and mitigating the effects of PAPR are critical for determining the techniques that yield optimal NOMA performance.

The concept of relaying, particularly in non-orthogonal coordinated transmissions between multiple small cells and a macro cell, holds promise for extending cell coverage and achieving enhanced capacity gains [114]. A hybrid scheme, as proposed by Ref. [115], combining topological interference management principles with NOMA, shows potential for improved sum rate performance. Further exploration of the MIMO form of this hybrid method, especially in dense network scenarios, presents a challenge in determining a fair power allocation method.

For NOMA to find its place in NG networks, it must demonstrate robustness in terms of system scalability, catering to the heterogeneous traffic and diverse radio environments inherent in NG networks. While existing work has largely focused on theoretical performance verification, practical demonstrations are crucial. Although Xiong *et al.* [116] introduced an SDR NOMA prototype, conducting more over-the-air experiments is essential to showcase NOMA's potential in real-world mobile network scenarios.

6.7 Implementation Issues

NOMA implementation includes designing and operating the system. Much of what will be attempted at this stage will rely upon a particular application and environment. This section provides some generic points, including some on computational complexity and error propagation that can be critical to assessing NOMA performance:

- **Decoding complexity:** Signal decoding using SIC requires additional implementation complexity compared to orthogonal schemes, as the receiver has to decode other users' information prior to decoding its own information [40]. Also, this complexity increases as the number of users in the cell of interest increases. However, users can be clustered into a number of groups, where each cluster contains a small number of users with bad channels. Single carrier (SC)/SIC can then be executed within each group. This group-wise SC and SIC operation basically provides a trade-off between performance gain and implementation complexity.

- **Error propagation:** It is intuitive that once an error occurs in SIC, all other user information will likely be decoded erroneously. However, the effect of error propagation can be compensated by using a stronger code (e.g., increasing the block length) when the number of users is reasonably small [3]. In the case of degradation in the performance for some users, non-linear detection techniques can also be considered to suppress error propagation. Based on computer simulations, it is shown [2] that error propagation can have a marginal impact on the NOMA performance. The reason is that a user with a bad channel gain is assigned to another user with a good channel gain during NOMA scheduling. This finding has been reported under the conditions of a worst-case model, which assumes that the decoding of the strong user at the second stage is always unsuccessful whenever the decoding of the weak user is unsuccessful at the first stage of the strong user receiver. Although there exist works that analytically study the SIC error propagation in basic MIMO systems [117, 118], there is no prominent research that provides a mathematical understanding of the effect of imperfect SIC on NOMA schemes. Therefore, a mathematical analysis of the impact of imperfect SIC on NOMA performances represents an interesting research direction.
- **Quantization error:** When the received signal strengths of users are disparate, the analog-to-digital (A/D) converter needs to support a very large full-scale input voltage range and requires high resolution to accurately quantize a weak signal. The more levels the ADC uses for quantization, the lower its quantization noise power. However, there is a limitation placed on arbitrarily high-resolution ADC due to its cost, conversion time, and hardware complexity. This constraint eventually leads to a trade-off between the quantization error and SIC gain.
- **PA complexity:** The achievable throughput of a user is affected by the transmit PA to that user. This PA also affects the achievable capacity of other users, as the basis of NOMA is power-domain user multiplexing. To achieve the best throughput performance in NOMA, a brute-force search over the possible user pairs with dynamic PA is required. However, this kind of exhaustive search is computationally expensive.

- **Residual timing offset:** Synchronous transmission has been considered in NOMA research, and this consideration is reasonable for the downlink scenario because the BS controls transmission for all users. However, perfect synchronization among NOMA users is impractical on the uplink since users are spatially distributed, and the mobile communication channel is usually dynamic in nature. In asynchronous communications, OFDM symbols from the superposition-coded users are time-misaligned. Thus, NOMA users' performance substantially depends on the relative time offset between interfering users [119]. It is important to further investigate NOMA in asynchronous communications. In practice, the asynchronous scheme requires information on multiple symbols for detection and interference cancellation. Otherwise, performance degrades when complete information on interfering signals is unknown.
- **Signaling and processing overhead:** There are several sources of additional signaling and processing overhead in NOMA compared to its orthogonal counterparts. For example, to collect the CSI from different receivers and to inform the receivers of the SIC order, some time slots need to elapse. This causes data rate degradation in NOMA. Also, with dynamic PA, and encoding and decoding for single carrier (SC) and SIC, NOMA signal processing requires additional energy overhead.
- **Limited number of user pairs:** NOMA exploits power-domain multiplexing to offer its benefits. More specifically, about an 8 dB difference in propagation loss is required to pair a cell-edge user with a cell-center user [121]. Therefore, the possible number of pairs of users in a typical NOMA scheme is limited, which eventually reduces the capacity gain of NOMA. Finding the appropriate signal detection and decoding strategy that increases the number of pairs of users is an important issue.
- **NOMA deploying environment:** Since the power domain is exploited, and channel gains are naturally adequately large in macrocells, researchers commonly consider macrocell deployment. In addition to macrocells, small cells are also becoming important and are being studied for future networks. Fortunately, the NOMA gain is still achievable, even in a small cell where NOMA offers a higher performance gain compared to OMA [122]. It is also important to note that NOMA gain in terms of cell throughput in small cells is even larger than in macrocells.

Therefore, NOMA can be equally deployed in both macrocell and small-cell environments.

- **Standardization status:** There are ongoing standardization efforts focused on NOMA technologies. The following is a summary of the current status of the ongoing standardization of NOMA. The 3GPP LTE Release 13 approved the study item (SI) of downlink MUST [123]. The primary objective of MUST was to identify promising enhancements of downlink multiuser transmission schemes within one cell. To achieve this objective, the SI focused on the evaluation of system-level gain and complexity–performance trade-off under practical deployment scenarios and traffic models [124–130]. The core objective of this work item is to identify necessary techniques to enable LTE to support downlink intra-cell MUST for the physical downlink shared channel.

6.8 Summary

This chapter provides a comprehensive overview of NOMA, focusing on its applications and benefits for NG wireless networks. It covers the fundamentals of NOMA, emphasizing power-domain and code-domain variations in both downlink and uplink scenarios. Power-domain NOMA is highlighted for its suitability in uplink environments, particularly for serving numerous users efficiently by leveraging BS-side decoding.

The key advantages of NOMA for 5G and beyond are discussed, including high SE, enhanced user fairness through optimal PA, and support for massive connectivity with low latency. The integration of NOMA with MIMO technologies, such as cluster-based and beamforming-based MIMO-NOMA in downlink settings, is explored. While beamforming-based MIMO-NOMA offers superior performance, it introduces additional implementation complexity compared to cluster-based approaches.

Resource allocation strategies, a critical aspect of NOMA, are detailed, covering user clustering, PA, and channel gain-based methods. Various C-NOMA forms are introduced, showcasing their superiority over conventional OMA approaches.

The chapter also examines NOMA's compatibility with emerging communication paradigms, such as heterogeneous networks, and millimeter-wave communications. It identifies open challenges and opportunities to fully realize NOMA's potential across diverse applications within the evolving wireless landscape. This chapter serves as a valuable resource for researchers interested in NOMA and its implications for future wireless communication systems.

References

[1] P. Wang, J. Xiao, and L. Ping, Comparison of orthogonal and non-orthogonal approaches to future wireless cellular systems, *IEEE Veh. Technol. Mag.*, 1(3), 4–11, Sep 2006.

[2] A. Benjebbour *et al.*, Concept and practical considerations of non-orthogonal multiple access (NOMA) for future radio access, in *Proc. Int. Symp. Intell. Signal Process. Commun. Syst. (ISPACS)*, Naha, Japan, 770–774, Nov 2013.

[3] D. Tse and P. Viswanath, *Fundamentals of Wireless Communication*. Cambridge, UK: Cambridge Univ. Press, 2005.

[4] R. Hoshyar, F. P. Wathan, and R. Tafazolli, Novel low-density signature for synchronous CDMA systems over AWGN channel, *IEEE Trans. Signal Process.*, 56(4), 1616–1626, Apr 2008.

[5] R. Razavi, R. Hoshyar, M. A. Imran, and Y. Wang, Information theoretic analysis of LDS scheme, *IEEE Commun. Lett.*, 15(8), 798–800, Aug 2011.

[6] M. Al-Imari, P. Xiao, M. A. Imran, and R. Tafazolli, Uplink non-orthogonal multiple access for 5G wireless networks, in *Proc. ISWCS*, 781–785, Aug 2014.

[7] M. Al-Imari, M. A. Imran, and R. Tafazolli, Low density spreading for next generation multicarrier cellular systems, in *Proc. IEEE ICFCN*, Baghdad, Iraq, 52–57, Apr 2012.

[8] H. Nikopour and H. Baligh, Sparse code multiple access, in *Proc. IEEE PIMRC*, London, U.K., 332–336, Sep 2013.

[9] H. Nikopour *et al.*, SCMA for downlink multiple access of 5G wireless networks, in *Proc. IEEE GLOBECOM*, Austin, TX, 3940–3945, Dec 2014.

[10] S. Kang, X. Dai, and B. Ren, Pattern division multiple access for 5G, *Telecommun. Netw. Technol.*, 5(5), 43–47, May 2015.

[11] L. Hanzo, M. Münster, B. J. Choi, and T. Keller, *OFDM and MC-CDMA for Broadband Multi-User Communications, WLANs and Broadcasting*. Piscataway, NJ, USA: IEEE Press, 2003.

[12] T. Abe and T. Matsumoto, Space-time turbo equalization in frequency-selective MIMO channels, *IEEE Trans. Veh. Technol.*, 52(3), 469–475, May 2003.

[13] M. Y. Alias, S. Chen, and L. Hanzo, Multiple-antenna-aided OFDM employing genetic-algorithm-assisted minimum bit error rate multiuser detection, *IEEE Trans. Veh. Technol.*, 54(5), 1713–1721, Sep 2005.

[14] S. M. Riazul Islam, Nurilla Avazov, Octavia A. Dobre and Kyung Sup Kwak, Power-Domain Non-Orthogonal Multiple Access (NOMA) in 5G Systems: Potentials and Challenges, *IEEE Comm. Surveys and Tutorials*, 19(2), 721–742, Oct. 2016.

[15] L. Dai *et al.*, Non-orthogonal multiple access for 5G: Solutions, challenges, opportunities, and future research trends, *IEEE Commun. Mag.*, 53(9), 74–81, Sep 2015.

[16] K. Higuchi and A. Benjebbour, Non-Orthogonal Multiple Access (NOMA) with Successive Interference Cancellation, *IEICE Transa. Commun.*, E98-B(3), 403–414, Mar 2015.

[17] L. Dai, B. Wang, Z. Ding, Z. Wang, S. Chen, and L. Hanzo, A Survey of Non-Orthogonal Multiple Access for 5G, *IEEE Communications Surveys & Tutorials*, 20(3), 2294–2323, 2018.

[18] Study on Downlink Multiuser Superposition Transmission for LTE, Document, 3rd Generation Partnership Project (3GPP), Jun. 2015.

[19] H. Tabassum, M. S. Ali, E. Hossain, M. J. Hossain, and D. I. Kim, "Uplink Vs. Downlink NOMA in cellular networks: Challenges and research directions, in *Proc. IEEE VTC Spring*, Sydney, Australia, 1–7, 2017.

[20] SMR Islam, M. Zeng, and Octavia A. Dobre, NOMA in 5G systems: exciting possibilities for enhancing spectral efficiency, *IEEE 5G Tech Focus*, 1(2), Jun 2017.

[21] R. Hoshyar, F. P. Wathan, and R. Tafazolli, Novel low-density signature for synchronous CDMA systems over AWGN channel, *IEEE Trans. Sig. Process.*, 56(4), 616–1626, Apr 2008.

[22] R. Hoshyar, R. Razavi, and M. Al-Imari, LDS-OFDM an efficient multiple access technique, in *Proc. IEEE VTC Spring*, Taipei, Taiwan, 1–5, 2010.

[23] H. Nikopour and H. Baligh, Sparse code multiple access, in *Proc. IEEE International Symposium on Personal Indoor Mobile Radio Communications (IEEE PIMRC)*, London, U.K., 332–336, 2013.

[24] M. Mohammadkarimi, M. A. Raza, and O. A. Dobre, Signature-based non-orthogonal massive multiple access for future wireless networks: Uplink massive connectivity for machine-type communications, *IEEE Veh. Technol. Mag.*, 13(4), 40–50, Dec 2018.

[25] T. M. Cover, Broadcast channels, *IEEE Trans. Inf. Theory*, 18(1), 2–14, Jan 1972.

[26] S. Vanka *et al.*, Superposition coding strategies: Design and experimental evaluation, *IEEE Trans. Wireless Commun.*, 11(7), 2628–2639, Jul 2012.

[27] L. Salaun, C. S. Chen, and M. Coupechoux, Optimal joint subcarrier and power allocation in NOMA is strongly NP-hard, in *Proc. IEEE ICC*, Kansas City, MO, 1–7, May 2018.

[28] L. Lei, D. Yuan, C. K. Ho, and S. Sun, power and channel allocation for non-orthogonal multiple access in 5G systems: Tractability and computation, *IEEE Trans. Wireless Commun.*, 15(12), 8580–8594, Dec 2016.

[29] B. Di, L. Song, and Y. Li, Sub-channel assignment, power allocation, and user scheduling for non-orthogonal multiple access networks, *IEEE Trans. on Wireless Commun.*, 15(11), 7686–7698, Nov 2016.

[30] Z. Ding, P. Fan, and H. V. Poor, Impact of user pairing on 5G Non-orthogonal multiple access downlink transmissions, *IEEE Trans. Veh. Tech.*, 65(8), 6010–6023, Aug 2016.

[31] S. M. R. Islam, M. Zeng, O. A. Dobre, and K. Kwak, Resource allocation for downlink NOMA systems: Key techniques and open issues, *IEEE Wireless Commun. Mag.*, 25(2), 40–47, Apr 2018.

[32] L. Zhu, J. Zhang, Z. Xiao, X. Cao, and D. O. Wu, Optimal user pairing for downlink non- orthogonal multiple access (NOMA), *IEEE Wireless Commun. Lett.*, 8(2), 328–331, Apr 2019.

[33] Y. Sun, D. W. K. Ng, Z. Ding and R. Schober, Optimal joint power and subcarrier allocation for MC-NOMA systems, in *Proc. IEEE GLOBECOM*, Washington, DC, 1–6, Dec 2016.

[34] K. Wang, Z. Ding and W. Liang, A game theory approach for user grouping in hybrid non- orthogonal multiple access systems, in *Proc. IEEE International Symposium on Wireless Communication Systems (ISWCS)*, Poznan, Poland, 643–647, Sep 2016.

[35] K. Wang, J. Cui, Z. Ding and P. Fan, A stackelberg game approach for NOMA in mmwave systems, in *Proc. IEEE GLOBECOM*, Abu Dhabi, United Arab Emirates, 1–6, Dec 2018.

[36] J. Ding, J. Cai and C. Yi, An improved coalition game approach for MIMO-NOMA clustering integrating beamforming and power allocation, *IEEE Trans. Veh. Tech.*, 68(2), 1672–1687, Feb 2019.

[37] Y. Gu, W. Saad, M. Bennis, M. Debbah and Z. Han, Matching theory for future wireless networks: Fundamentals and applications, *IEEE Commun. Mag.*, 53(5), 52–59, May 2015.

[38] M. Zeng, A. Yadav, O. A. Dobre and H. V. Poor, Energy-efficient joint user-RB association and power allocation for uplink hybrid NOMA-OMA, *IEEE Internet Things J.*, 1–1, 2019, doi: 10.1109/JIOT.2019.2896946.

[39] W. Liang, Z. Ding, Y. Li, and L. Song, User pairing for downlink non-orthogonal multiple access networks using matching algorithm, *IEEE Trans. Commun.*, 65(12), 5319–5332, Dec 2017.

[40] A. Benjebbovu, A. Li, Y. Saito, Y. Kishiyama, A. Harada, and T. Nakamura, System-level performance of downlink NOMA for future LTE enhancements, in *Proc. IEEE GLOBECOM*, Atlanta, 66–70, 2013.

[41] M. B. Shahab, M. F. Kader, and S. Y. Shin, A virtual user pairing scheme to optimally utilize the spectrum of unpaired users in non-orthogonal multiple access, *IEEE Sig. Process. Lett.*, 23(12), 1766–1770, Dec 2016.

[42] M. Zeng, G. I. Tsiropoulos, O. A. Dobre, and M. H. Ahmed, Power allocation for cognitive radio networks employing non-orthogonal multiple access, in *Proc. IEEE GLOBECOM*, Washington DC, 1–5, Dec 2016.

[43] M. Zeng, A. Yadav, O. A. Dobre, and H. V. Poor, Energy-efficient power allocation for MIMO- NOMA with multiple users in a cluster, *IEEE Access*, 6, 5170–5181, Feb 2018.

[44] S. Timotheou and I. Krikidis, Fairness for non-orthogonal multiple access in 5G systems, *IEEE Sig. Process. Lett.*, 22(10), 1647–1651, Oct 2015.

[45] M. Zeng, G. I. Tsiropoulos, A. Yadav, O. A. Dobre, and M. H. Ahmed, A two-phase power allocation scheme for CRNs employing NOMA," in *Proc. IEEE Global Communications Conference (GLOBECOM)*, Singapore, 1–6, Dec 2017.

[46] W. Hao, M. Zeng, Z. Chu, and S. Yang, Energy-efficient power allocation in millimeter wave massive MIMO with non-orthogonal multiple access, *IEEE Wireless Commun. Lett.*, 6(6), 782–785, Dec 2017.

[47] M. Zeng, A. Yadav, O. A. Dobre and H. V. Poor, Energy-efficient power allocation for uplink NOMA, in *Proc. IEEE GLOBECOM*, Abu Dhabi, UAE, 1–6, Dec 2018.

[48] M. Zeng, N. P. Nguyen, O. A. Dobre, and H. V. Poor, Securing downlink massive MIMO-NOMA networks with artificial noise, *IEEE JSTSP*, 1–1, Feb 2019, doi: 10.1109/JSTSP.2019.2901170.

[49] S. Shi, L. Yang, and H. Zhu, Outage balancing in downlink non-orthogonal multiple access with statistical channel state information, *IEEE Trans. Wireless Commun.*, 15(7), 4718–4731, Jul 2016.

[50] Z. Ding, M. Peng, and H. V. Poor, Cooperative non-orthogonal multiple access in 5G systems, *IEEE Commun. Lett.*, 19(8), 1462–1465, Aug 2015.

[51] Z. Ding, F. Adachi, and H. V. Poor, The application of MIMO to non-orthogonal multiple access, *IEEE Trans. Wireless Commun.*, 15(1), 537–552, Jan 2016.

[52] M. Zeng, A. Yadav, O. A. Dobre, and H. V. Poor, A fair individual rate comparison between MIMO-NOMA and MIMO-OMA, in *Proc IEEE GLOBECOM Wkshps*, Singapore, 1–5, Dec 2017.

[53] M. Zeng, A. Yadav, O. A. Dobre, and H. V. Poor, Energy-efficient power allocation for hybrid multiple access systems, in *Proc. IEEE International Conference on Communications Workshops (ICC Workshops)*, Kansas City, MO, 1–5, May 2018.

[54] M. Zeng, N. P. Nguyen, O. A. Dobre, Z. Ding, and H. V. Poor, Spectral and energy efficient resource allocation for multi-carrier uplink NOMA systems, *IEEE Trans. Veh. Tech.*, submitted.

[55] Z. Wei, D. W. K. Ng, and J. Yuan, Power-efficient resource allocation for MC-NOMA with statistical channel state information, in *Proc. IEEE Global Communications Conference (GLOBECOM)*, Washington, DC, 1–7, Dec 2016.

[56] G. I. Tsiropoulos, A. Yadav, M. Zeng, and O. A. Dobre, Cooperation in 5G Het-Nets: Advanced spectrum access and D2D assisted communications, *IEEE Wireless Commun. Mag.*, 24(5), 110–117, Oct 2017.

[57] Y. Sun, D.W.K. Ng, Z. Ding, and R. Schober, Optimal joint power and subcarrier allocation for full-duplex multicarrier non-orthogonal multiple access systems, *IEEE Trans. Commun.*, 65(3), 1077–1091, Mar 2017.

[58] S. M. Riazul Islam, Nurilla Avazov, Octavia A. Dobre, and Kyung-Sup Kwak, Power-domain non-orthogonal multiple access (NOMA) in 5G systems: Potentials and challenges, *IEEE Commun. Surv. Tutor.*, 19(2), 721–742, 2017.

[59] A. Li, A. Harada, and H. Kayama, A novel low computational complexity power allocation method for non-orthogonal multiple access systems, *IEICE Trans. Fundamentals Electron. Commun. Comput. Sci.*, E97-A(1), 57–67, Jan 2014.

[60] M. F. Kader, M. B. Shahab, and S. Y. Shin, Exploiting non-orthogonal multiple access in cooperative relay sharing, *IEEE Commun. Lett.*, 21(5), 1159–1162, May 2017.

[61] M. F. Kader, M. B. Uddin, SMR Islam, and S. Y. Shin, Capacity and outage analysis of a dual-hop decode-and-forward relay-aided NOMA scheme, *DSP*, 88, 138–148, May 2019.

[62] J. Kim and I. Lee, Non-orthogonal multiple access in coordinated direct and relay transmission, *IEEE Commun. Lett.*, 19(11), 2037–2040, Nov 2015.

[63] J. Men and J. Ge, Non-orthogonal multiple access for multiple-antenna relaying networks, *IEEE Commun. Lett.*, 19(10), 1686–1689, Oct 2015.

[64] Y. Jing and H. Jafarkhani, Single and multiple relay selection schemes and their achievable diversity orders, *IEEE Trans. Wireless Commun.*, 8(3), 1414–1423, Mar 2009.

[65] Z. Ding, H. Dai, and H. V. Poor, Relay selection for cooperative NOMA, in *IEEE Wireless Commun. Lett.*, 5(4), 416–419, Aug 2016.
[66] J. Choi, Non-orthogonal multiple access in downlink coordinated two-point systems, *IEEE Commun. Lett.*, 18(2), 313–316, Feb 2014.
[67] S. M. Alamouti, "A simple transmit diversity technique for wireless communications, *IEEE J. Sel. Areas Commun.*, 16(8), 1451–1458, Oct 1998.
[68] S. Han, C.-L. I, Z. Xu, and Q. Sun, Energy efficiency and spectrum efficiency co-design: From NOMA to network NOMA, *IEEE Commun. Soc. MMTC E-Lett.*, 9(5), 21–24, Sep 2014.
[69] W. Shin *et al.*, Coordinated beamforming for multi-cell MIMO-NOMA, *IEEE Commun. Lett.*, doi: 10.1109/LCOMM.2016.2615097.
[70] W. Shin, M. Vaezi, J. Lee, and H. V. Poor, On the number of users served in MIMO-NOMA cellular networks, in *Proc. ISWCS*, Poznan, Poland, 638–642, Sep 2016.
[71] Z. Chen, Z. Ding, P. Xu, and X. Dai, Optimal precoding for a QoS optimization problem in two-user MISO-NOMA downlink, *IEEE Commun. Lett.*, 20(6), 1263–1266, Jun 2016.
[72] Z. Ding, R. Schober, and H. V. Poor, A general MIMO framework for NOMA downlink and uplink transmission based on signal alignment, *IEEE Trans. Wireless Commun.*, 15(6), 4438–4454, Jun 2016.
[73] M. Zeng, A. Yadav, O. A. Dobre, G. I. Tsiropoulos, and H. V. Poor, Capacity comparison between MIMO-NOMA and MIMO-OMA with multiple users in a cluster, *IEEE JSAC*, 35(10), 2413–2424, Oct 2017.
[74] M. Zeng, A. Yadav, O. A. Dobre, G. I. Tsiropoulos, and H. V. Poor, On the sum rate of MIMO-NOMA and MIMO-OMA systems, *IEEE Wireless Commun. Lett.*, 6(4), 534–537, Aug 2017.
[75] S. M. Riazul Islam, Ming Zeng, Octavia A. Dobre, Kyung-Sup Kwak, Non-Orthogonal Multiple Access (NOMA): How It Meets 5G and Beyond, *Wiley 5G Ref: The Essential 5G Reference Online*, John Wiley & Sons, Ltd. Oct 2019.
[76] B. Kimy *et al.*, Non-orthogonal multiple access in a downlink multiuser beamforming system, in *Proc. IEEE Mili. Commun. Conf. (MILCOM)*, San Diego, CA, 1278–1283, Nov 2013.
[77] K. Higuchi and Y. Kishiyama, Non-orthogonal access with random beamforming and intra-beam SIC for cellular MIMO downlink, in *Proc. IEEE Veh. Technol. Conf.*, Las Vegas, NV, 1–5, Sep 2013.
[78] Q. Sun, S. Han, I. Chin-Lin, and Z. Pan, On the ergodic capacity of MIMO NOMA systems, *IEEE Wireless Commun. Lett.*, 4(4), 405–408, Dec 2015.
[79] J. Choi, On the power allocation for MIMO-NOMA systems with layered transmissions, *IEEE Trans. Wireless Commun.*, 15(5), 3226–3237, May 2016.
[80] M. F. Hanif, Z. Ding, T. Ratnarajah, and G. K. Karagiannidis, "A minorization-maximization method for optimizing sum rate in the downlink of non-orthogonal multiple access systems, *IEEE Trans. Sig. Process.*, 64(1), 76–88, Jan 2016.
[81] J. Choi, H-ARQ based non-orthogonal multiple access with successive interference cancellation, in *Proc. IEEE Glob. Telecommun. Conf.*, New Orleans, LA, 1–5, Nov/Dec 2008.

[82] A. Li, A. Benjebbour, X. Chen, H. Jiang, and H. Kayama, Investigation on hybrid automatic repeat request (HARQ) design for NOMA with SU-MIMO," in *Proc. IEEE 26th Annu. Int. Symp. Pers.*, Indoor Mobile.

[83] A. Shokrollahi, Raptor codes, *IEEE Trans. Inf. Theory*, 52(6), 2551–2567, Jun 2006.

[84] H. Liu, N.I. Miridakis, T. A. Tsiftsis, K.J. Kim, and K.S. Kwak, Coordinated Uplink Transmission for Cooperative NOMA Systems, *2018 IEEE Global Communications Conference (GLOBECOM)*, Dec. 2018.

[85] S. Park and D.-H. Cho, Random linear network coding based on non-orthogonal multiple access in wireless networks, *IEEE Commun. Lett.*, 19(7), 1273–1276, Jul 2015.

[86] Y. Liu, Z. Ding, M. Eikashlan, and H. V. Poor, Cooperative non-orthogonal multiple access in 5G systems with SWIPT, in *Proc. 23rd Eur. Signal Process. Conf. (EUSIPCO)*, Nice, France, 1999–2003, Aug/Sep 2015.

[87] Z. Ding, P. Fan, and H. V. Poor, Impact of user pairing on 5G nonorthogonal multiple-access downlink transmissions, *IEEE Trans. Veh. Technol.*, 65(8), 6010–6023, Aug 2016.

[88] B. Clerckx, *et al.*, Is NOMA efficient in multi-antenna networks? A critical look at next generation multiple access techniques, *IEEE Open J. Commun. Soc.*, 2, 1310–1343, 2021.

[89] T. Seyama, T. Dateki, and H. Seki, Efficient selection of user sets for downlink non-orthogonal multiple access, in *Proc. IEEE PIMRC*, Hong Kong, 1062–1066, Sep 2015.

[90] S. Choudhury and J. D. Gibson, Information transmission over fading channels, in *Proc. IEEE GLOBECOM*, Washington, DC, 3316–3321, Nov 2007.

[91] S. M. R. Islam and K. S. Kwak, Outage capacity and source distortion analysis for NOMA users in 5G systems, *Electron. Lett.*, 52(15), 1344–1345, Jul 2016.

[92] Z. Ding, M. Peng, and H. V. Poor, Cooperative non-orthogonal multiple access in 5G systems, *IEEE Commun. Lett.*, 19(8), 1462–1465, Aug 2015.

[93] B. G. Lee, D. Park, and H. Seo, *Wireless Communications Resource Management.* Hoboken, NJ, USA: Wiley, 2009.

[94] L. Lei, D. Yuan, C. K. Ho, and S. Sun, Joint optimization of power and channel allocation with non-orthogonal multiple access for 5G cellular systems, in *Proc. IEEE Glob. Telecommun. Conf. (GLOBECOM)*, San Diego, CA, 1–6, Dec 2015.

[95] M. Zeng, G. I. Tsiropoulos, O. A. Dobre, and M. H. Ahmed, Power allocation for cognitive radio networks employing non-orthogonal multiple access, in *Proc. IEEE Glob. Telecommun. Conf. (GLOBECOM)*, Washington, DC, Dec 2016.

[96] X. Chen, A. Beiijebbour, A. Li, H. Jiang, and H. Kayama, Consideration on successive interference canceller (SIC) receiver at cell-edge users for non-orthogonal multiple access (NOMA) with SU-MIMO, in *Proc. IEEE 26th Annu. Int. Symp. Pers. Indoor. Mobile Radio Commun. (PIMRC)*, Hong Kong, 522–526, Aug/Sep 2015.

[97] M. Zeng, W. Hao, O. A. Dobre, and H. V. Poor, "Energy-efficient power allocation in uplink mmWave massive MIMO with NOMA, *IEEE Trans. Veh. Tech.*, 68(3), 3000–3004, Mar 2019.

[98] K. Saito, A. Benjebbour, Y. Kishiyama, Y. Okumura, and T. Nakamura, Performance and design of SIC receiver for downlink NOMA with open-loop SU-MIMO, in *Proc. IEEE ICCW*, London, U.K., 1161–1165, Jun 2015.

[99] Y. Endo, Y. Kishiyama, and K. Higuchi, Uplink non-orthogonal access with MMSE-SIC in the presence of inter-cell interference, in *Proc. ISWCS*, Paris, France, 261–265, Aug 2012.

[100] Z. Ding, F. Adachi, and H. V. Poor, The application of MIMO to non-orthogonal multiple access, *IEEE Trans. Wireless Commun.*, 15(1), 537–552, Jan 2016.

[101] X. Chen, A. Benjebbour, Y. Lan, A. Li, and H. Jiang, Impact of rank optimization on downlink non-orthogonal multiple access (NOMA) with SU-MIMO, in *Proc. IEEE Int. Conf. Commun. Syst. (ICCS)*, 233–237, Nov 2014.

[102] J. Kim, J. Koh, J. Kang, K. Lee, and J. Kang, Design of user clustering and pre-coding for downlink non-orthogonal multiple access (NOMA), in *Proc. IEEE Milit. Commun. Conf. (MILCOM)*, Tampa, FL, 1170–1175, Oct 2015.

[103] R. Q. Hu and Y. Qian, *Heterogeneous Cellular Networks*. Chichester, UK: Wiley, 2013.

[104] R. Q. Hu and Y. Qian, An energy efficient and spectrum efficient wireless heterogeneous network framework for 5G systems, *IEEE Commun. Mag.*, 52(5), 94–101, May 2014.

[105] B. Kim *et al.*, Non-orthogonal multiple access in a downlink multiuser beamforming system, in *Proc. IEEE Milit. Commun. Conf.*, San Diego, CA, 1278–1283, Nov 2013.

[106] R. C. Kizilirmak, C. R. Rowell, and M. Uysal, Non-orthogonal multiple access (NOMA) for indoor visible light communications, in *Proc. 4th Int. Workshop Opt. Wireless Commun. (IWOW)*, Istanbul, Turkey, 98–101, Sep 2015.

[107] M. J. Hagh and M. R. Soleymani, Raptor coding for non-orthogonal multiple access channels, in *Proc. IEEE Int. Conf. Commun. (ICC)*, 1–6, Jun 2011.

[108] Z. Ding, Z. Yang, P. Fan, and H. V. Poor, On the performance of non-orthogonal multiple access in 5G systems with randomly deployed users, *IEEE Sig. Process. Lett.*, 21(12), 1501–1505, Dec 2014.

[109] S. Rangan, T.S. Rappaport, and E. Erkip, Millimeter-wave cellular wireless networks: Potentials and challenges, *Proc. IEEE*, 102(3), 366–385, Mar 2014.

[110] T. S. Rappaport *et al.*, Millimeter wave mobile communications for 5G cellular: It will work! *IEEE Access*, 1, 335–349, 2013.

[111] S. Timotheou and I. Krikidis, Fairness for non-orthogonal multiple access in 5G systems, *IEEE Sig. Process. Lett.*, 22(10), 1647–1651, Oct 2015.

[112] G. Yuan, X. Zhang, W. Wang, and Y. Yang, Carrier aggregation for LTE-advanced mobile communication systems, *IEEE Commun. Mag.*, 48(2), 88–93, Feb 2010.

[113] S. Liu and C. Zhang, Downlink non-orthogonal multiple access system with limited feedback channel, *in Proc. IEEE 7th Int. Conf. Wireless Commun. Signal Process. (WCSP)*, Nanjing, China, 1–5, Oct 2015.

[114] J.-B. Kim and I.-H. Lee, Non-orthogonal multiple access in coordinated direct and relay transmission, *IEEE Commun. Lett.*, 19(11), 2037–2040, Nov 2015.

[115] V. Kalokidou, O. Johnson, and R. Piechocki, A hybrid TIM-NOMA scheme for the SISO broadcast channel, in *Proc. IEEE ICCW*, London, U.K., 387–392, Jun 2015.

[116] X. Xiong, W. Xiang, K. Zheng, H. Shen, and X. Wei, An open source SDR-based NOMA system for 5G networks, *IEEE Wireless Commun.*, 22(6), 24–32, Dec 2015.

[117] P. Li, R. C. de Lamare, and R. Fa, Multiple feedback successive interference cancellation detection for multiuser MIMO systems, *IEEE Trans. Wireless Commun.*, 10(8), 2434–2439, Aug 2011.

[118] J. Kim, D. Kim, and S. K. Yun, Mitigating error propagation in successive interference cancellation, *IEICE Trans. Commun.*, E89-B(10), 2956–2960, Oct 2006.

[119] H. Haci and H. Zhu, Performance of non-orthogonal multiple access with a novel interference cancellation method, in *Proc. IEEE ICC*, London, U.K., 2912–2917, Jun 2015.

[120] Tao Han, Jie Gong, Xiong Liu, S. M. Riazul Islam, Qiang Li, Zhiquan Bai and Kyung Sup Kwak, On Downlink NOMA in Heterogeneous Networks with Non-Uniform Small Cell Deployment, *IEEE Access*, Vol. 6, (s): 31099–31109, June 2018.

[121] T. Yazaki and Y. Sanada, Effect of joint detection and decoding in non-orthogonal multiple access, in *Proc. SPACS*, Kuching, Malaysia, 245–250, Dec 2014.

[122] Y. Saito, A. Benjebbour, Y. Kishiyama, and T. Nakamura, System-level performance of downlink non-orthogonal multiple access (NOMA) under various environments, in *Proc. IEEE VTC*, Glasgow, U.K., 1–5, May 2015.

[123] MediaTek Inc., Study on downlink multiuser superposition transmission for LTE, 3GPP TSG RAN Meeting #68, Malmö, Sweden, *Tech. Rep.* RP-151100, Jun 2015.

[124] A. Benjebbour *et al.*, NOMA: From concept to standardization, in *Proc. IEEE Conf. Standards Commun. Netw. (CSCN)*, Tokyo, Japan, 18–23, May 2015.

[125] NTT DOCOMO, Deployment scenarios for downlink multiuser superposition transmissions, 3GPP TSG RAN WG1 Meeting #80bis, Belgrade, Serbia, *Tech. Rep.* R1-152062, Apr. 2015.

[126] NTT DOCOMO, Evaluation methodologies for downlink multiuser superposition transmissions, 3GPP TSG RAN WG1 Meeting #81, Fukuoka, Japan, *Tech. Rep.* R1-153332, May 2015.

[127] MediaTek Inc., "Candidate non-orthogonal multiple access," 3GPP TSG RAN WG1 Meeting #81, Fukuoka, Japan, *Tech. Rep.* R1-153335, May 2015.

[128] NTT DOCOMO, System-level evaluation results for downlink multiuser superposition transmissions, 3GPP TSG RAN WG1 Meeting #82, Beijing, China, *Tech. Rep.* R1-154536, Aug 2015.

[129] NTT DOCOMO, Link-level evaluation results for downlink multiuser superposition transmissions, 3GPP TSG RAN WG1 Meeting #82, Beijing, China, *Tech. Rep.* R1-154537, Aug 2015.

[130] MediaTek Inc. and CMCC, Downlink multiuser superposition transmissions for LTE, 3GPP TSG RAN Meeting #71, Gothenburg, Sweden, Tech. Rep. RP-160680, Mar. 2016.

Chapter 7

Enabling Applications with NOMA Systems

7.1 Coordinated Transmission for Cooperative NOMA System

This chapter explores a coordinated direct and relay transmission scheme for the uplink of a cooperative non-orthogonal multiple access (NOMA) system. It involves two users communicating with a base station (BS) aided by a decode-and-forward (DF) relay. By integrating relay and direct links, the scheme forms an orthogonal structure at the BS receiver side, allowing linear combinations for signal recovery without successive interference cancellation. The outage probability under the Nakagami-m fading channels is derived, with asymptotic analysis revealing equal asymptotic outage performance for both users. The results demonstrate that this scheme outperforms a DF relay without direct links, offering improved outage performance and higher target rates, as validated by Monte Carlo simulations.

7.1.1 *Introduction*

NOMA emerges as a promising multiple access technique for fifth-generation (5G) networks, aiming to deliver spectral-efficient massive connectivity [1]. To alleviate the burden on time, frequency, and code resources, power-domain NOMA has gained significant interest by overlaying multiple users' signals in the power domain [1]. In the context of 3GPP long-term evolution (LTE), the concept of multiple user superposition transmission (MUST) has been proposed, representing a specific instance of power-domain NOMA [2]. Incorporating relaying into cooperative NOMA systems enhances both end-to-end system performance and user fairness through collaborative transmissions. In certain studies [1], a nearby user with superior channel conditions acts as a partially functioning relay to enhance reception reliability for a distant user. Additionally, dedicated amplify-and-forward (AF) relay and DF relay strategies have been introduced in Refs [3, 4], respectively, to assist transmissions from a BS to NOMA distant users. Furthermore, for cooperative NOMA setups featuring multiple AF relay nodes, various relay selection (RS) schemes, such as two-stage RS and partial RS, have been proposed to mitigate outage probability and achieve spatial diversity [5, 6]. Addressing physical-layer security concerns, investigations have been conducted in Ref. [7] for NOMA cognitive radio networks incorporating simultaneous wireless information and power transfer (SWIPT).

By incorporating non-negligible direct links (DLs) into coordinated direct and relay transmissions (CDRTs), studies have demonstrated the potential for enhancing outage performance and ergodic sum rate in cooperative NOMA systems [8–11]. Given the prevalence of direct links in various communication scenarios, including device-to-device (D2D) communications, Internet of Things (IoT) networks, and overlapping macro, micro, and small cells [6, 8], it becomes imperative to consider the utilization of direct links for CDRTs in cooperative NOMA systems. However, the acquisition of side information is essential for CDRTs to mitigate inter-user interference, leading to substantial overheads associated with managing the extensive connectivity in cooperative NOMA setups [8–11].

While considerable attention has been devoted to downlink designs in cooperative NOMA systems, there remains a critical need to enhance system performance for uplink scenarios [12]. In addressing this, a hybrid downlink–uplink transmission scheme was proposed in Ref. [13], where a nearby user served as a relay for a distant user. In another study [14], a cooperative direct and relay transmission (CDRT) scheme was introduced for uplink NOMA systems, leveraging the direct link from a nearby user to the BS in the presence of a dedicated relay.

Furthermore, performance improvements can be achieved by exploiting non-negligible direct links from multiple users to the BS [15]. For scenarios involving multiple antennas at relays and BSs, a two-stage interference cancellation and decoupling method was proposed for uplink cellular NOMA systems [15]. Additionally, investigations into compute-and-forward relay and generalized analog network coding were conducted in Refs. [16, 17], respectively, considering direct links from two users to the BS. Furthermore, noisy network coding was implemented at a relay for two-user relay-aided uplink NOMA systems [18], although these approaches necessitate intricate network cooperation and encoding/decoding procedures [15–18].

In this chapter, a straightforward yet efficient approach termed coordinated direct and relay transmission for uplink systems is provided within a cooperative NOMA framework, considering significant direct links between two users and the BS. The method employs a dedicated DF relay to assist transmissions from both users to the BS. The primary objectives are outlined in the following section.

A coordinated direct and relay transmission scheme is provided for the cooperative NOMA system, considering direct links from both users to the BS. This simplifies decoding with linear combining at the BS, eliminating the need for SIC and reducing the decoding complexity compared to prior approaches. Analytically, closed-form expressions for outage probabilities under independent but not identically distributed Nakagami-m fading channels are derived for the two users. Additionally, asymptotic expressions for outage probabilities in the high SNR regime show identical asymptotic performance for both users. Leveraging coordinated direct link transmissions, the coordinated direct and relay transmission scheme demonstrates improved outage performance in small and moderate SNR regimes [18].

7.1.2 *System model*

In the cooperative NOMA uplink system depicted in Fig. 7.1, two users, denoted as U_1 and U_2, aim to communicate with a BS. The direct links from both users to the BS are considered non-negligible. To assist with transmissions from the users to the BS, particularly in cases of long-distance

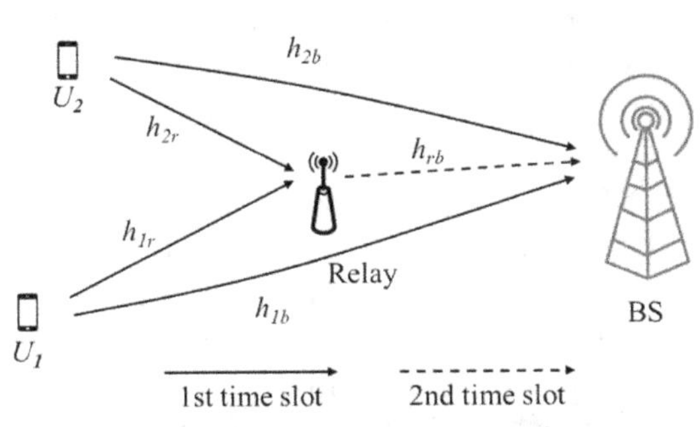

Fig. 7.1. Uplink model of the cooperative NOMA system.

or non-line-of-sight (NLoS) propagation, a DF relay is employed. It is assumed that all nodes operate with a single antenna and function in half-duplex mode.

The channels from U_i to the BS, from U_i to the relay, and from the relay to the BS are denoted by h_{ib}, h_{ir}, and h_{rb}, respectively, with $i = 1$ and 2. For each link, the channel coefficient is denoted by $h_j = \sqrt{L_j}\tilde{h}_j$, where $j \in \{ir, ib, rb\}$, L_j is the large-scale fading, and $\tilde{h}_j$ is the small-scale fading.

The channel, $|h_j|$ as i.i.d., is modeled by the Nakagami fading with the unit mean and fading shape factor m_j. In addition, the small-scale channel gain $\tilde{g}_j \equiv |\tilde{h}_j|^2$ is distributed according to the gamma distribution with the shape factor m_j and the scale factor $\widetilde{\theta}_j \equiv E[|\tilde{h}_j|^2]/m_j$. For the overall channel gain $g_j \equiv L_j\widetilde{g}_j$, its PDF and CDF can be expressed, respectively, as

$$f_{gj}(x) = \frac{{m_j}^{m_j}}{{\theta_j}^{m_j}\Gamma(m_j)}x^{m_j-1}e^{-\frac{m_jx}{\theta_j}} \tag{7.1}$$

and

$$F_{gj}(x) = 1 - \frac{\Gamma(m_j, m_jx/\theta_j)}{\Gamma(m_j)} \tag{7.2}$$

where $\theta_j \equiv L_j\widetilde{\theta}_j$. All the channels are assumed to be static block fading, i.e., a channel keeps constant during one transmission block and can vary from one transmission block to another. In addition, for implementing the coordinated direct and relay transmission scheme and correct decoding at the relay and BS, we assume that the relay knows perfect channel state information (CSI) of h_{ir} and h_{ib} and the BS knows the perfect CSI of h_{ib} and h_{rb} [12–15, 18]. The additive white Gaussian noises (AWGNs) at the relay and BS are assumed to have zero mean and variance σ^2, while the information signal transmitted by U_i is denoted by x_i satisfying $\mathrm{E}\{x_i\} = 0$ and $\mathrm{E}\{|x_i|^2\} = 1$.

In cooperative NOMA systems, users can be grouped based on their channel qualities or quality of service (QoS) requirements [4, 5]. When QoS requirements are used for grouping, two users can be paired even if their channel qualities are statistically similar. Additionally, applying QoS-based user pairing can significantly reduce the need for CSI acquisition at the transmitters [5]. In this study, QoS-based user pairing is employed, assuming that user U_1 requires a low target rate but necessitates timely service, while user U_2 can tolerate more delay but requires a higher throughput.

In the cooperative NOMA system under consideration, each uplink transmission block consists of two time slots. In the first time slot, users U_1 and U_2 transmit their signals $\sqrt{P_u \alpha_1} x_1$ and $\sqrt{P_u \alpha_2} x_2$ to the relay and BS, respectively, where P_u is the transmit peak power for each user and α_i is the power allocation for x_i, satisfying $\sum_{i=1}^{2} \alpha_i = 1$ and $\alpha_i > 0$. The received signal at the relay and BS during the first time slot can be expressed, respectively, as

$$y_r = \sqrt{\alpha_1 P_u} h_{1r} x_1 + \sqrt{\alpha_2 P_u} h_{2r} x_2 + n_r \tag{7.3}$$

and

$$y_b{}^{(1)} = \sqrt{\alpha_1 P_u} h_{1b} x_1 + \sqrt{\alpha_2 P_u} h_{2b} x_2 + n_b{}^{(1)} \tag{7.4}$$

where n_r and $n_b{}^{(1)}$ are the AWGNs at the relay and BS, respectively. Since the QoS requirements of U_1 are assumed to be much less demanding than those of U_2, the decoding order at the relay is always from U_1 to U_2 [5]. Based on (7.3), the received signal-to-interference-plus-noise ratio (SINR) and SNR at the relay for decoding x_1 and x_2 can be expressed, respectively, as

$$\gamma_r{}^{x_1} = \frac{\alpha_1 \rho g_{1r}}{\alpha_2 \rho g_{2r} + 1} \quad \text{and} \quad \gamma_r{}^{x_2} = \alpha_2 \rho g_{2r} \tag{7.5}$$

where $\rho \equiv P_u / \sigma^2$ is the transmit SNR.

7.1.2.1 *DF relay without direct links*

In the scenario where the direct links from the two users to the BS are unavailable, the relay performs the DF operation and forwards the superposition $x_r = \sqrt{\alpha_1 P_r} x_1 + \sqrt{\alpha_2 P_r} x_2$ to the BS, where P_r is the relay transmit power. Accordingly, in the second time slot, the received signal at the BS is

$$y_b{}^{(2)} = \sqrt{\alpha_1 P_r} h_{rb} x_1 + \sqrt{\alpha_2 P_r} h_{rb} x_2 + n_b{}^{(2)} \tag{7.6}$$

where $n_b{}^{(2)}$ is the AWGN at the BS in the second time slot. Then, the BS detects x_1 and x_2 using the conventional NOMA detecting procedure involving SIC. The received SINR and SNR at the BS for decoding x_1 and

x_2 are given, respectively, by

$$\gamma_b{}^{x_1} = \frac{\alpha_1 \rho g_{1b}}{\alpha_2 \rho g_{2b} + 1} \quad \text{and} \quad \gamma_b{}^{x_2} = \alpha_2 \rho g_{rb} \tag{7.7}$$

Considering the DF operation at the relay, the end-to-end SNR for the BS to detect x_i can be expressed as

$$\gamma_i = \min\{\gamma_r{}^{x_i}, \gamma_b{}^{x_i}\} \tag{7.8}$$

for $i = 1$ and 2.

7.1.2.2 *Coordinated direct and relay transmission scheme*

When the non-negligible direct links from two users to the BS are available, the coordinated direct and relay transmission (CDRT) scheme can enhance the system performance. In the first time slot, the BS does not decode its received signal and waits for the relay transmission. In the second time slot, the relay forwards the following new superposition:

$$x_r = \frac{1}{\sqrt{\beta}}\left(\sqrt{\alpha_2 P_r} h_{2b}^* x_1 - \sqrt{\alpha_1 P_r} h_{1b}^* x_2\right) \tag{7.9}$$

where $h_j{}^*$ is the conjugate of h_j. Then, the received signal at the BS in the second time slot can be expressed as

$$y_b{}^{(2)} = \frac{h_{rb}}{\sqrt{\beta}}\left(\sqrt{\alpha_2 P_r} h_{2b}^* x_1 - \sqrt{\alpha_1 P_r} h_{1b}^* x_2\right) + n_b{}^{(2)} \tag{7.10}$$

where

$$\beta \triangleq \alpha_1 g_{1b} + \alpha_2 g_{2b} \tag{7.11}$$

is the power normalization factor at the relay and $n_b{}^{(2)}$ is AWGN at the BS receiver in the second time slot. With the BS receiving signals from both the first and second time slots, the signals can be rewritten in matrix form as

$$\begin{bmatrix} \dfrac{y_b{}^{(1)}}{\sqrt{P_u}} \\ \dfrac{y_b{}^{(2)}\sqrt{\beta}}{h_{rb}\sqrt{P_r}} \end{bmatrix} = \underbrace{\begin{bmatrix} \sqrt{\alpha_1} h_{1b} & \sqrt{\alpha_2} h_{2b} \\ \sqrt{\alpha_2} h_{2b}^* & -\sqrt{\alpha_1} h_{1b}^* \end{bmatrix}}_{\mathcal{P}} \begin{bmatrix} x_1 \\ x_2 \end{bmatrix} + \begin{bmatrix} \dfrac{n_b{}^{(1)}}{\sqrt{P_u}} \\ \dfrac{n_b{}^{(2)}\sqrt{\beta}}{h_{rb}\sqrt{P_r}} \end{bmatrix} \tag{7.12}$$

where $\mathcal{P}$ is the equivalent superposing matrix in the power domain. It can be shown that

$$\mathcal{P}^H \mathcal{P} = \beta I_2 \tag{7.13}$$

where I_2 is the 2×2 identity matrix. With the aid of the orthogonality provided by (7.13), the BS can recover the information signal transmitted by each user using a simple linear combining. Specifically, the estimations for x_1 and x_2 can be obtained as

$$\hat{x}_1 = \frac{{y_b}^{(1)} h_{1b}^* \sqrt{\alpha_1}}{\sqrt{P_u}} + \frac{{y_b}^{(2)} h_{2b} \sqrt{\alpha_2 \beta}}{h_{rb}\sqrt{P_r}} = \beta x_1 + \tilde{n}_b^{(1)} \tag{7.14}$$

$$\hat{x}_2 = \frac{{y_b}^{(1)} h_{2b}^* \sqrt{\alpha_2}}{\sqrt{P_u}} - \frac{{y_b}^{(2)} h_{1b} \sqrt{\alpha_1 \beta}}{h_{rb}\sqrt{P_r}} = \beta x_2 + \tilde{n}_b^{(2)} \tag{7.15}$$

where

$$\tilde{n}_b^{(1)} \triangleq \frac{{n_b}^{(1)} h_{1b}^* \sqrt{\alpha_1}}{\sqrt{P_u}} + \frac{{n_b}^{(2)} h_{2b}^* \sqrt{\alpha_2 \beta}}{h_{rb}\sqrt{P_r}} \quad \text{and}$$

$$\tilde{n}_b^{(2)} \triangleq \frac{{n_b}^{(1)} h_{2b}^* \sqrt{\alpha_2}}{\sqrt{P_u}} - \frac{{n_b}^{(2)} h_{1b}^* \sqrt{\alpha_1 \beta}}{h_{rb}\sqrt{P_r}} \quad \text{are the equivalent noises.}$$

As seen in (7.14) and (7.15), the estimations for x_1 and x_2 are free of inter-user interference. Furthermore, the BS receiver does not perform successive interference cancellation as done by a conventional NOMA BS receiver. Hereinafter, we assume $P_u = P_r = P$ for mathematical tractability. Based on (7.14) and (7.15), the received SNRs for the BS to decode x_1 and x_2 can be, respectively, expressed as

$$\gamma_b^{x_1} = \frac{\beta \rho g_{rb}}{\alpha_1 g_{1b} g_{rb} + \alpha_2 \beta g_{2b}} \tag{7.16}$$

and

$$\gamma_b^{x_2} = \frac{\beta \rho g_{rb}}{\alpha_1 g_{1b} + \alpha_2 g_{2b} g_{rb}} \tag{7.17}$$

7.1.3 *Outage probability*

In this subsection, the outage probabilities achieved by the coordinated direct and relay transmission scheme for the BS to detect x_1 and x_2 are analyzed. Denoting the targeted transmit rate for U_i by R_i, an outage event occurs for the BS to detect x_1 if $\gamma_1 < \gamma_{th,1}$, where $\gamma_{th,1} \equiv 2^{2R_1} - 1$. Moreover, an outage event occurs for the BS to detect x_2 under two conditions: (i) ${\gamma_1}^{x_1} < \gamma_{th,1}$; (ii) ${\gamma_1}^{x_1} \geq \gamma_{th,1}$ and $\gamma_2 < \gamma_{th,2}$, where $\gamma_{th,2} \equiv 2^{2R_2} - 1$.

The outage probability for the BS to detect x_1 can be written as

$$P_{\text{out},1} = P_r\{\gamma_1 < \gamma_{th,1}\} \tag{7.18}$$

Theorem 1. *The outage probability for the BS to detect x_1 is given by*

$$P_{\text{out},1} = 1 - I_1 I_2 (1 - I_3) \tag{7.19}$$

where

$$I_1 = \frac{m_{2r}^{m_{2r}}}{\theta_{2r}^{m_{2r}} \Gamma(m_{2r})} e^{-\frac{m_{1r}\gamma_{th,1}}{\alpha_1 \rho \theta_{1r}}} \sum_{k=0}^{m_{1r}-1} \sum_{l=0}^{k} \frac{1}{k!} \left(\frac{m_{1r}\alpha_2 \gamma_{th,1}}{\alpha_1 \theta_{1r}} \right)^k \binom{k}{l}$$
$$\times (\alpha_2 \rho)^{l-k} \left(\frac{m_{2r}}{\theta_{2r}} + \frac{m_{1r}\alpha_2 \gamma_{th,1}}{\alpha_1 \theta_{1r}} \right)^{-m_{2r}-l} \Gamma(m_{2r} + l) \tag{7.20}$$

$I_2 = 1$ if $\gamma_{th,1} < \rho$; otherwise

$$I_2 = \frac{m_{1b}^{m_{1b}}}{\theta_{1b}^{m_{1b}} \Gamma(m_{1b})} \sum_{k=0}^{m_{1b}-1} \frac{1}{k!} \left(\frac{m_{2b}\alpha_1 (\gamma_{th,1} - \rho)}{\alpha_2 \rho \theta_{2b}} \right)^k$$
$$\times \left(\frac{m_{1b}}{\theta_{1b}} + \frac{m_{2b}\alpha_1 (\gamma_{th,1} - \rho)}{\alpha_2 \rho \theta_{2b}} \right)^{-m_{1b}-k} \Gamma(m_{1b} + k) \tag{7.21}$$

and

$$I_3 = \frac{m_{rb}^{m_{rb}}}{\theta_{rb}^{m_{rb}} \Gamma(m_{rb})} \sum_{k=0}^{m_{2b}-1} \frac{1}{k!} \left(\frac{m_{2b}\rho}{\alpha_2 \gamma_{th,1} \theta_{2b}} \right)^k$$
$$\times \left(\frac{m_{rb}}{\theta_{rb}} + \frac{m_{2b}\rho}{\alpha_2 \gamma_{th,1} \theta_{2b}} \right)^{-m_{rb}-k} \Gamma(m_{rb} + k) \tag{7.22}$$

Proof. See Appendix A.7.1.

The closed-form expression in (7.19) consists of the finite summations so that it can be easily evaluated via computer computation. Note that (7.19) is valid for $\alpha_1/\alpha_2 > \gamma_{th,1}$. While $\alpha_1/\alpha_2 \le \gamma_{th,1}$, we have $P_{\text{out},1} = 1$. As $\to \infty$, we have $\gamma_r^{x_1} \to \alpha_1 g_{1r}/\alpha_2 g_{r2}$ and $\gamma_b{}^{x_1} \to \rho g_{rb}/\alpha_2 g_{2b}$ so that the asymptotic outage probability has the form of $P_{\text{out},1}{}^{\infty} = P_r\{\gamma_r{}^{x_1} < \gamma_{th.1}\} = P_r\{g_{1r} < \alpha_2\gamma_{th,1}g_{2r}/\alpha_1\}$. After some mathematical manipulation, we can express the asymptotic outage probability as

$$P_{\text{out},1}^{\infty} = 1 - \frac{m_{2r}^{m_{2r}}}{\theta_{2r}^{m_{2r}}\Gamma(m_{2r})}\sum_{k=0}^{m_{1r}-1}\frac{1}{k!}\left(\frac{m_{1r}\alpha_2\gamma_{th,1}}{\alpha_1\theta_{1r}}\right)^k \times \left(\frac{m_{2r}}{\theta_{2r}} + \frac{m_{1r}\alpha_2\gamma_{th,1}}{\alpha_1\theta_{1r}}\right)^{-m_{2r}-k}\Gamma(m_{2r}+k) \tag{7.23}$$

which results in an outage probability floor being determined by Nakagami-m fading parameters.

The outage probability for the BS to detect x_2 can be expressed as

$$P_{\text{out},2} = P_r\{\gamma_r^{x_1} < \gamma_{th,1}\} + P_r\{\gamma_r^{x_1} \ge \gamma_{th,1}\}P_r\{r_2 < \gamma_{th,2}\} \tag{7.24}$$

□

Theorem 2. *The outage probability for the BS to detect x_2 is given by*

$$P_{\text{out},2} = 1 - I_1\tilde{I}_1\tilde{I}_2\tilde{I}_3 \tag{7.25}$$

where I_1 is given by (7.20),

$$\tilde{I}_1 = e^{-\frac{\gamma_{th,2}}{\alpha_2\rho\theta_{2r}}}\sum_{k=0}^{m_{2r}-1}\frac{1}{k!}\left(\frac{\gamma_{th,2}}{\alpha_2\rho\theta_{2r}}\right)^k$$

$\tilde{I}_2 = 1$ *if* $\gamma_{th,2} < \rho$; *otherwise*

$$\tilde{I}_2 = \frac{m_{2b}^{m_{2b}}}{\theta_{2b}^{m_{2b}}\Gamma(m_{2b})}\sum_{k=0}^{m_{1b}-1}\frac{1}{k!}\left(\frac{m_{1b}\alpha_2(\gamma_{th,2}-\rho)}{\alpha_1\rho\theta_{1b}}\right)^k \times \left(\frac{m_{2b}}{\theta_{2b}} + \frac{m_{1b}\alpha_2(\gamma_{th,2}-\rho)}{\alpha_1\rho\theta_{1b}}\right)^{-m_{2b}-k}\Gamma(m_{2b}+k) \tag{7.26}$$

and

$$\tilde{I}_3 = 1 - \frac{m_{rb}^{m_{rb}}}{\theta_{rb}^{m_{rb}}\Gamma(m_{rb})}\sum_{k=0}^{m_{1b}-1}\frac{1}{k!}\left(\frac{m_{1b}\rho}{\alpha_1\gamma_{th,2}\theta_{1b}}\right)^k \times \left(\frac{m_{rb}}{\theta_{rb}} + \frac{m_{1b}\rho}{\alpha_1\gamma_{th,2}\theta_{1b}}\right)^{-m_{rb}-k}\Gamma(m_{rb}+k) \tag{7.27}$$

Proof. See Appendix B.7.1.

The closed-form expression in (7.25) characterizes the outage probability in terms of the transmit SNR and Nakagami-m fading parameters of both the relay and direct links, facilitating straightforward evaluation through computer computation. Also, when $\alpha_1/\alpha_2 \leq \gamma_{th,1}$, $P_{\text{out},2} = 1$. Moreover, as $\rightarrow\infty$, $P_r\{\gamma_2 < \gamma_{th.2}\} = 0$. With respect to (7.24), the asymptotic outage probability becomes $P_{\text{out},2}{}^{\infty} = P_r\{\gamma_r{}^{x_1} < \gamma_{th.1}\}$, which has its closed-form expression given by (7.23). Thus, U_1 and U_2 achieve the same asymptotic outage floor in the high SNR regime. Defining the diversity order achieved by U_i as

$$d^i = -\lim_{\rho\rightarrow\infty} \frac{\log(P_{\text{out},i})}{\log(\rho)} \tag{7.28}$$

It can be shown that the diversity orders of U_1 and U_2 achieved by the coordinated direct and relay transmission scheme are the same as zero due to the DF operation at the relay. □

7.1.4 *Performance analysis*

In this section, Monte Carlo simulation results are provided to verify the system performance achieved by the coordinated direct and relay transmission scheme. For all the channels, the large-scale fading is set as $\mathcal{L}_{\text{j}} = (d_j/d_0)^{-\epsilon}$, where the path loss exponent is set as $\epsilon = 3$, d_j is the distance associated with the channel h_j for $j \in \{ir, ib, rb\}$, and the reference distance is set as $d_0 = 50$ m [8]. Throughout the simulations, we set $\mathcal{L}_{rb} = \mathcal{L}_{1r} = \mathcal{L}_{2r} = 0.5$, $\mathcal{L}_{1b} = \mathcal{L}_{2b} = 0.3$, and $m_{1b} = m_{2b} = 1$. Unless otherwise stated, the targeted rates for U_1 and U_2 are set as $R_1 = 0.2$ bps/Hz and $R_2 = 1$ bps/Hz, respectively, and the corresponding power allocation coefficients are set as $\alpha_1 = 0.75$ and $\alpha_2 = 0.25$, respectively. For the comparison purpose, the outage performance achieved by the DF w/o DLs is also provided.

The outage probability versus transmit SNR is investigated in Fig. 7.2, considering two sets of shape factors for Nakagami-m fading: $\{m_{rb} = m_{ir} = 2\}$ and $\{m_{rb} = m_{ir} = 4\}$. The simulation results verify the correctness of the analytical outage probability. It is observed that the coordinated direct and relay transmission scheme achieves better outage performance for both U_1 and U_2 compared to the DF w/o DLs by incorporating direct

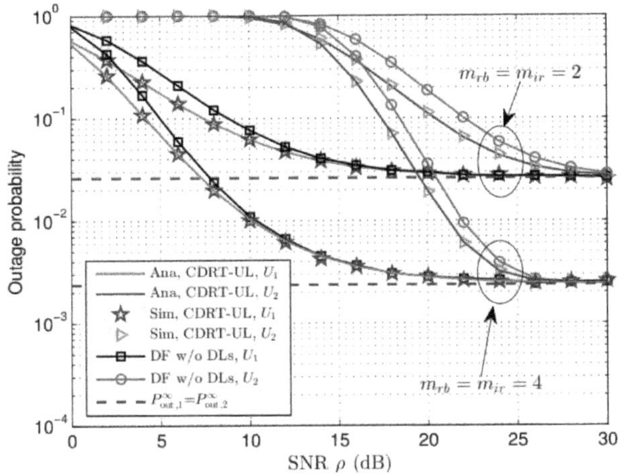

Fig. 7.2. Outage probability versus transmit SNR ρ.

link transmissions. For U_1, the coordinated direct and relay transmission scheme exhibits smaller outage probability than the DF w/o DLs in the small and middle SNR regimes. In the high SNR regime, both schemes approach the same outage probability floor beyond a certain large value of ρ. Similarly, for U_2, the coordinated direct and relay transmission scheme outperforms the DF w/o DLs in the middle and high SNR regimes, with convergence to the same outage probability floor for large ρ. Additionally, increasing the shape factors from $\{m_{rb} = m_{ir} = 2\}$ to $\{m_{rb} = m_{ir} = 4\}$ leads to more shape factor diversity, while smaller shape factor sets result in larger outage probability floors in the high SNR regime.

In Fig. 7.3, the impact of targeted rates on the outage probability is investigated, with $\rho = 20$ dB. The targeted rates R_2 are varied from 0.2 to 3 bps/Hz, while R_1 is fixed at 0.2 bps/Hz. It is observed that the coordinated direct and relay transmission scheme achieves smaller outage probability compared to the DF w/o DLs for U_2 in the considered R_2 regime. Moreover, the outage probability for U_2 increases with higher values of R_2 and tends to approach 1 in the high R_2 regime. Increasing the shape factor set from $\{m_{rb} = m_{ir} = 2\}$ to $\{m_{rb} = m_{ir} = 4\}$ results in decreased outage probabilities for both the coordinated direct and relay transmission and DF w/o DLs. Therefore, the coordinated direct and relay transmission scheme can support larger targeted rates compared to the DF w/o DLs.

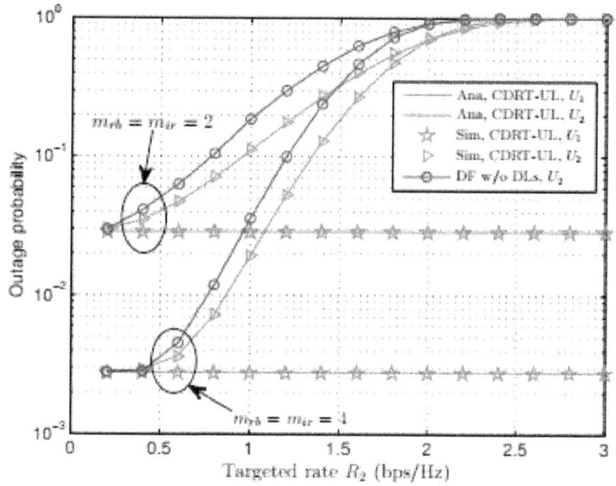

Fig. 7.3. Outage probability versus targeted rate R_2.

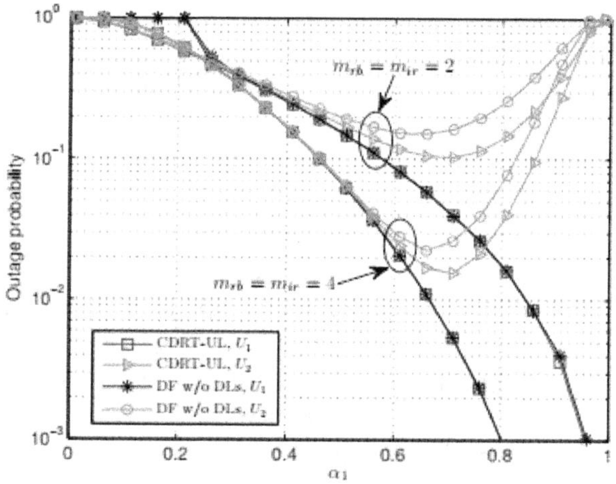

Fig. 7.4. Outage probability versus α_1.

Figure 7.4 illustrates the impact of power allocation on outage performance, with ρ set to 20 dB and varying α_1 from 0.01 to 0.99. For U_1, both the coordinated direct and relay transmission and DF w/o DLs exhibit monotonically decreasing outage probabilities with increasing α_1 once $\alpha_1/\alpha_2 > \gamma_{th,1}$. However, for U_2, the outage probabilities initially decrease and then increase with rising α_1. Additionally, the coordinated

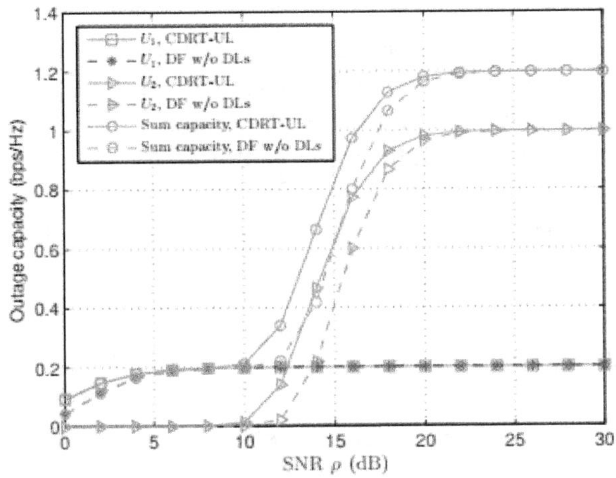

Fig. 7.5. Outage capacity versus transmit SNR ρ.

direct and relay transmission scheme outperforms the DF w/o DLs for U_2. Moreover, a larger shape factor set leads to improved outage performance for both U_1 and U_2.

Figure 7.5 investigates the outage capacity versus transmit SNR, with shape factors set to $\{m_{rb} = m_{ir} = 2\}$. The outage capacity of U_i is given by $C_i = R_i(1 - P_{\text{out},i})$, and the sum capacity is given by $C_{\text{sum}} = C_1 + C_2$. In the small SNR regime, the coordinated direct and relay transmission scheme achieves a higher outage capacity compared to the DF w/o DLs. As ρ increases, both schemes approach the same capacity ceiling in the middle and high SNR regimes. This convergence occurs because both schemes achieve the same outage probability floor in these regimes for U_1. For U_2, both schemes exhibit almost zero outage capacities in the small SNR regime. However, in the middle SNR regime, the coordinated direct and relay transmission scheme outperforms the DF w/o DLs, resulting in a higher outage capacity. Consequently, the coordinated direct and relay transmission scheme achieves a higher sum capacity than the DF w/o DLs in the small and middle SNR regimes.

7.1.5 *Summary*

The coordinated direct and relay transmission scheme in this chapter addresses cooperative NOMA systems with non-negligible direct links from two users to the BS. By integrating direct and relay links, this scheme establishes an orthogonal structure at the BS receiver, facilitating linear combining decoding without SIC. Analytical outage probabilities for both users under i.i.d Nakagami-m fading channels, along with asymptotic expressions, have been derived. Both users exhibit identical asymptotic outage performance in the high SNR regime. The simulation results validate the improved system performance of the coordinated direct and relay transmission scheme compared to DF w/o DLs.

Appendix A.7.1. Proof of Theorem 1

The outage probability for the BS to detect x_1 can be rewritten as

$$P_{\text{out}}^{x_1} = 1 - P_r\{\gamma_r^{x_1} > \gamma_{th,1}\}P_r\{\gamma_b^{x_1} > \gamma_{th,1}\} \triangleq 1 - I_1 I_2(1 - I_3) \quad \text{(A.1)}$$

where

$$I_1 = P_r\left\{g_{1r} > \frac{\gamma_{th,1}}{\alpha_1\rho} + \frac{\alpha_2\gamma_{th,1}g_{2r}}{\alpha_1}\right\},$$

$$I_2 = P_r\left\{g_{2b} > \frac{\alpha_1 g_{1b}(\gamma_{th,1} - \rho)}{\alpha_2\rho}\right\}, \quad \text{and}$$

$$I_3 = P_r\left\{g_{2b} > \frac{\rho g_{rb}}{\alpha_2\gamma_{th,1}}\right\}$$

By using Refs. [19], the term I_1 can be evaluated as

$$\begin{aligned} I_1 &= \frac{1}{\Gamma(m_{1r})}\int_0^\infty \Gamma\left(m_{1r}, \frac{m_{1r}\gamma_{th,1}}{\alpha_1\rho\theta_{1r}} + \frac{m_{1r}\alpha_2\gamma_{th,1}x}{\alpha_1\theta_{1r}}\right) f_{g_{2r}}(x)dx \\ &= \frac{m_{2r}^{m_{2r}}}{\theta_{2r}^{m_{2r}}\Gamma(m_{2r})} e^{-\frac{m_{1r}\gamma_{th,1}}{\alpha_1\rho\theta_{1r}}} \sum_{k=0}^{m_{1r}-1}\sum_{l=0}^{k}\frac{1}{k!}\left(\frac{m_{1r}\alpha_2\gamma_{th,1}}{\alpha_1\theta_{1r}}\right)^k \binom{k}{l} \\ &\times (\alpha_2)^{l-k}\int_0^\infty x^{m_{2r}+l-1} e^{-\left(\frac{m_{2r}}{\theta_{2r}} + \frac{m_{1r}\alpha_2\gamma_{th,1}}{\alpha_1\theta_{1r}}\right)x} dx \end{aligned} \quad \text{(A.2)}$$

which can be further expressed as (7.20). When $\gamma_{th,1} < \rho$, we have $I_2 = 1$ due to the non-negative value of g_{2b}. When $\gamma_{th,1} \geq \rho$, the term I_2 can be evaluated as (7.21) following a similar procedure as that for (A.2). As such, the term I_3 is derived as (7.22), and finally leads to (7.19).

Appendix B.7.1. Proof of Theorem 2

The outage probability for the BS to detect x_2 can be rewritten as

$$\begin{aligned} P_{\text{out},2} &= P_r\{\gamma_r^{x_1} < \gamma_{th,1}\} + P_r\{\gamma_r^{x_1} \geq \gamma_{th,1}\} \\ &\quad \times (1 - P_r\{\gamma_r^{x_2} > \gamma_{th,2}\} P_r\{\gamma_b^{x_2} > \gamma_{th,2}\}) \\ &= 1 - I_1 + I_1(1 - \tilde{I}_1 P_r\{\gamma_b^{x_2} > \gamma_{th,2}\}) \end{aligned} \tag{B.1}$$

where $I_1 = P_r\{g_{1r} > \frac{\gamma_{th,1}}{\alpha_1\rho} + \frac{\alpha_2\,\gamma_{th,1}g_{2r}}{\alpha_1}\}$ and $\tilde{I}_1 = P_r\{g_{2r} > \frac{\gamma_{th,2}}{\alpha_2\rho}\}$. Note that I_1 has been evaluated as (7.20), while $\tilde{I}_1$ can be evaluated as

$$\tilde{I}_1 = e^{-\frac{\gamma_{th,2}}{\alpha_2\rho\theta_{2r}}} \sum_{k=0}^{m_{2r}-1} \frac{1}{k!}\left(\frac{\gamma_{th,2}}{\alpha_2\rho\theta_{2r}}\right)^k \tag{B.2}$$

The term $P_r\{{\gamma_b}^{x_2} > \gamma_{th,2}\}$ in (B.1) can be further expressed as

$$P_r\{{\gamma_b}^{x_2} > \gamma_{th,2}\} = \tilde{I}_2\tilde{I}_3 \tag{B.3}$$

where $\tilde{I}_2 = P_r\{g_{1b} > \frac{\alpha_2 g_{2r}(\gamma_{th,2}-\rho)}{\alpha_1\rho}\}$ and $\tilde{I}_3 = P_r\{g_{1b} > \frac{\rho g_{rb}}{\alpha_1\gamma_{th,2}}\}$. Straightforwardly, the outage probability can be expressed as

$$P_{\text{out},2} = 1 - I_1\tilde{I}_1\tilde{I}_2\tilde{I}_3 \tag{B.4}$$

Following a similar procedure as that for (A.2), $\tilde{I}_2$ and $\tilde{I}_3$ can be evaluated as (7.26) and (7.27), respectively.

7.2 Heterogeneous Networks for NOMA System

In this section, the non-orthogonal multiple access (NOMA) is investigated in a two-tier heterogeneous network (HetNet) with non-uniform small-cell deployment, focusing on the key performance metrics of coverage probability and achievable rate. Initially, a NOMA-based HetNet model is established, with users paired using a proposed scheme. Following that, the distribution of order statistics for the distances between NOMA users and serving base stations (BSs) is examined, taking into account channel qualities. Analytical demonstrations illustrate the influence of various network parameters on NOMA user coverage probability and achievable rate, encompassing the signal-to-interference-plus-noise ratio (SINR) threshold and BS density. Additionally, an analysis is provided to offer insights into the energy efficiency (EE) of the system. Finally, extensive simulations and comparisons are conducted to validate the superiority of NOMA over orthogonal multiple access (OMA) in the HetNet environment [33].

7.2.1 *Introduction*

NOMA is increasingly recognized as a promising multiple access solution in next-generation wireless communication systems [20, 21]. Studies have shown that applying NOMA to users with distinct channel conditions can lead to higher data rates [23, 24]. Typically, users with better channel conditions are allocated less power by the BS, ensuring fairness in terms of data rate across users [25]. NOMA has gained significant attention in next-generation wireless networks due to its promising benefits, particularly in addressing massive access demands [26–28]. It has been shown to be compatible with various key enabling techniques in fifth-generation (5G) systems, including cooperative communications, relay networks, and cognitive radios. However, challenges persist, such as integrating NOMA with multi-cell networks.

In this section, we focus on the performance of NOMA-based HetNet with non-uniform small-cell deployment, and the main results are as follows: A NOMA scheme is applied to a non-uniform HetNet, wherein BSs are distributed independently according to Poisson point processes (PPPs) with varying densities across tiers. Users associated with specific BSs, determined by the maximum average received power, are segmented into distinct groups based on their access method. Four distinct types of typical users are considered and organized into two NOMA groups, while OMA is implemented across these groups to enable intergroup multiple access [29–31].

The coverage probability and the average achievable rate of NOMA users are derived using stochastic geometry, assuming perfect successive interference cancellation (SIC) on the user side [22, 32–34]. The analysis includes the distribution of order statistics for distances between NOMA users and their associated BSs. Moreover, the impact of macro base station (MBS)-prescribed distances on the coverage probability and the power allocation factor's effect on the sum achievable rate of NOMA groups are examined. A user-pairing scheme is proposed based on the NOMA framework and compared with a random pairing scheme. Performance enhancements are demonstrated with the proposed pairing scheme. Finally, the performance simulations of NOMA and OMA in a two-tier HetNet illustrate the substantial performance gains for NOMA over OMA.

7.2.2 *System model*

7.2.2.1 *Two-tier HetNet*

A two-tier HetNet is considered, comprising BSs with distinct characteristics in terms of transmit powers and densities. The BSs in the kth tier, where $k \in \{1, 2\}$, are modeled as a homogeneous PPP Φ_k with density λ_k and transmit power P_k where $P_1 \geq P_2$ (P_1 and P_2 represent the transmit power of MBS and S BS, respectively). The MBSs form the first tier, while the small base stations (SBSs) constitute the second tier. The inner region is defined as the collection of locations within a predefined distance D from the MBSs. Conversely, the outer region encompasses locations where the distance to any MBS exceeds D. To maintain a balance between coverage and EE, the SBSs within the inner region A_{inner} remain inactive, resulting in an approximate PPP distribution for the SBSs. Meanwhile, users are distributed according to a homogeneous PPP Φ_u with density λ_u. NOMA is employed for user communication with both tiers of BSs. Given that the user density typically exceeds that of the BSs, there are ample users associated with any given BS to facilitate NOMA transmission. All channels between users and BSs experience independent and identically distributed Rayleigh fading. The path loss exponent for links between any BS and users is denoted by α, with uniform noise power σ^2. The channel fading coefficient between the ith NOMA user $\{i \in (1, 2)\}$ and the associated BS in the kth tier is denoted as $h_{k,i}$.

7.2.2.2 *User association and user pairing*

In this model, users located in the inner region A_{inner} are associated with the nearest MBS. This means that when the distance between a user and its nearest MBS is less than D, the user is associated with the MBS. For users in the outer region A_{outer}, a biased largest long-term average receive power scheme is implemented. Specifically, users in A_{outer} are associated with the nearest MBS if $P_1 R_1{}^{-\alpha} \geq \beta P_2 R_2{}^{-\alpha}$, otherwise they are associated with the nearest SBS. Here, R_i denotes the distance between the user and the nearest BS in the ith BS tier, and β represents the transmit power biased factor. The corresponding power allocation factor of the ith NOMA user in the kth tier is denoted as $a_{k,i}$. For this factor, $\sum_{i=1}^{N_k} a_{k,i} = 1$ is assumed with

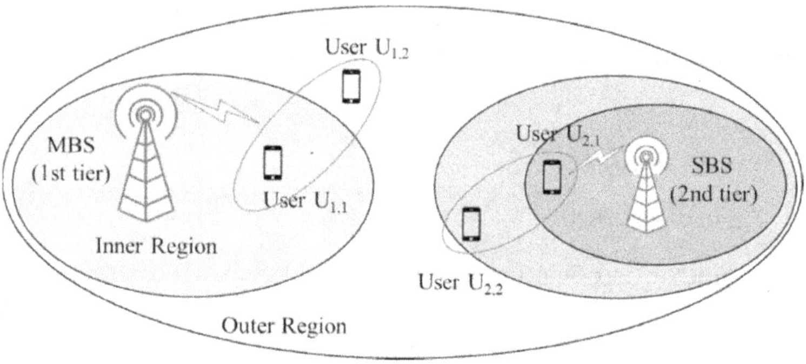

Fig. 7.6. The user-pairing system of two-tier HetNet.

$N_k = 2$, where N_k denotes the number of users in the kth NOMA group. Furthermore, adhering to the characteristics of NOMA, it is stipulated that $a_{k,1} < a_{k,2}$. Figure 7.6 illustrates the proposed user-pairing strategy in NOMA groups.

In the user association and pairing scheme, there are four types of users categorized by their access methods and two types of NOMA groups. These groups consist of the near user $\{U_{k,1}\}$ and the far user $\{U_{k,2}\}$ associated with the kth tier BSs in the network. The scheme can be summarized as follows:

$$U_{k.i} = \begin{cases} U_{1,1} & R_1 \leq D \\ U_{1,2} & \dfrac{P_1 R_1^{-\alpha}}{P_2 R_2^{-\alpha}} > \beta, \qquad R_1 > D \\ U_{2,1} & \dfrac{P_1 R_1^{-\alpha}}{P_2 R_2^{-\alpha}} \leq 1, \qquad R_1 > D \\ U_{2,2} & 1 < \dfrac{P_1 R_1^{-\alpha}}{P_2 R_2^{-\alpha}} \leq \beta, \quad R_1 > D \end{cases} \tag{7.29}$$

It's evident that $U_{1,1}$ and $U_{1,2}$ are associated with MBSs, while $U_{2,1}$ and $U_{2,2}$ are associated with SBSs.

7.2.2.3 *Signal-to-interference-plus-noise*

In this HetNet employing NOMA, the received interference at a NOMA user consists of three components: intra-tier interference from the BSs in the same tier to the associated user, inter-tier interference from BSs in the other tier, and intra-NOMA interference from signals that have not been decoded within the same NOMA group. It is noteworthy that the NOMA user with the best channel condition can mitigate interference from signals of all other users in the same NOMA group [34, 35].

Now, the receiving SINR is defined to be $\gamma_{k,i}$ at the ith NOMA user $U_{k,i}$ in the kth BS tier. If the SIC is perfect at the user side to decode signals intended for other users whose channel qualities are worse in the same group, then the SINR is as follows for the pairing strategy in NOMA groups illustrated in Fig. 7.7.

$$\gamma_{k,i} = \frac{|h_{k,i}|^2 a_{k,i} P_k r_{k,i}^{-\alpha}}{\sum_{m=1,2} I_{m,k,i} + \sum_{n=i+1}^{N_k} |h_{k,i}|^2 a_{k,n} P_k r_{k,i}^{-\alpha} + \sigma^2} \tag{7.30}$$

where $r_{k,i}$ is the distance between $U_{k,i}$ and the associated BS, $a_{k,n}$ is the power allocation factor of the nth NOMA user in the kth tier, and $I_{m,k,i}$ is the aggregate interference from all mth tier BSs to $U_{k,i}$, which can be represented as

$$I_{m,k,i} = \sum_{j \in (\Phi_m \backslash O_{k,i})} |h_{k,i,m,j}|^2 P_m r_{k,i,m,j}^{-\alpha} \tag{7.31}$$

where $h_{k,i,m,j}$ and $r_{k,i,m,j}$ denote the channel coefficient and the distance between $U_{k,i}$ and the jth BS in the mth BS tier, respectively. Φ_m represents some modeled locations among {MBSs, SBSs, and users} and $O_{k,i}$ is the association of the BS to $U_{k,i}$. The intra-NOMA interference of $U_{k,1}$ from the signal of $U_{k,2}$ is $\sum_{n=i+1}^{N_k} |h_{k,i}|^2 a_{k,n} P_k r_{ki}{}^{-\alpha}$, which is not canceled in the NOMA group, and the interference is none for $U_{k,2}$.

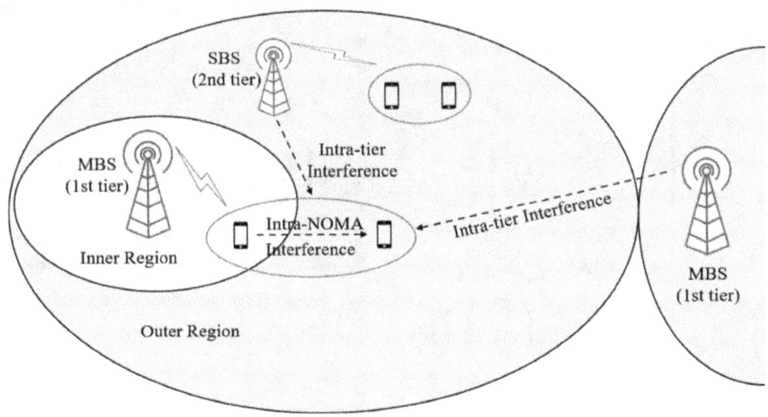

Fig. 7.7. Interference in a NOMA group of two-tier HetNet.

7.2.3 *Coverage probability*

7.2.3.1 *Distance distribution*

The distribution of $r_{k,i}$, the distance $r_{k,i}$ from the typical user $U_{k,i}$ to the associated BS, plays a key role in the analysis of coverage probability. So, the probability density function (PDF) of $r_{k,i}$ can be obtained in the following.

Lemma 1. *The PDF of* $\mathrm{r}_{1,1}$, *the distance between the typical user* $U_{1,1}$ *and the associated BS, is*

$$f_{r_{1,1}}(r) = \frac{2\pi\lambda_1 r \cdot e^{-\pi\lambda_1 r^2}}{1 - e^{-\pi\lambda_1 D^2}} \tag{7.32}$$

Proof. See Appendix A.7.2. □

Lemma 2. *The PDF of* $\mathrm{r}_{1,2}$, *the distance between the typical user* $U_{1,2}$ *and the associated BS, is given by*

$$f_{r_{1,2}}(r) = 2\pi\Delta_1 r \cdot e^{-\pi\Delta_1(r^2 - D^2)} \tag{7.33}$$

where

$$\Delta_1 = \lambda_1 + \lambda_2\left(\frac{\beta P_2}{P_1}\right)^{\frac{2}{\alpha}}$$

Proof. See Appendix A.7.2. □

Lemma 3. *The PDF of* $\mathrm{r}_{2,1}$, *the distance between the typical user* $U_{2,1}$ *and the associated BS, is*

$$f_{r_{2,1}}(r) = \begin{cases} 2\pi\varpi\lambda_2 r \cdot e^{-\pi\lambda_2 r} & r \le R_d \\ \dfrac{2\pi\varpi\lambda_2 r}{e^{-\pi\lambda_2 D^2}}\exp\left\{-\pi\left[\lambda_2 + \lambda_1\left(\dfrac{P_1}{P_2}\right)^{\frac{2}{\alpha}}\right]r^2\right\} & r > R_d \end{cases} \tag{7.34}$$

where

$$R_d = \left(\frac{P_2}{P_1}\right)^{\frac{1}{\alpha}} D,$$

$$\varpi = \frac{\Delta_2}{\Delta_2 - \lambda_1 e^{-\pi(\lambda_1 - \Delta_2)D^2}} e^{-\pi\lambda_1 D^2} \tag{7.35}$$

and

$$\Delta_2 = \lambda_1 + \lambda_2 \left(\frac{P_2}{P_1}\right)^{\frac{2}{\alpha}}$$

Proof. See Appendix A.7.2. □

Lemma 4. *The PDF of $r_{2,2}$, the distance between the typical user $U_{2,2}$ and the associated BS, is given by*

$$f_{r_{2,2}}(r) = \begin{cases} \frac{1}{\mathcal{Z}}\left[-2\pi\left(\Delta_2-\lambda_1\right)\left(\frac{P_1}{P_2}\right)^{\frac{2}{\alpha}} r e^{-\pi\Delta_2\left(\frac{P_1}{P_2}\right)^{\frac{2}{\alpha}} r^2} \right. \\ \left. +2\pi\lambda_2 r e^{-\pi\lambda_1 D^2} e^{-\pi\lambda_2 r^2}\right], & R_d \le r \le R_p \\ \frac{1}{\mathcal{Z}}\left[-2\pi\left(\Delta_2-\lambda_1\right)\left(\frac{P_1}{P_2}\right)^{\frac{2}{\alpha}} r e^{-\pi\Delta_2\left(\frac{P_1}{P_2}\right)^{\frac{2}{\alpha}} r^2} \right. \\ \left. +2\pi\left(\Delta_1-\lambda_1\right)\left(\frac{P_1}{\beta P_2}\right)^{\frac{2}{\alpha}} r e^{-\pi\Delta_1\left(\frac{P_1}{\beta P_2}\right)^{\frac{2}{\alpha}} r^2}\right], & r > R_p \end{cases} \tag{7.36}$$

where

$$R_p = \left(\frac{\beta P_2}{P_1}\right)^{\frac{1}{\alpha}} D$$

and

$$\mathcal{Z} = \frac{\lambda_1}{\Delta_2} e^{-\pi\Delta_2 D^2} - \frac{\lambda_1}{\Delta_1} e^{-\pi\Delta_1 D^2} \tag{7.37}$$

Proof. See Appendix A.7.2. □

7.2.3.2 *Coverage probability*

The coverage probability of a typical user $U_{k,i}$ is defined as the probability that the received SINR $\gamma_{k,i}$ is greater than a predefined threshold. According to the definition of the coverage probability, $\rho_{k.i}$, as a general result for the four types of typical users, the coverage probability can be expressed as

$$\rho_{k,i} = P_r\{\gamma_{k,i} > \varepsilon\} \tag{7.38}$$

where ε is the SINR threshold of the users.

Theorem 1. *Considering all four kinds of typical users, the coverage probability of user $U_{k.i}$ is given by*

$$\rho_{k,i} = \int_{r>0} e^{-\frac{\theta_{k,i} r^{\alpha} \sigma^2}{P_k}} \prod_{m=1}^{2} \mathcal{L}_{I_{k,i,m}}\left(\frac{\theta_{k,i} r^{\alpha}}{P_k}\right) f_{r_{k,i}}(r)dr \tag{7.39}$$

where $\mathcal{L}_{I_{k,i,m}}(s)$ is the Laplace transform of the random variable $I_{k,i,m}$ that can be expressed as

$$\mathcal{L}_{I_{k,i,m}}(s) = \mathbb{E}_{I_{k,i,m}}[e^{-sI_{k,i,m}}] \tag{7.40}$$

and

$$\theta_{k,i} = \frac{\varepsilon}{a_{k,i} - \varepsilon \sum_{n=i+1}^{N_k} a_{k,n}}$$

Note that the above result is obtained based on $\theta_{k,i} > 0$, so it must satisfy

$$a_{k,i} > \varepsilon \sum_{n=i+1}^{N_k} a_{k,n} \tag{7.41}$$

Otherwise, the coverage probability would always be zero.

Proof. See Appendix B.7.2. □

7.2.4 *Average achievable rate*

7.2.4.1 *Single-user case*

The general result of the achievable rate for a typical NOMA user in the downlink is provided as follows:

Theorem 2. *The achievable rate of the ith NOMA user in the kth tier can be expressed as*

$$R_{k,i} = \ln(1+\varepsilon) + \frac{1}{\rho_{k,i}} \int_{\ln(1+\varepsilon)}^{\infty} \int_{0}^{\infty} e^{-\mu \frac{\tau_{k,i,t} r_{k,i}^{\alpha} \sigma^2}{P_k}} \times \prod_{m=1}^{K} \mathcal{L}_{I_{k,i,m}} \left(\mu \frac{\tau_{k,i,t} r_{k,i}^{\alpha}}{P_k} \right) f_{r_{k,i}}(r) dr dt \quad (7.42)$$

where

$$\tau_{k,i,t} = \frac{e^t - 1}{a_{k,i} - (e^t - 1) \sum_{n=i+1}^{N_k} a_{k,n}}$$

Proof: *From* $E[X] = \int_{t>0} P[X > t]dt$, *we can have*

$$\begin{aligned} R_{k,i} &= \int_{t>0} \mathbb{P}[\ln(1+\gamma_{k,i}) > t | \gamma_{k,i} > \varepsilon] dt \\ &= \int_{t>0} \mathbb{P}[\gamma_{k,i} > (e^t - 1) | \gamma_{k,i} > \varepsilon] dt \\ &= \int_{t>0} \frac{\mathbb{P}[\gamma_{k,i} > (e^t - 1), \gamma_{k,i} > \varepsilon]}{\mathbb{P}[\gamma_{k,i} > \varepsilon]} dt = \ln(1+\varepsilon) \\ &+ \frac{1}{\rho_{k,i}} \int_{\ln(1+\varepsilon)}^{\infty} \mathbb{P}[\gamma_{k,i} > (e^t - 1)] dt \end{aligned} \quad (7.43)$$

Substituting ε *with* $(e^t - 1)$ *in the expression of* $\theta_{k,i}$, $P[\gamma_{k,i} > (e^t - 1)]$ *can be obtained according to* (7.39). *Then Eq. (7.42) can be obtained.* □

7.2.4.2 *Energy efficiency*

The EE metric serves as a measure for quantifying the energy consumption associated with information transmission at the access station level. In this context, EE is defined as the attainable rate per unit of transmit power. Mathematically, this can be expressed using the rate in equation (7.42) as:

$$\eta_k = \frac{\sum_{i=1}^{N_k} R_{k,i}}{P_k} \tag{7.44}$$

It is evident that, with a fixed P_k, the EE is inversely proportional to the sum of the average achievable rates of a typical user group. In the subsequent section, a comparison is conducted between the EE of a NOMA group and that of a typical OMA group. While (7.44) does not yield a closed-form expression, it can be efficiently computed to derive numerical results. As demonstrated later in the performance analysis section, under a specific transmit power scenario, a higher sum of the average achievable rates within the same user group is achieved and thus corresponds to an increased EE.

7.2.5 *Performance analysis*

In this section, numerical results are provided on the coverage probability and the average achievable rate of NOMA users considering variations in the BS density. Comparative analyses are conducted, including comparisons with the OMA system and a random user-pairing strategy. It is noteworthy that all results presented herein stem from a two-tier HetNet, comprising MBS as the primary tier and SBS as the secondary tier. Additionally, to capture the realistic scenario, Monte Carlo simulations are employed.

In Fig. 7.8, the relationships between the coverage probability of typical users and the radius of the inner region D are depicted. As D increases, the coverage probability of typical users within the inner region decreases, whereas the coverage probability of users associated with SBS remains nearly unchanged. This phenomenon can be attributed to the reduction in the average distance between users and MBS as D increases, leading to a decrease in the coverage probability of users associated with MBS.

Figure 7.9 presents a comparison between the sum of the average achievable rates of a NOMA group in the proposed scheme and a random user-pairing strategy. In the random user-pairing strategy, users associated with the same BS are paired randomly to form a NOMA group. It is evident that the sum of the average achievable rates in the proposed scheme significantly surpasses that of the random user-pairing strategy. This observation aligns with the theoretical notion that the sum of the average achievable rates of two users in a NOMA group is greater when there is a larger discrepancy in channel conditions.

Furthermore, it is noted that the sum of the average achievable rates of a NOMA group decreases with a higher SBS density. However, for the random user-pairing strategy, the sum of the average achievable rates of users associated with MBS increases with a higher SBS density. This phenomenon arises because both users associated with MBS can be located in the inner region, thus mitigating the impact of interference.

When disregarding noise in the two-tier interference-limited HetNet, as per equation (7.42), the average achievable rate of a NOMA user is dependent upon its channel condition order, irrespective of the tier it is associated with.

The sum of the average achievable rates for both NOMA and OMA groups is presented in Fig. 7.10. Here, orthogonal frequency-division multiplexing (OFDM) technology represents OMA technology. When the transmit power of BSs and the frequency band are consistent, the achievable rate of users associated with MBS in the proposed strategy surpasses that in OMA. Conversely, the achievable rate of users associated with SBS in the strategy is comparable to that in OMA. This observation suggests that when the transmitting power of BSs is low, both the power allocation factor and the channel conditions significantly impact the achievable rate.

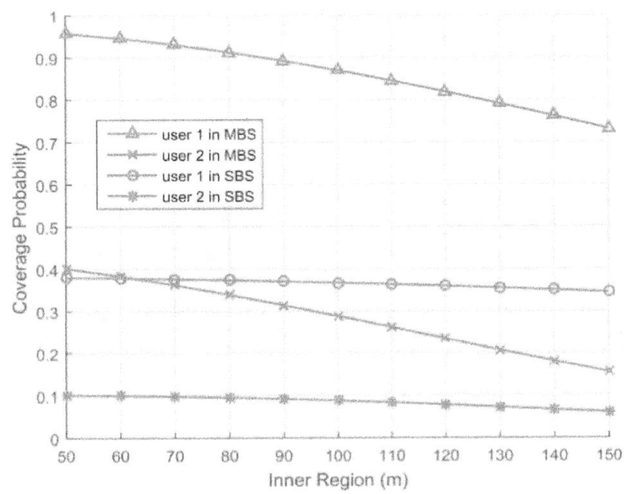

Fig. 7.8. Coverage probability with varying values of the inner region: $\{\lambda_2 = 10\lambda_1, \beta = 5, \{P_1, P_2\} = \{10, 1\}\text{dBm}, \{a_{1,1}, a_{1,2}\} = \{a_{2,1}, a_{2,2}\} = \{0.3, 0.7\}\}$.

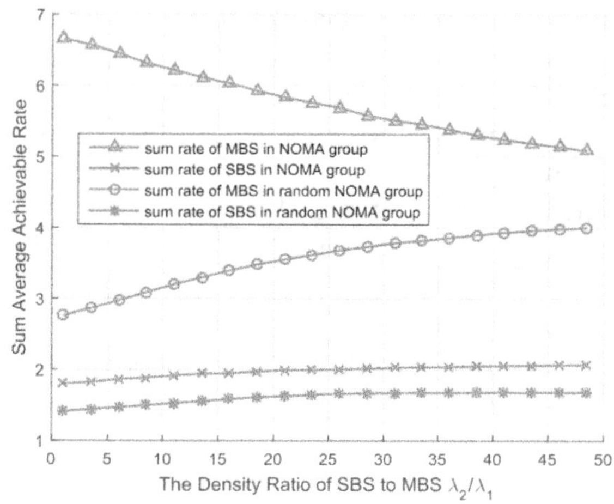

Fig. 7.9. Achievable rate in the NOMA group compared to the random matching strategy: $\{D = 100\text{m}, \beta = 5, \{P_1, P_2\} = \{10, 1\}\text{dBm}, \{a_{1,1}, a_{1,2}\} = \{a_{2,1}, a_{2,2}\} = \{0.3, 0.7\}\}$.

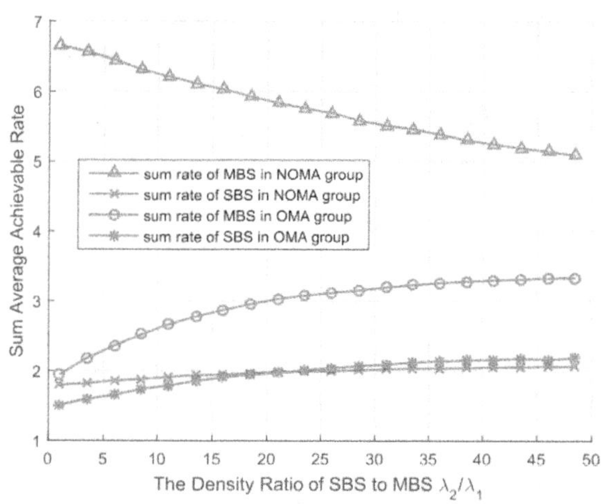

Fig. 7.10. Average achievable rate in the NOMA group compared to the OMA group: $\{D = 100\text{m}, \beta = 5, \{P_1, P_2\} = \{10, 1\}\text{dBm}, \{a_{1,1}, a_{1,2}\} = \{a_{2,1}, a_{2,2}\} = \{0.3, 0.7\}\}$.

Furthermore, as depicted in Fig. 7.11, it can be deduced that, compared with the OMA system, higher EE can be attained for users associated with SBSs in the considered NOMA system, while the EE of users associated with MBSs exhibits minimal variation.

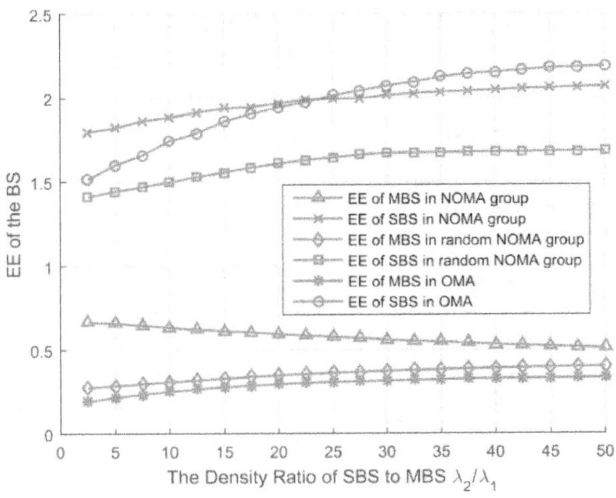

Fig. 7.11. EE of the BS in the NOMA group compared to the random NOMA and OMA groups: $\{D = 100\text{m}, \beta = 5, \{P_1, P_2\} = \{10, 1\}\text{dBm}, \{a_{1,1}, a_{1,2}\} = \{a_{2,1}, a_{2,2}\} = \{0.3, 0.7\}\}$.

7.2.6 *Summary*

This chapter presents a comprehensive model for analyzing the performance of NOMA in a two-tier HetNet with non-uniform small-cell deployment. Users associated with the BS providing the maximum average received power are differentiated within the same NOMA group based on their distances from the associated BS. Leveraging the proposed NOMA user-pairing scheme, analytical expressions for the coverage probability and the average achievable rate are derived for a NOMA group comprising two users. The impacts of network parameters such as BS density and power allocation factor are thoroughly investigated. The findings reveal that the average achievable rate of a NOMA group exceeds that of a randomly paired NOMA group, and the deployment of NOMA in HetNet leads to enhanced EE compared to OMA.

Appendix A.7.2. Proof of Lemmas 1–4

Proof of Lemma 1. Obviously,

$$P_r\{r_{1,1} < r\} = P_r\{R_1 < r | R_1 \leq D\} \tag{A.1}$$

This means that when the user is in the inner region, there is at least one MBS located within a radius r of user $U_{1,1}$. Since

$$P_r\{R_1 < r|R_1 \leq D\} = \frac{P_r\{R_1 < r\}}{P_r\{R_1 \leq D\}} = \frac{1 - \lambda_1 \cdot e^{-\pi \lambda_1 r^2}}{1 - e^{-\pi \lambda_1 D^2}} \tag{A.2}$$

then it can be expressed as

$$f_{r_{1,1}}(r) = \frac{2\pi \lambda_1 r \cdot e^{-\pi \lambda_1 r^2}}{1 - e^{-\pi \lambda_1 D^2}} \tag{A.3}$$

□

Proof of Lemma 2. When the distance between a user and its nearest MBS exceeds D but $P_1 R_1 \geq \beta P_2 R_2$, the user is still considered associated with this MBS. For $r_{1,2}$, with $\delta_1 = (\beta P_2 / P_1)^{\alpha^{-1}}$,

$$\begin{aligned} P_r\{r_{1,2} < r\} &= P_r\{R_1 < r|P_1 R_1^{-\alpha} \geq \beta P_2 R_2^{-\alpha}, R_1 > D\} \\ &= \frac{P_r\{R_1 < r, R_1 < \delta_1 R_2 | R_1 > D\} \cdot P_r\{R_1 > D\}}{P_r\{R_1 < \delta_1 R_2, R_1 > D\}} \end{aligned} \tag{A.4}$$

where

$$\begin{aligned} &P_r\{R_1 < \delta_1 R_2, R_1 > D\} \\ &= \int_{R_p}^{\infty} P_r \left\{ D < R_1 < \left(\frac{\beta P_2}{P_1}\right)^{\frac{1}{\alpha}} x \right\} f_{R_2}(x) dx \\ &= \int_{R_p}^{\infty} \left[e^{-\pi \lambda_1 D^2} - e^{-\pi \lambda_1 \left(\frac{P_1}{\beta P_2}\right)^{\frac{2}{\alpha}} x^2} \right] \cdot 2\pi \lambda_2 x e^{-\pi \lambda_2 x^2} dx \\ &= \frac{\lambda_1}{\Delta_1} \cdot e^{-\pi \Delta_1 D^2} \end{aligned} \tag{A.5}$$

and

$$\begin{aligned} &P_r\{R_1 < r, R_1 < \delta_1 R_2 | R_1 > D\} \\ &= \int_{D}^{r} P_r \left\{ R_2 > \left(\frac{P_1}{\beta P_2}\right)^{\frac{1}{\alpha}} x \right\} f_{R_1|R_1 > D}(x) dx \\ &= \frac{\lambda_1}{\Delta_1 \cdot e^{-\pi \lambda_1 D^2}} (e^{-\pi \Delta_1 D^2} - e^{-\pi \Delta_1 r^2}) \end{aligned} \tag{A.6}$$

Then

$$f_{r_{1,2}}(r) = 2\pi\Delta_1 r \cdot e^{-\pi\Delta_1(r^2-D^2)} \tag{A.7}$$

where

$$\Delta_1 = \lambda_1 + \lambda_2\left(\frac{\beta P_2}{P_1}\right)^{\frac{2}{\alpha}}$$

Proof of Lemma 3. For this case, the cumulative distribution function (CDF) of $r_{2,1}$ is as follows:

$$\begin{aligned} P_r\{r_{2,1} < r\} &= P_r\left\{R_2 < r | R_2 < \left(\frac{P_2}{P_1}\right)^{\frac{1}{\alpha}} R_1, R_1 > D\right\} \\ &= \frac{P_r\{R_2 < r, R_2 < (\frac{P_2}{P_1})^{\frac{1}{\alpha}} R_1 | R_1 > D\} \cdot P_r\{R_1 > D\}}{P_r\{R_2 < (\frac{P_2}{P_1})^{\frac{1}{\alpha}} R_1, R_1 > D\}} \end{aligned} \tag{A.8}$$

then the corresponding PDF with $R_d = (P_2/P_1)^{\alpha^{-1}} D$ can be expressed as

$$f_{r_{2,1}}(r) = \begin{cases} \varpi \cdot 2\pi\lambda_2 r \cdot e^{-\pi\lambda_2 r^2}, & r \le R_d \\ \varpi \cdot \dfrac{2\pi\lambda_2 r}{e^{-\pi\lambda_1 D^2}} \cdot \exp\left\{-\pi\left[\lambda_2 + \lambda_1\left(\dfrac{P_1}{P_2}\right)^{\frac{2}{\alpha}}\right] r^2\right\}, & r > R_d \end{cases} \tag{A.9}$$

where

$$\varpi = \frac{\Delta_2}{\Delta_2 - \lambda_1 e^{-\pi(\lambda_1 - \Delta_2)D^2}} e^{-\pi\lambda_1 D^2} \tag{A.10}$$

Proof of Lemma 4. For the case of $P_1 R_1{}^{-\alpha} < P_2 R_2{}^{-\alpha}$, with $\delta_2 = (P_2/P_1)^{\alpha^{-1}}$, the CDF of $r_{2,2}$ is

$$\begin{aligned} P_r\{r_{2,2} \le r\} &= P_r\{R_2 < r | \delta_2 R_1 < R_2 < \delta_1 R_1, R_1 > D\} \\ &= \frac{P_r\{R_2 < r, \delta_2 R_1 < R_2 < \delta_1 R_1 | R_1 > D\} P_r\{R_1 > D\}}{P_r\{\delta_2 R_1 < R_2 < \delta_1 R_1, R_1 > D\}} \end{aligned} \tag{A.11}$$

where

$$P_r\{R_2 < r, \delta_2 R_1 < R_2 < \delta_1 R_1 | R_1 > D\}$$
$$= \begin{cases} \frac{1}{e^{-\pi D^2}} \left[\left(1 - \frac{\lambda_1}{\Delta_2}\right) e^{-\pi \Delta_2 (\frac{P_1}{P_2})^{\frac{2}{\alpha}} r^2} \right. \\ \left. + \frac{\lambda_1}{\Delta_1} e^{-\pi \Delta_2 D^2} - e^{-\pi \lambda_1 D^2} e^{-\pi \lambda_2 r^2} \right], \quad R \le r \le R_p \\ \frac{1}{e^{-\pi D^2}} \left[\left(1 - \frac{\lambda_1}{\Delta_2}\right) e^{-\pi \Delta_2 (\frac{P_1}{P_2})^{\frac{2}{\alpha}} r^2} \right. \\ \left. - \left(1 - \frac{\lambda_1}{\Delta_1}\right) e^{-\pi \Delta_1 (\frac{P_1}{\beta P_2})^{\frac{2}{\alpha}} r^2} + \mathcal{Z} \right], \quad r > R_p \end{cases} \tag{A.12}$$

Following similar derivations as those outlined above, equation (7.37) can be obtained.

Appendix B.7.2. Proof of Theorem 1

From the definition of coverage probability, one can obtain,

$$\rho_{k,i} = \mathbb{E}_{r_{k,i}}[\mathbb{P}[\gamma_{k,i} > \varepsilon | r_{k,i}]] = \int_{r>0} \mathbb{P}[\gamma_{k,i} > \varepsilon | r_{k,i}] f_{r_{k,i}}(r) dr \tag{B.1}$$

Here, $P[\gamma_{ki,} > \varepsilon | r_{k,i}]$ is the coverage probability conditioned on the nearest kth tier BS being at a distance r from the user. From equation (7.30), one can have

$$\mathbb{P}[\gamma_{k,i} > \varepsilon | r_{k,i}] = \mathbb{P}\left[|h_{k,i}|^2 > \frac{(\sum_{m=0}^{K} I_{k,i,m} + \sigma^2)\varepsilon}{\sum_{n=i+1}^{N_k} (\beta_{k,i} - \varepsilon \beta_{k,n}) P_k r_{k,i}^{-\alpha}} | r_{k,i} \right] \tag{B.2}$$

Since the Rayleigh fading channel is considered, $|h_{k,i}|^2$ follows an exponential distribution with mean μ. Then we can obtain

$$\mathbb{P}[\gamma_{k,i} > \varepsilon | r] = \mathbb{E}\left\{ \exp\left[-\frac{(\sum_{m=0}^{K} I_{k,i,m} + \sigma^2)\varepsilon}{\theta_{k,i} P_k r_{k,i}^{-\alpha}} \right] \right\}$$

$$= e^{-\frac{\mu\theta_{k,i} r^{\alpha}\sigma^2}{P_k}} \prod_{m=1}^{K} \mathbb{E}[e^{-\frac{\mu\theta_{k,i} r^{\alpha}}{P_k} I_{k,i,m}}]$$

$$= e^{-\frac{\mu\theta_{k,i} r^{\alpha}\sigma^2}{P_k}} \prod_{m=1}^{K} \mathcal{L}_{I_{k,i,m}}\left(\frac{\mu\theta_{k,i} r^{\alpha}}{P_k}\right) \tag{B.3}$$

where

$$\theta_{k,i} = \frac{\varepsilon}{a_{k,i} - \varepsilon \sum_{n=i+1}^{N_k} a_{k,n}}$$

Here, the Laplace transform of the random variable $I_{k,i,m}$ is denoted by $\mathcal{L}_{I_{k,i,m}}(s)$, which can be derived as

$$\begin{aligned}
\mathcal{L}_{I_{k,i,m}}(s) &= \mathbb{E}_{I_{k,i,m}}[e^{-sI_{k,i,m}}] \\
&= \mathbb{E}_{I_{k,i,m}}\left[\exp\left\{s \sum_{j\in(\Phi_m\setminus\mathcal{O}_{k,i})} h_{k,i,m,j}^2 P_m r_{k,i,m,j}^{-\alpha}\right\}\right] \\
&= \mathbb{E}\left[\prod_{j\in(\Phi_m\setminus\mathcal{O}_{k,i})} \exp\left\{-sh_{k,i,m,j}^2 P_m r_{k,i,m,j}^{-\alpha}\right\}\right] \\
&= \exp\left(-2\pi\lambda_m \int_{z_{k,i,m}}^{\infty} (1-\mathbb{E}[-sh^2 v^{-\alpha}])v dv\right) \\
&= \exp\left(-2\pi\lambda_m \int_{z_{k,i,m}}^{\infty} \left(1-\frac{\mu}{\mu+sP_m v^{-\alpha}}\right) v dv\right)
\end{aligned} \tag{B.4}$$

where

$$z_{k,i,m} = r(P_m/P_k)^{\alpha^{-1}}.$$

References

[1] Z. Ding, M. Peng, and H. V. Poor, Cooperative non-orthogonal multiple access in 5G systems, *IEEE Commun. Lett.*, 19(8), 1462–1465, Aug 2015.

[2] 3rd Generation Partnership Project (3GPP), Study on downlink multiuser superposition transmission for LTE, TSG RAN meeting 67, *Tech. Rep.* RP-150496 Mar 2015.

[3] J. Men and J. Ge, Non-orthogonal multiple access for multiple-antenna relaying networks, *IEEE Commun. Lett.*, 19(10), 1686–1689, Oct 2015.

[4] Z. Ding, H. Dai, and H. V. Poor, Relay selection for cooperative NOMA, *IEEE Wireless Commun. Lett.*, 5(4), 416–419, Aug 2016.

[5] Z. Yang, Z. Ding, Y. Wu, and P. Fan, Novel relay selection strategies for cooperative NOMA, *IEEE Trans. Veh. Technol.*, 66(11), 10 114–10 123, Nov 2017.

[6] S. Lee, D. B. da Costa, and T. Q. Duong, Outage probability of nonorthogonal multiple access schemes with partial relay selection, in Proc. IEEE 26th PIMRC, Valencia, Spain, 1–6, 4–8 Sep 2016.

[7] F. Zhou, Z. Chu, H. Sun, R. Q. Hu, and L. Hanzo, Artificial noise aided secure cognitive beamforming for cooperative MISO-NOMA using SWIPT, *IEEE J. Sel. Areas Commun.*, 36(4), 918–931, Apr 2018.

[8] J. B. Kim and I. H. Lee, Non-orthogonal multiple access in coordinated direct and relay transmission, *IEEE Trans. Veh. Technol.*, 19(11), 2037–2040, Nov 2015.

[9] C. Zhong and Z. Zhang, Non-orthogonal multiple access with cooperative full-duplex relaying, *IEEE Commun. Lett.*, 20(12), 2478–2481, Dec 2016.

[10] H. Liu, Z. Bai, H. Lei, G. Pan, K. J. Kim and T. A. Tsiftsis, A new rate splitting strategy for uplink CR-NOMA Systems, *IEEE Trans. Veh. Tech.*, 71 (7), 7947–7951, Jul 2022.

[11] J. B. Kim, I. H. Lee, and J. Lee, Capacity scaling for D2D aided cooperative relaying systems using NOMA, *IEEE Commun. Lett.*, 7(1), 42–45, 2018.

[12] M. Al-Imari, P. Xiao, M. A. Imran, and R. Tafazolli, Uplink non-orthogonal multiple access for 5G wireless networks, in *Proc. 2014 11th International Symposium on Wireless Communications Systems (ISWCS)*, Barcelona, Spain, 781–785, 26–29 Aug 2014.

[13] Z. Wei, L. Dai, D. W. K. Ng, and J. Yuan, Performance analysis of a hybrid downlink-uplink cooperative NOMA scheme, in *Proc. 2017 IEEE 85th Vehicular Technology Conference (VTC Spring)*, Sydney, NSW, Australia, 1–7, 4–7 Jun 2017.

[14] M. F. Kader and S. Y. Shin, Coordinated direct and relay transmission using uplink NOMA, *IEEE Wireless Commun. Lett.*, 7(3), 400–403, Jun. 2018.

[15] W. Shin, H. Yang, M. Vaezi, J. Lee, and H. V. Poor, Relay-aided NOMA in uplink cellular networks, *IEEE Sig. Process. Lett.*, 24(12), 1842–1846, Dec 2017.

[16] M. E. Soussi, A. Zaidi, and L. Vandendorpe, Compute-and-forward on a multiaccess relay channel: Coding and symmetric-rate optimization, *IEEE Trans. Wireless Commun.*, 13(4), 1932–1947, Apr 2014.

[17] S. Wei, J. Li, W. Chen, L. Zheng, and H. Su, Design of generalized analog network coding for a multiple-access relay channel, *IEEE Trans. Commun.*, 63(1), 170–185, Jan 2015.

[18] H. Liu, Z. Ding, K. J. Kim, K. S. Kwak, H. V. Poor, Decode-and-forward relaying for cooperative non-orthogonal multiple access with direct links, *IEEE Transactions on Wireless Communications*, 17(12), 8077–8093, Dec. 2018.

[19] I. S. Gradshteyn and I. M. Ryzhik, *Table of Integrals, Series, and Products*. New York: Academic Press, 2007.

[20] Y. Saito, Y. Kishiyama, A. Benjebbour, T. Nakamura, A. Li, and K. Higuchi, Non-orthogonal multiple access (NOMA) for cellular future radio access, in *Proc. 77th IEEE VTC-Spring*, Dresden, Germany, pp. 1–5, Jun 2013.

[21] S. Park and D.-H. Cho, Random linear network coding based on non-orthogonal multiple access in wireless networks, *IEEE Commun. Lett.*, 19(7), 1273–1276, Jul 2015.

[22] G. Zhang and M. Haenggi, The performance of successive interference cancellation in random wireless networks, *IEEE Trans. Inf. Theory*, 60(10), 6368–6388, Oct 2014.

[23] Z. Ding *et al.*, Impact of user pairing on 5G nonorthogonal multiple access downlink transmissions, *IEEE Trans. Veh. Technol.*, 65(8), 6010–6023, Aug 2016.

[24] Y. Hayashi, Y. Kishiyama, and K. Higuchi, Investigations on power allocation among beams in non-orthogonal access with random beamforming and intra-beam SIC for cellular MIMO downlink, in *Proc. IEEE Veh. Technol. Conf. (VTC Fall)*, 1–5, Sep 2013.

[25] J. Choi, Power allocation for max-sum rate and max-min rate proportional fairness in NOMA, *IEEE Commun. Lett.*, 20(10), 2055–2058, Oct 2016.

[26] Z. Ding, M. Peng, and H. V. Poor, Cooperative non-orthogonal multiple access in 5G systems, *IEEE Commun. Lett.*, 19(8), 1462–1465, Aug 2015.

[27] X. Xiong, W. Xiang, K. Zheng, H. Shen, and X. Wei, An open source SDR-based NOMA system for 5G networks, *IEEE Wireless Commun.*, 22(6), 24–32, Dec 2015.

[28] S. M. R. Islam, N. Avazov, O. A. Dobre, and K.-S. Kwak, Power-domain non-orthogonal multiple access (NOMA) in 5G systems: Potentials and challenges, *IEEE Commun. Surveys Tuts.*, 19(2), 721–742, 2nd Quart., 2017.

[29] Y. Liu, Z. Qin, M. Elkashlan, A. Nallanathan, and J. A. McCann, Non-orthogonal multiple access in large-scale heterogeneous networks, *IEEE J. Sel. Areas Commun.*, 35(12), 2667–2680, Dec 2017.

[30] T. X. Tran and K. C. Teh, Performance analysis of two-tier HetNets with massive MIMO and nonuniformly small cell deployment, *IEEE Trans. Veh. Technol.*, 66(11), 10044–10054, Nov 2017.

[31] H. Wang, X. Zhou, and M. C. Reed, Coverage and throughput analysis with a non-uniform small cell deployment, *IEEE Trans. Wireless Commun.*, 13(4), 2047–2059, Apr 2014.

[32] F. Muhammad, Z. H. Abbas, and F. Y. Li, Cell association with load balancing in nonuniform heterogeneous cellular networks: Coverage probability and rate analysis, *IEEE Trans. Veh. Technol.*, 66(6), 5241–5255, Jun 2017.

[33] T. Han, J. Gong, X. Liu, S. M. R. Islam, Q. Li, Z. Bai and K. S. Kwak, On Downlink NOMA in Heterogeneous Networks with Non-Uniform Small Cell Deployment, *IEEE Access*, Vol. 6, 31099–31109, June 2018.

[34] H. S. Dhillon, R. K. Ganti, F. Baccelli, and J. G. Andrews, Modeling and analysis of K-tier downlink heterogeneous cellular networks, *IEEE J. Sel. Areas Commun.*, 30(3), 550–560, Apr 2012.

[35] Y. Liu, X. Li, F. R. Yu, H. Ji, H. Zhang, and V. C. M. Leung, Grouping and cooperating among access points in user-centric ultra-dense networks with non-orthogonal multiple access, *IEEE J. Sel. Areas Commun.*, 35(10), 2295–2311, Oct 2017.

Chapter 8

Rate-Splitting Multiple Access Systems

8.1 Introduction

The emergence of the sixth-generation (6G) mobile communication system has garnered considerable attention within academia and industry, promising to redefine connectivity standards. Positioned as the linchpin of the Internet of Everything, 6G seeks to usher in an era of unparalleled throughput, ultra-reliable connectivity, heterogeneous quality of service (QoS), extensive connectivity, and the convergence of diverse technologies encompassing communications, sensing, localization, computing, and control. Consequently, the development of 6G necessitates a more effective utilization of wireless resources and the deployment of advanced interference management techniques, prompting a reassessment and refinement of the physical (PHY) layer and multiple access (MA) strategies for wireless communication systems [1].

Throughout the past few decades, the evolution of cellular networks, spanning from the first-generation (1G) to the fifth-generation (5G), has coincided with the emergence of novel MA techniques. Across this progression, the overarching objective of MA has revolved around the allocation of orthogonal radio resources to users, aiming to mitigate multi-user interference. Orthogonal multiple access (OMA) methodologies have been widely embraced, where users are scheduled within orthogonal domains, such as the time domain (e.g., time-division TDMA), frequency domain (e.g., FDMA), code domain (e.g., CDMA), or through a fusion of time and frequency domains (e.g., orthogonal frequency-division multiple access (OFDMA)) [2].

In 4G and 5G networks, multiple-input multiple-output (MIMO) communication and space-division multiple access (SDMA) have addressed increasing capacity demands. MIMO networks introduce a spatial dimension, enabling the application of SDMA [3–5]. By leveraging spatial resources and multi-antenna processing at the transmitter, SDMA can serve multiple users in the same time–frequency resource, allowing each user to directly decode the intended data streams while treating residual interference as noise.

To further enhance spectral efficiency (SE), MA has advanced toward NOMA. In NOMA, users are superposed in the same time–frequency resources via the power domain (e.g., power-domain NOMA) or code domain (e.g., sparse code multiple access). Power-domain NOMA relies on superposition coding (SC) at the transmitter to combine user messages in the power domain and successive interference cancellation (SIC) at the receivers (also known as SC-SIC) [6, 7]. This approach manages multi-user interference by ensuring that at least one user successfully decodes messages and removes interference from other users.

Rate-splitting multiple access (RSMA) is a new way to manage interference and handle multiple users in 6G networks. It works by splitting user messages into two parts: common and private. This helps deal with interference better compared to other methods like SDMA and NOMA. RSMA is flexible and can adjust to different levels of interference. Whether the interference is weak or strong, RSMA can adapt and work like SDMA or NOMA. It acts as a bridge between these approaches, making the transition smooth without sudden changes [8].

8.2 Basics of OMA/SDMA/NOMA

Interference poses a significant challenge in modern wireless communication networks, owing to the broadcast nature of wireless signals. Existing MA schemes have tackled interference differently: some allocate separate radio resources to each user (OMA), others treat interference as noise (SDMA), and some decode interference (NOMA). Looking ahead, the next generation of MA schemes aims to handle interference more effectively by adjusting their approach based on the level of interference: fully treating it as noise when interference is low or high, and partially decoding it when interference is moderate.

RSMA emerges as a promising solution to interference management and accommodating multiple users in future wireless networks. Built on the concept of rate-splitting (RS), RSMA divides each user's message into two parts, allowing for partial decoding of interference and partial treatment of interference as noise. This approach offers a flexible and efficient means of interference management, catering to the evolving demands of future networks. RSMA, also known as RS in the literature, forms the foundational block of the RSMA framework for multi-antenna broadcast channels (BCs), with each user's message divided into at most two parts.

To clarify the distinctions among the MA schemes mentioned, let's consider a scenario with two users sharing a single radio resource, as depicted in Fig. 8.1. In OMA, the radio resource is exclusively allocated to one user (user 1), while the other user (user 2) must wait for its turn to access the resource. In contrast, in SDMA, both users' messages are individually encoded and linearly precoded at the transmitter. Each user then directly decodes its intended message, treating any remaining interference from the other user's message as noise. NOMA involves encoding each user's message independently into streams, which are then combined at the transmitter and transmitted to both users simultaneously. In NOMA, one user's message (user 2) must be decoded by both users to retrieve the intended information.

In RSMA, each user message m_k ($k = 1, 2$) by the transmitter is divided into a common part $m_{c,k}$ and a private part $m_{p,k}$. Now $m_{c,1}$ and $m_{c,2}$ are combined to form a common message m_c, and similarly, $m_{p,1}$ and $m_{p,2}$ are combined to form a private message m_p. The resulting three messages $m_c, m_{p,1}$, and $m_{p,2}$ derived from m_1 and m_2 are independently encoded and linearly precoded at the transmitter. Each user initially decodes the common stream s_c, treating all private streams as noise. After removing the decoded common stream from the received signal, each user proceeds to decode its intended private stream s_k ($k = 1, 2$), treating the other user's private stream as noise. This approach allows for flexible interference management by partially decoding interference through the common stream and treating interference as noise when decoding the intended private stream at each user [8].

RSMA simplifies SDMA (depicted in Fig. 8.1(b)); the common stream s_c is deactivated, implying that m_k is directly encoded into s_k, where all interference between different users' private streams is treated as noise. Similarly, RSMA simplifies NOMA (Fig. 8.1(c)) when s_2 is deactivated, m_2 encodes s_c, and m_1 encodes s_1. RSMA therefore unifies the existing MA schemes and represents a promising PHY-layer transmission paradigm for non-orthogonal transmission, interference management, and multi-user communications [9, 30].

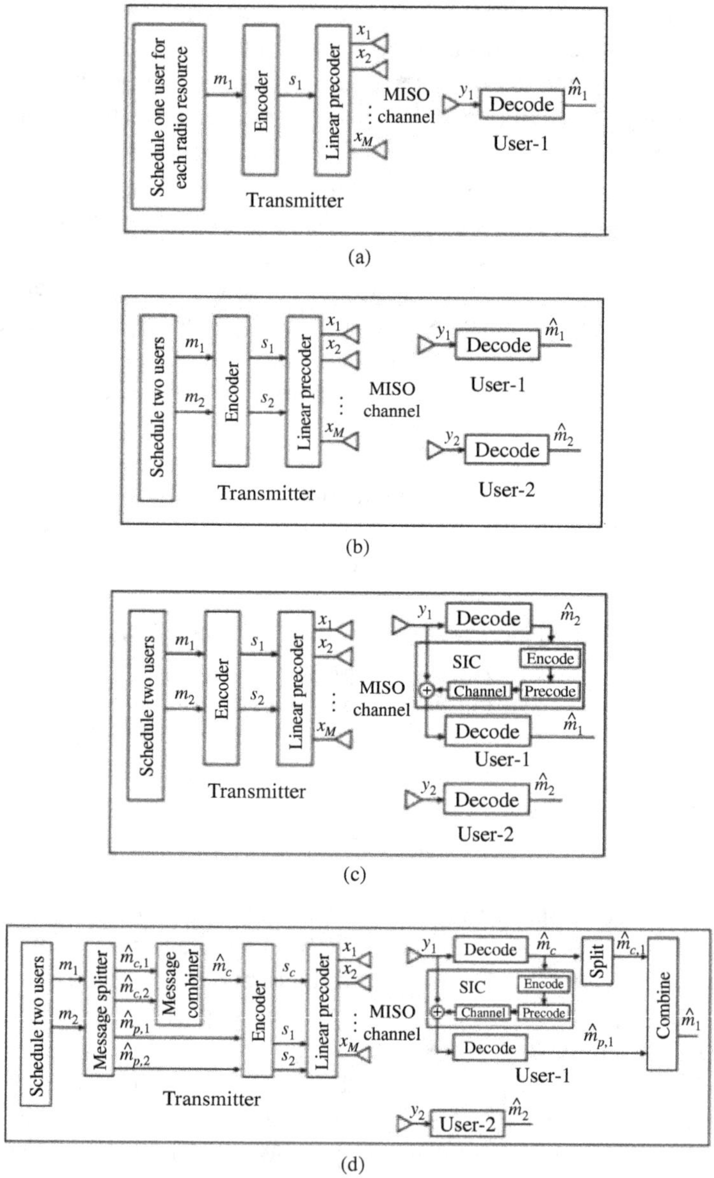

Fig. 8.1. Two-user OMA, SDMA, NOMA, and RSMA for MISO BC: (a) OMA, (b) SDMA, (c) NOMA (SC-SIC) and (d) RSMA (one-layer RS).

Next, the discussion will address the primary PHY-layer issues associated with MA schemes and emphasize the importance of adopting this new paradigm in the design of present and future wireless networks.

8.2.1 *Orthogonal multiple access*

OMA encompasses several techniques used in earlier generations of cellular networks to allocate radio resources to users. In 1G, FDMA divided the spectrum into non-overlapping frequency bands for individual users. Second-generation (2G) systems utilized TDMA, allocating time slots to users. Third-generation (3G) networks introduced CDMA, where user-specific codes allowed simultaneous transmission in the same time–frequency resources. In 4G, OFDMA partitioned frequency and time resources into subcarriers and slots, offering efficient resource allocation. OMA's simplicity and interference avoidance are notable advantages. However, dedicating orthogonal resources to individual users limits SE and may lead to inefficient spectrum use, especially when low-rate users occupy the entire resource blocks. Moreover, effective user scheduling is necessary to maintain system performance, adding to signaling overhead [9].

8.2.2 *Space-division multiple access*

SDMA plays a crucial role in modern wireless networks, especially with the widespread adoption of MIMO technology across various wireless standards like 5G NR and LTE. SDMA utilizes linear precoding techniques at the transmitter to handle multi-user interference, treating it as noise at the receivers. This approach is particularly beneficial in scenarios such as multi-user MIMO, networked MIMO, and massive MIMO, contributing to improved SE and reliability [10].

SDMA offers several advantages, including its ability to suppress multi-user interference when the channel state information (CSI) at the transmitter is perfect, especially in underloaded systems. It achieves the maximum degrees of freedom (DoF) in underloaded multi-antenna BC with low transmitter and receiver complexities. However, SDMA has limitations, notably its sensitivity to network load, user deployment, and CSI at the transmitter (CSIT) accuracy. In overloaded networks, SDMA's performance suffers, necessitating complex user grouping and scheduling strategies to mitigate interference effectively. Moreover, SDMA requires users to have nearly orthogonal channels and similar channel strengths, posing challenges for scheduler design and increasing complexity in maximizing SE under imperfect CSIT conditions.

8.2.3 *Non-orthogonal multiple access*

NOMA, a concept rooted in information theory and wireless communications, has deep roots dating back several decades, notably in the study of the SISO BC and multiple access channel (MAC) [11]. In these contexts, SC-SIC stands out as the capacity-achieving strategy for the SISO BC, achieving capacity with time-sharing for the SISO MAC. Recognizing the performance benefits of NOMA compared to OMA in single-antenna networks [12], recent years have seen significant efforts to integrate NOMA into multi-antenna networks. Within the multi-antenna BC, MIMO/MISO NOMA combines SDMA and NOMA [13]. This approach involves segregating K users into $G(1 \leq G \leq K)$ distinct groups. Within each group, SC-SIC is employed to manage interference from other users, while interference from users in different groups is addressed through SDMA. In cases where $G = 1$, akin to single-antenna NOMA (SC-SIC), users are ordered based on their effective channel strengths, facilitating successful message decoding. Conversely, when $G > 1$, linear precoding is utilized to decompose the multi-antenna BC into G single-antenna NOMA channels, effectively mitigating interference between these channels given adequate transmit antennas and perfect CSIT knowledge.

In the multi-antenna MAC, NOMA employing minimum mean square error (MMSE)–SIC receivers achieves the vertices at the boundary of the capacity region, yet time-sharing is necessary to reach all points along the line segment between the adjacent vertices: NOMA shows considerable potential for enhancing SE in scenarios with high levels of overload by efficiently serving users with closely aligned channels and diverse channel strengths within the same time–frequency resources [14]. However, increasing the number of users leads to heightened receiver complexity (i.e., more SIC layers) and exacerbates SIC error propagation. In a K-user SISO BC, the user with the strongest channel requires $K - 1$ layers of SIC to decode and eliminate interference from the $K - 1$ messages of all other co-scheduled users before accessing its intended message. To address this issue, one approach is to limit the number of SIC layers per user by grouping users into smaller clusters, applying SC-SIC within each group, and scheduling user groups via OMA. However, this method may compromise performance and increase latency. While compelling a user to decode another user's entire message is effective in the degraded SISO BC, it is inefficient in most scenarios in multi-antenna networks due to the suboptimal utilization of multiple antennas and SIC receivers [15].

Comparing various MA schemes, OMA allocates one user per radio resource, while SDMA and NOMA can handle multiple users concurrently. SDMA allows overlapping of users within the same resource, but the crucial aspect is how they handle multi-user interference, distinguishing OMA, SDMA, and NOMA. OMA assigns orthogonal resources to users to mitigate interference, whereas SDMA treats residual multi-user interference as noise through multi-user linear precoding (MU-LP), and NOMA utilizes SIC to decode and eliminate interference.

The challenges of current MA schemes revolve around interference management strategies. When multi-user interference is weak, treating it solely as noise is effective, whereas fully decoding interference is preferable under strong interference. However, for moderate interference levels, partially decoding interference can significantly improve performance. This indicates the need for future MA schemes to dynamically adjust how interference is handled based on its level. Unfortunately, SDMA and NOMA are optimized for weak and strong interference, respectively. To address this, one approach is to dynamically switch between SDMA and NOMA for adaptive interference management. However, this method may not be ideal for moderate interference levels, as it solely supports one of the two approaches, lacking flexibility.

8.3 Fundamentals of RSMA

In this section, the fundamentals of interference management are presented. To illustrate these concepts, a symmetric two-user interference channel (IC) is considered, representing the simplest scenario for capturing the essence of interference management. This straightforward example is leveraged to derive key insights applicable to MA design [9, 29].

8.3.1 *Two-user interference channel*

Consider a two-user symmetric IC depicted in Fig. 8.2 with four nodes, including two transmitters (Tx) and two receivers (Rx) each equipped with

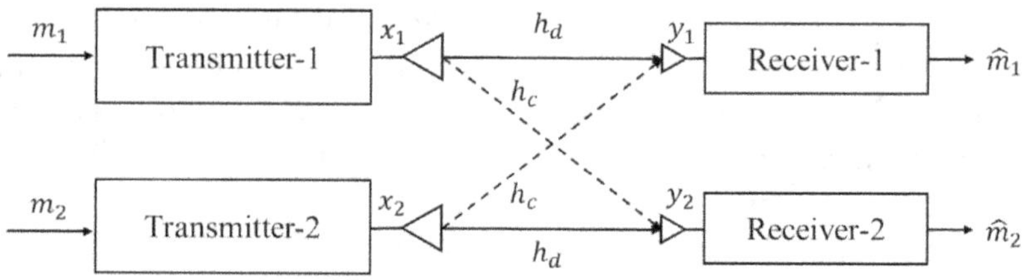

Fig. 8.2. A two-user symmetric Gaussian interference channel.

a single antenna; Tx-1 (transmitter 1) aims to transmit message m_1 to Rx-1 (receiver 1), while Tx-2 (transmitter 2) aims to transmit message m_2 to Rx-2 (receiver 2); for simplicity, assume symmetry where the direct channel from a transmitter to its intended receiver (Tx-1 $\rightarrow$ Rx-1, Tx-2 $\rightarrow$ Rx-2) is denoted by h_d and the cross channel (Tx-1 $\rightarrow$ Rx-2, Tx-2 $\rightarrow$ Rx-1) is denoted by h_c, such that the stronger the h_d compared to h_c, the stronger the interference from Tx-k to Rx-j (where $j \neq k$).

Various strategies can be considered for managing interference in this scenario: orthogonalizing transmissions between Tx-1 and Tx-2 to ensure that they do not interfere with each other, which is suboptimal as it does not account for the strength of the interference; treating interference as additive white Gaussian noise (AWGN), which is effective when the interference is weak compared to the signal but overlooks the potential information carried by the interference; and decoding interference fully at the receivers, which is particularly effective and optimal when the interference is strong enough relative to the signal strength, resulting in better reception of unintended signals than the intended ones [17].

However, more advanced interference management strategies can be achieved by adopting the RS approach [9, 18, 19]. RS involves dividing the transmitted information into two parts: a common part decoded by both receivers and a private part decoded only by the intended receiver. The key benefit of RS is that it allows a receiver to decode the common part, effectively canceling off a portion of the interference, while treating the remaining private part from the other transmitter as noise. By properly adjusting the power allocated to the common and private parts, RS can partially decode interference and partially treat the remaining interference as noise. This approach bridges and unifies the two extremes of treating interference as noise and decoding interference, ultimately outperforming both strategies.

8.3.2 *Rate-splitting for two-user systems*

A RS strategy involves splitting each transmitter message m_k into two parts: a common part $m_{c,k}$ and a private part $m_{p,k}$, as depicted in Fig. 8.1(d) and Fig. 8.3 [18, 19]. A shared codebook between both transmitters is utilized to encode the common parts $m_{c,k}(k = 1, 2)$ into the common stream $s_{c,k}(k = 1, 2)$, allowing them to be decodable by both receivers. Meanwhile, the private parts $m_{p,k}(k = 1, 2)$ are encoded into private streams s_k, i.e., $s_{p,k}(k = 1, 2)$, constructed from independent codebooks. Each transmitter Tx-k then transmits the superposition of its common stream $s_{c,k}$ and private stream $s_{p,k}$ with proper power allocation to both streams as follows:

$$x_k = \sqrt{P_c} s_{c,k} + \sqrt{P_k} s_k \tag{8.1}$$

By defining $s_k = [s_{c,k}, s_{p,k}]^T$ and assuming that $E[s_k {s_k}^H] = \mathrm{I}$, the average transmit (sum) power constraint at each transmitter is P, with P_c and P_k ($P = P_k + P_c$) being the power allocated to the common and private streams, respectively. Let $\gamma = P_c/P$ denote the fraction of the transmit power allocated to the common stream.

The received signals at both receivers can be written as

$$y_1 = h_d x_1 + h_c x_2 + n_1 \tag{8.2}$$

$$y_2 = h_c x_1 + h_d x_2 + n_2 \tag{8.3}$$

where $n_k \sim \mathcal{CN}(0, {\sigma_{n,k}}^2)$ is the AWGN. Without loss of generality, the variance is ${\sigma_{n,k}}^2 = 1$. Also, perfect CSI at the transmitter and receiver is assumed at the receivers.

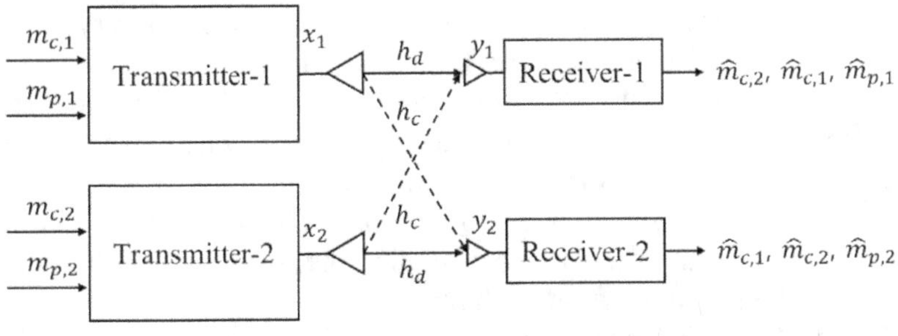

Fig. 8.3. RS for two-user SISO interference channel.

Each receiver (Rx) decodes the common streams and its intended private stream, as depicted in Fig. 8.3. To achieve this, each Rx jointly decodes the common streams s_c into messages $\hat{m}_{c,1}$ and $\hat{m}_{c,2}$ by treating the private streams as noise. Subsequently, Rx-k cancels the common streams from the received signal and decodes the intended private stream $s_{p,k}$ into message $\hat{m}_{p,k}$ by treating the other private stream $s_{p,j}$, $j \neq k$, as noise. Finally, Rx-k reconstructs the original message by combining $\hat{m}_{p,k}$ and $\hat{m}_{c,k}$ (the message intended for the other receiver) into $\hat{m}_k$, which matches the original message m_k in the absence of any decoding error. Through this process, Rx-k partially decodes the message intended for Rx-j (namely $\hat{m}_{c,j}$), effectively canceling off part of the interference.

Due to symmetry, the rates of the common streams $s_{c,1}$ and $s_{c,2}$ are equal and denoted simply as R_c. Similarly, the rates of the private streams $s_{p,1}$ and $s_{p,2}$ are also equal and denoted as R_p. Assuming Gaussian signaling and infinite block length, R_c must satisfy the following inequalities at Rx-1 (and equivalently at Rx-2 due to symmetry):

$$R_c \leq \log_2\left(1 + \frac{\gamma P|h_d|^2}{1+I}\right) \tag{8.4}$$

$$R_c \leq \log_2\left(1 + \frac{\gamma P|h_c|^2}{1+I}\right) \tag{8.5}$$

$$2R_c \leq \log_2\left(1 + \frac{\gamma P[|h_d|^2 + |h_c|^2]}{1+I}\right) \tag{8.6}$$

where $I = (1-\gamma)P|h_d|^2 + |h_c|^2$. The inequalities (8.4), (8.5), and (8.6) stem from the joint decoding of the common streams at each receiver, where the private streams are treated as noise. These inequalities are derived by imposing rate constraints on a two-user MAC formed by two virtual transmitters sending $s_{c,1}$ and $s_{c,2}$ to Rx-1 (or 2 due to symmetry), subject to the additional noise power I created by the private streams $s_{p,1}$ and $s_{p,2}$.

Specifically, inequality (8.4) represents the individual rate constraint for decoding $s_{c,1}$ at Rx-1 while treating $s_{p,1}$ and $s_{p,2}$ as noise. Similarly, inequality (8.5) represents the individual rate constraint for decoding $s_{c,2}$ at Rx-2 while treating $s_{p,1}$ and $s_{p,2}$ as noise. The sum-rate constraint is expressed by inequality (8.6), encapsulating the overall rate constraint for jointly decoding s_c and $s_{p,j}$ at each receiver.

The rate R_p of the private streams in (8.7) is determined by decoding a private stream after canceling the common streams and treating the other interfering private stream as noise. This rate can be written as

$$R_p \leq \log_2\left(1+\frac{(1-\gamma)P|h_d|^2}{1+(1-\gamma)P|h_c|^2}\right) \tag{8.7}$$

where $(1-\gamma)P|h_c|^2$ expresses the interference from the private stream of Tx-k owed to Rx-j private stream.

Combining inequalities (8.4)–(8.6) and (8.7), the achievable rate in this two-user IC is expressed as the sum of the private and common rates, denoted as $R_{sym} = R_p + R_c$ in equation (8.8). The concept of RS becomes evident here, where R_{sym} is split into two parts: the private rate (R_p) and the common rate (R_c).

The benefit of this RS architecture lies in the ability to adjust the power allocation to the common and private streams through the parameter γ to maximize R_{sym} depending on the channel strengths of h_d and h_c.

$$\begin{aligned} R_{sym} &= \log_2\left(1+\frac{(1-\gamma)P|h_d|^2}{1+(1-\gamma)P|h_c|^2}\right) \\ &+\min\left\{\begin{array}{l} \log_2\left(1+\dfrac{\gamma P|h_d|^2}{1+I}\right), \quad \log_2\left(1+\dfrac{\gamma P|h_c|^2}{1+I}\right), \\ \dfrac{1}{2}\log_2\left(1+\dfrac{\gamma P[|h_d|^2+|h_c|^2]}{1+I}\right) \end{array}\right\} \end{aligned} \tag{8.8}$$

8.4 Transceiver Model of RSMA System

Expanding on the RS scheme for a two-user IC, insights into designing RSMA for two-user scenarios in both downlink (BC) and uplink (MAC) settings will be discussed in the following sections. Furthermore, connections between the discussed interference management techniques and conventional multiple access designs such as OMA, SDMA, and NOMA will be highlighted, demonstrating a direct relationship or analogy [9, 29].

8.4.1 *Two-user downlink RSMA*

Consider a scenario with a single transmitter potentially equipped with M antennas serving two users. In this setup, suppose the two transmit antennas depicted in Fig. 8.2 cooperate by exchanging CSI and messages, effectively operating as part of the same transmitter, as illustrated in Fig. 8.4.

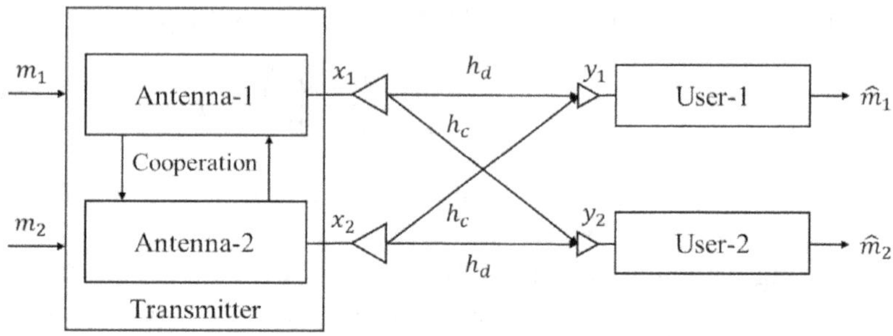

Fig. 8.4. Two-user MISO downlink (Gaussian broadcast channel).

Consequently, the two-user interference channel transforms into a two-user multiple-input single-output (MISO) BC with $M = 2$ transmit antennas, where the base station transmits to two active receiving users with channel responses represented by $\boldsymbol{h}_1$ and $\boldsymbol{h}_2$, respectively.

In the context of Fig. 8.4, where $\boldsymbol{h}_1 = [h_d,\ h_c]$ and $\boldsymbol{h}_2 = [h_c,\ h_d]$, considering a transmit signal vector $\boldsymbol{x} = [x_1,\ x_2]$ spanning across the transmit antennas, the received signal at user k can be expressed as

$$y_k = \boldsymbol{h}_k \boldsymbol{x} + n_k \tag{8.9}$$

where $n_k \sim \mathrm{CN}(0, {\sigma_{n,k}}^2)$ is assumed.

A multi-antenna BC serves as a fundamental component of modern downlink communication systems. Assuming perfect CSI at both the transmitter and receiver the capacity of this channel is well established and can be achieved using dirty paper coding (DPC) [21, 22]. However, DPC is known for its complexity due to its nonlinear encoding/precoding mechanism. RSMA has emerged in recent years as an appealing strategy to achieve performance close to DPC while maintaining the lower complexity of linear precoding [8]. In scenarios with imperfect CSIT, the capacity and the optimal strategy for achieving the capacity remain unknown. Nevertheless, it is recognized that RS plays a central role in achieving optimal DoF [22, 23], and RSMA has been shown to outperform DPC [24].

In the subsequent sections, the key design principles and schemes of RSMA in the two-user downlink scenario will be progressively delineated.

8.4.1.1 *MISO RSMA model*

An initial architecture [9] is developed by splitting the two messages m_1 and m_2 into common and private components and encoding each component into respective streams, leading to $m_{c,k} \rightarrow s_{c,k}$ and $m_{p,k} \rightarrow s_{p,k}$, $k = 1, 2$. With multiple antennas available, precoding or beamforming can be implemented across the two antennas, and the transmit signal model for each message can be represented as

$$x = \sum_{k=1}^{2} \boldsymbol{p}_{c,k} s_{c,k} + \boldsymbol{p}_k s_k \tag{8.10}$$

where $\boldsymbol{p}_{c,k}$ is the precoder of the common stream k and $\boldsymbol{p}_k$ is the precoder of the private stream k.

Defining $\boldsymbol{s} = [s_{c,1}, s_1, s_{c,2}, s_2]^T$ and assuming that $E[\boldsymbol{s}\,\boldsymbol{s}^H] = I$, where I is the identity matrix, the average transmit power constraint at the transmitter is $\sum_{k=1}^{2} P_{c,k} + P_k \leq P$, with $P_{c,k} = \|\boldsymbol{p}_{c,k}\|^2$ and $P_k = \|\boldsymbol{p}_k\|^2$ representing the power allocated to the common stream k and the private stream k, respectively.

In this downlink setting where both $s_{c,1}$ and $s_{c,2}$ are transmitted from the same transmitter and decoded by both users, one can choose $p_{c,1} = p_{c,2} = p_c$ to use a single common precoder, and $x = p_c(s_{c,1} + s_{c,2}) + \sum_{k=1}^{2} p_k s_k$. User k can decode using SIC or joint decoding of the common streams $s_{c,1}$ and $s_{c,2}$, followed by the private stream s_k. In the SIC approach, each receiver k would need to be equipped with two SIC layers to decode the three streams $s_{c,j}$, $s_{c,k}$, and s_k, $j \neq k$. In summary, this architecture generates four streams from two messages, requiring two SIC layers to recover the original messages. This RSMA architecture is illustrated in Fig. 8.5.

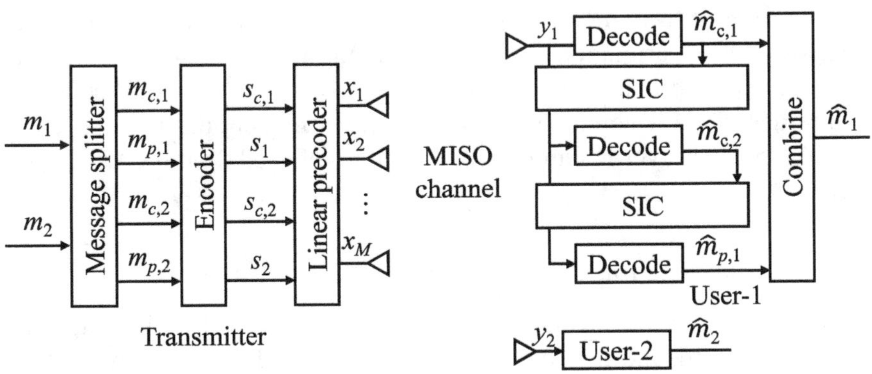

Fig. 8.5. Two-user downlink MISO RSMA without message combiner.

In a second MISO RSMA model for the downlink, an additional advantage is recognized where instead of encoding each common part into separate common streams $m_{c,k} \rightarrow s_{c,k}$, the common parts are combined into a single common message $m_c = \{m_{c,1}, m_{c,2}\}$, which is then encoded into a unified common stream $m_c \rightarrow s_c$ such that

$$x = p_c s_c + \sum_{k=1}^{2} p_k s_k \tag{8.11}$$

By defining $s = [s_c, s_1, s_2]^T$ and assuming $E[s\, s^H] = I$, where I is the identity matrix, the average transmit power constraint at the transmitter is $P_c + \sum_{k=1}^{2} P_k \leq P$, with $P_c = \|p_{c,k}\|^2$ being the power allocated to the unique common stream.

The received signal at user k, $k = 1, 2$, $j \neq k$, is expressed as

$$y_k = h_k p_c s_c + h_k p_k s_k + h_k p_j s_j + n_k \tag{8.12}$$

This architecture is known as the one-layer RS architecture of RSMA because it utilizes a single common stream, simplifying both the encoding and decoding complexities compared to previous architectures. In this setup, each receiver employs a single SIC layer instead of two, reducing the overall complexity. Both users decode the single common stream s_c into $\hat{m}_c$ by treating interference from all private streams as noise. Subsequently, each user retrieves the estimate $\hat{m}_{c,k}$ from $\hat{m}_c$. Using SIC, $\hat{m}_c$ is then re-encoded, precoded, and subtracted from the received signal to decode its intended private stream s_k from $\hat{m}_{p,k}$ by treating the remaining interference from the other private streams as noise. Finally, each user reconstructs their original message m_k by combining $\hat{m}_{c,k}$ and $\hat{m}_{p,k}$ into the message $\hat{m}_k$, assuming that no decoding errors occur. The simplified architecture described is depicted in Fig. 8.6. In subsequent discussions about RSMA, the one-layer RS architecture will be referred to unless otherwise specified.

8.4.1.2 *Achievable rate of two-user MISO RSMA*

Under the assumption of Gaussian signaling and infinite block length, along with perfect CSI, the instantaneous rates for decoding the common and

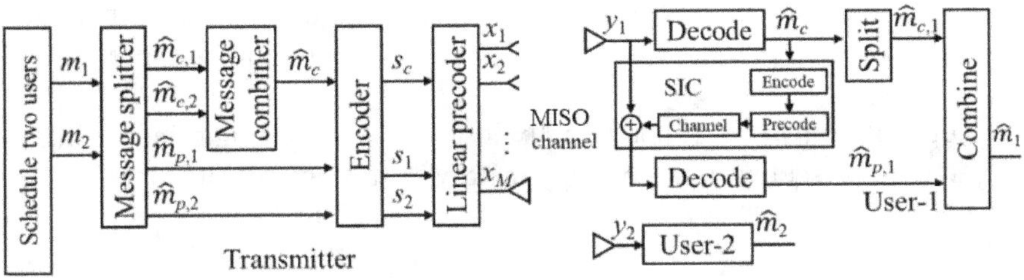

Fig. 8.6. Two-user downlink MISO RSMA with message combiner (one-layer RS).

private streams at user k can be expressed as follows [9]:

$$R_{c,k} = \log_2\left(1 + \frac{|h_k p_c|^2}{|h_k p_1|^2 + |h_k p_2|^2 + 1}\right)$$

$$R_k = \log_2\left(1 + \frac{|h_k p_k|^2}{|h_k p_j|^2 + 1}\right) \tag{8.13}$$

where the noise variance is normalized at ${\sigma_{n,k}}^2 = 1$. To ensure that the common stream s_c is successfully decoded by both users, its rate R_c cannot exceed

$$R_c = \min\{R_{c,1}, R_{c,2}\} \tag{8.14}$$

Given that s_c contains sub-messages $m_{c,1}$ and $m_{c,2}$ from the two users, the rate distribution of R_c among the users is adapted based on the amount of sub-messages contributed by each user. Let C_k denote the portion of rate R_c allocated to user k for $m_{c,k}$. Then, R_c can be expressed as

$$C_1 + C_2 = R_c \tag{8.15}$$

The overall achievable rate $R_{k,\text{tot}}$ for user k can be expressed as

$$R_{k,\text{tot}} = C_k + R_k \tag{8.16}$$

It is noted that the rate of each user can be divided into two components: the rate from the private rate and a portion of the rate from the common rate.

Note that for $M = 1$, specifically in the downlink single-input single-output (SISO) scenario, the one-layer RSMA strategy results in

$$x = \sqrt{P_c} s_c + \sqrt{P_1} s_1 + \sqrt{P_2} s_2 \tag{8.17}$$

The common and private rates can be expressed for $k = 1, 2$ as

$$\begin{aligned} R_{c,k} &= \log_2\left(1 + \frac{|h_k|^2 P_c}{|h_k|^2 P_1 + |h_k|^2 P_2 + 1}\right) \\ R_k &= \log_2\left(1 + \frac{|h_k|^2 P_k}{|h_k|^2 P_j + 1}\right) \end{aligned} \tag{8.18}$$

Without loss of generality, let $|h_2| \leq |h_1|$ such that

$$R_c = \log_2\left(1 + \frac{|h_2|^2 P_c}{|h_2|^2 P_1 + |h_2|^2 P_2 + 1}\right) \tag{8.19}$$

Since the capacity of the Gaussian SISO BC with perfect CSI is achieved through SC with SIC in the SISO NOMA scheme, optimizing P_1, P_2, and P_c to maximize the weighted sum rate would involve setting $P_2 = 0$ so that the message of the weaker user (user 2) m_2 is entirely encoded in s_c, thereby transforming SISO RSMA into SISO NOMA $x = \sqrt{P_c} s_c + \sqrt{P_1} s_1$. The achievable rates in this setup would be

$$R_1 = \log_2(1 + |h_1|^2 P_1) \tag{8.20}$$

$$R_c = \log_2\left(1 + \frac{|h_2|^2 P_c}{|h_2|^2 P_1 + 1}\right) \tag{8.21}$$

This illustrates how RSMA can function similar to NOMA in the Gaussian SISO downlink scenario.

8.4.1.3 *MIMO RSMA model*

Let's consider a scenario involving a two-user MIMO downlink with M transmit antennas and N receive antennas at each user. The goal is to transmit two N-dimensional message vectors $\boldsymbol{m}_1$ and $\boldsymbol{m}_2$ to user 1 and user 2, respectively. To achieve this, each message vector for each user is split into a common part and a private part. The common parts of both users are combined and encoded into an N-dimensional vector of common streams $\boldsymbol{s}_c$, while the private parts are encoded into two N-dimensional vectors of private streams $\boldsymbol{s}_k$. Precoding operations, involving $M \times N$ matrices, are applied to the common and private stream vectors, resulting in the transmit signal expressed as

$$x = \boldsymbol{P}_c\boldsymbol{s}_c + \sum_{k=1,2} \boldsymbol{P}_k\boldsymbol{s}_k \tag{8.22}$$

Consider an example with $N = 2$ to demonstrate the framework's versatility. Four messages (two for each user) are to be transmitted: $\boldsymbol{m}_1 = [m_1^{(1)}, m_1^{(2)}]$ and $\boldsymbol{m}_2 = [m_2{}^{(1)}, m_2{}^{(2)}]$, where $m_i{}^{(j)}$ denotes the jth message of user i ($i = 1, 2$, $j = 1, 2$). These messages are split into common and private parts as follows:

$$m_1^{(1)} = \left\{m_{c,1}^{(1)}, m_{p,1}^{(1)}\right\} \tag{8.23}$$

$$m_1^{(2)} = \left\{m_{c,1}^{(2)}, m_{p,1}^{(2)}\right\} \tag{8.24}$$

$$m_2^{(1)} = \left\{m_{c,2}^{(1)}, m_{p,2}^{(1)}\right\} \tag{8.25}$$

$$m_2^{(2)} = \left\{m_{c,2}^{(2)}, m_{p,2}^{(2)}\right\} \tag{8.26}$$

The common parts are combined into $m_c{}^{(1)}$ and $m_c{}^{(2)}$, which are then encoded into common streams $s_c{}^{(1)}$ and $s_c{}^{(2)}$ to form the common stream vector $\boldsymbol{s}_c = [s_c{}^{(1)}, s_c{}^{(2)}]^T$

$$m_c^{(1)} = \left\{m_{c,1}^{(1)}, m_{c,2}^{(1)}\right\} \rightarrow s_c^{(1)} \tag{8.27}$$

$$m_c^{(2)} = \left\{m_{c,1}^{(2)}, m_{c,2}^{(2)}\right\} \rightarrow s_c^{(2)} \tag{8.28}$$

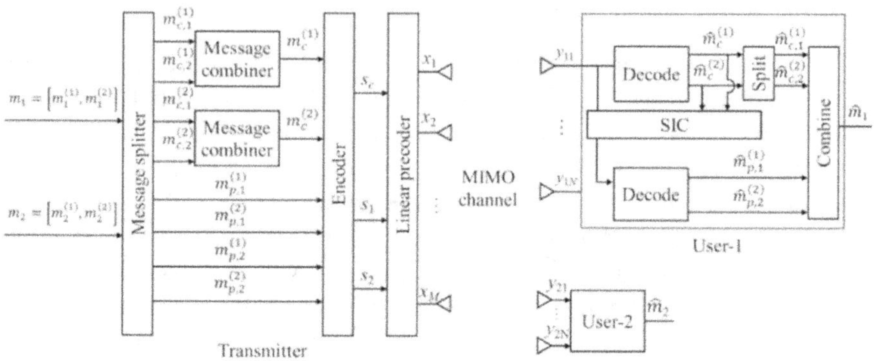

Fig. 8.7. Downlink MIMO RSMA.

The private parts are encoded into private streams to form two private stream vectors:

$$\mathbf{s_1} = [s_1{}^{(1)}, s_1{}^{(2)}]^T \quad \text{and} \quad \mathbf{s_2} = [s_2{}^{(1)}, s_2{}^{(2)}]^T$$

$$m_{p,1}^{(1)} \rightarrow s_1^{(1)}, \quad m_{p,1}^{(2)} \rightarrow s_1^{(2)} \tag{8.29}$$

$$m_{p,2}^{(1)} \rightarrow s_2^{(1)}, \quad m_{p,2}^{(2)} \rightarrow s_2^{(2)} \tag{8.30}$$

At the receivers, both users first decode the vector of the common streams using their two receive antennas, perform SIC, and then decode their respective private stream vectors while treating the private stream vector of the co-scheduled user as noise. Figure 8.7 illustrates the MIMO RSMA architecture.

The MIMO NOMA scheme proposed is also a subscheme of MIMO RSMA [34–37], which can switch among MU-MIMO, NOMA, or various RSMA configurations depending on different numbers of transmit antennas and receive antennas.

The optimization of MIMO RSMA, considering both perfect and imperfect CSI, can be conducted as described in Ref. [25]. Similar to the MISO case, RSMA demonstrates superior performance compared to MU-MIMO and NOMA in the MIMO scenario. It's important to note that RSMA is information theoretically optimal from a DoF perspective in MIMO BCs with imperfect CSI [26, 27]. In asymmetric scenarios where the number of receive antennas differs between users, employing RSMA with a multi-channel transmission is necessary to achieve optimality. For further details on such space–time RSMA schemes for asymmetric MIMO BCs, readers are directed to Ref. [27].

8.4.2 *Two-user uplink RSMA*

In the uplink, two single-antenna users simultaneously transmit their messages to a receiver equipped with M antennas. When $M = 1$, such a two-user MAC can be considered as a special case of the two-user IC, as illustrated in Fig. 8.2, where the two receivers are co-located and cooperatively decode the messages of the two users.

8.4.2.1 *Upper link SIMO RSMA model*

Inspired by the RSMA design in the IC, the uplink RSMA architecture was proposed in Ref. [28], where user 1's message m_1 is split into two parts, $m_{1,1}$ and $m_{1,2}$. Without loss of generality, this involves independently encoding these two parts into $s_{1,1}$ and $s_{1,2}$, respectively, allocating transmit power $P_{1,1}$ and $P_{1,2}$, and then superposing the two streams. The transmit signal at user 1 is thus given by

$$x_1 = \sqrt{P_{1,1}} s_{1,1} + \sqrt{P_{1,2}} s_{1,2} \tag{8.31}$$

At user 2, the message m_2 is directly encoded into s_2. By allocating a specific power P_2, the transmit signal at user 2 is $x_2 = \sqrt{P_2} s_2$. The received signal is then represented as

$$y = \boldsymbol{h}_1 x_1 + \boldsymbol{h}_2 x_2 + n = \sqrt{P_{1,1}} \boldsymbol{h}_1 s_{1,1} + \sqrt{P_{1,2}} h_1 s_{1,2} + \sqrt{P_2} \boldsymbol{h}_2 s_2 + n \tag{8.32}$$

where $\boldsymbol{h}_1, \boldsymbol{h}_2 \in \mathbb{C}^{M\times 1}$ are the channel vectors and $n \sim \mathcal{CN}(0, \mathbf{I}_M)$ is the AWGN vector. The receiver can employ SIC or joint decoding to decode the three streams $s_{1,1}$, $s_{1,2}$ and s_2. While using SIC, two layers of SIC are required at the receiver. This uplink RSMA architecture is illustrated in Fig. 8.8. In summary, two-user uplink RSMA creates two virtual users from the user that splits its message into two parts, resulting in a total of three streams sent to the receiver. Two SIC layers are required to recover the original messages of the two users.

8.4.2.2 *Upper link MIMO RSMA model*

An extension to the two-user uplink SIMO RSMA in Fig. 8.8 is the two-user uplink MIMO RSMA with N transmit antennas at each user and M

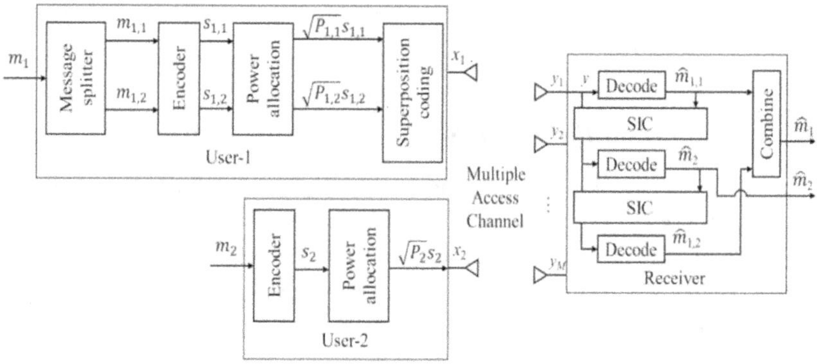

Fig. 8.8. Two-user uplink SIMO RSMA.

receive antennas at the receiver [9]. In this case, each user k transmits an N-dimensional vector of messages $\boldsymbol{m}_k = [m_k{}^{(1)}, \ldots, m_k{}^{(N)}]$ to the receiver. At user 1, each of its N messages $m_1{}^{(j)}$, $j \in \{1, 2, \ldots, N\}$ is split into two parts, $m_{1,1}{}^{(j)}$ and $m_{1,2}{}^{(j)}$, and encoded independently into $s_{1,1}^{(j)}$ and $s_{1,2}^{(j)}$. The two N-dimensional vectors $\boldsymbol{s}_{1,1} = [s_{1,1}^{(1)}, \ldots, s_{1,1}^{(N)}]^T$ and $\boldsymbol{s}_{1,2} = [s_{1,2}^{(1)}, \ldots, s_{1,2}^{(N)}]^T$ are linearly precoded by the precoders $P_{1,1}$ and $P_{1,2} \in \mathbb{C}^{M \times N}$, respectively, and superposed such that the transmit signal at user 1 is

$$x_1 = P_{1,1}\boldsymbol{s}_{1,1} + P_{1,2}\boldsymbol{s}_{1,2} \tag{8.33}$$

At user 2, the message vector $\boldsymbol{m}_2$ is encoded into an N-dimensional stream vector $\boldsymbol{s}_2$ and linearly precoded by P_2. The transmit signal at user 2 is $x_2 = P_2\boldsymbol{s}_2$.

An example of the two-user uplink MIMO RSMA when $N = 2$ is illustrated in Fig. 8.9. Each message in the message vector $\boldsymbol{m}_1$ of user 1 is split into two parts as

$$m_1^{(1)} = \{m_{1,1}^{(1)}, m_{1,2}^{(1)}\}, \quad m_1^{(2)} = \{m_{1,1}^{(2)}, m_{1,2}^{(2)}\} \tag{8.34}$$

The four sub-messages are, respectively, encoded into four streams, and two 2-dimensional stream vectors $\boldsymbol{s}_{1,1} = [s_{1,1}^{(1)}, s_{1,1}^{(2)}]^T$ and $\boldsymbol{s}_{1,2} = [s_{1,2}^{(1)}, s_{1,2}^{(2)}]^T$ are created. The two stream vectors are linearly precoded and superposed to form the transmit signal x_1. At user 2, the two messages in $\boldsymbol{m}_2$ are directly encoded into two streams in $\boldsymbol{s}_2 = [s_2{}^{(1)}, s_2{}^{(2)}]$ and linearly

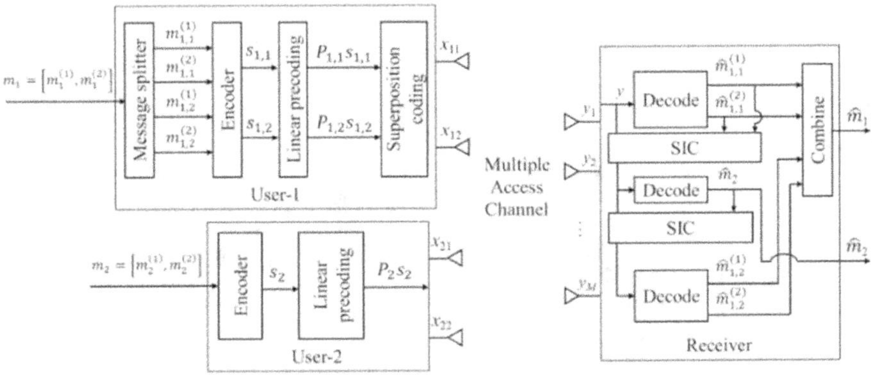

Fig. 8.9. Two-user uplink MIMO RSMA.

precoded to form the transmit signal x_2. Again, MIMO NOMA and OMA are the subschemes of MIMO RSMA in the uplink.

8.4.3 *K-user downlink RSMA*

The strategies discussed previously for managing interference in wireless communications can be extended from two-user scenarios to scenarios involving K users, with similar conclusions and insights applicable across different user counts. The K-user scenario also introduces new possibilities for RSMA schemes that incorporate varying levels of SIC layers. This subsection outlines the latest state-of-the-art K-user RSMA schemes, covering both downlink and uplink configurations [29].

8.4.3.1 *Rate-splitting scheme*

Beginning with a downlink symmetric MIMO BC scenario, a transmitter with M antennas serves K users, each with N receive antennas. Users are indexed by $\mathcal{K} = \{1, \ldots, K\}$ assuming that a Q_k-dimensional vector of messages $\boldsymbol{m}_k = [m_k{}^{(1)}, \ldots, m_k{}^{(Q_k)}]$ is transmitted to user k, where $Q_k \leq \min(M, N)$. Depending on the selected RSMA schemes, user messages $\{\boldsymbol{m}_k, k \in \mathcal{K}\}$ are split, combined, and encoded into different stream vectors. These stream vectors are then mapped to the transmit antennas using linear or non-linear precoders to generate the transmit signal x. The signal is transmitted through the MIMO BC, resulting in a received signal at user k represented as $y_k = H_k^H x + n_k$, where $H_k \in \mathbb{C}^{M \times N}$ denotes the channel

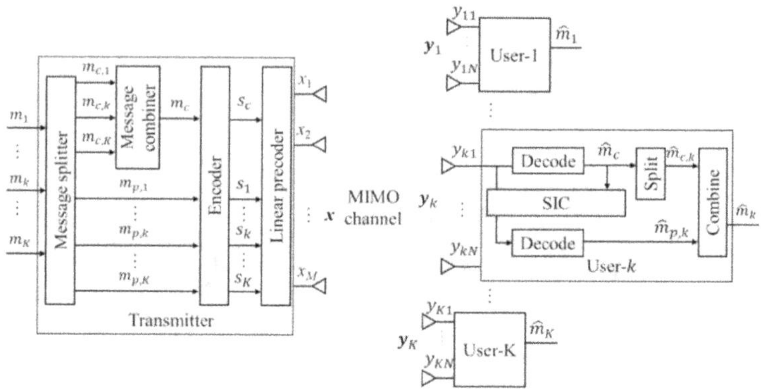

Fig. 8.10. K-user downlink one-layer MIMO RSMA.

matrix between the base station and user k, and n_k denotes the AWGN vector at user k. Equal noise variances are assumed for all users, i.e., $\sigma_{n,k}^2 = 1$, $\forall k \in \mathcal{K}$. Subsequently, detailed descriptions of transceiver architectures for various RSMA schemes in this symmetric MIMO BC setting will be provided.

8.4.3.1.1 One-layer RS

The system model for one-layer RS in a MIMO BC was introduced in Ref. [35]. Illustrated in Fig. 8.10, each message $m_k{}^{(i)}$ in $\boldsymbol{m}_k$ for user k is split into a common sub-message and a private sub-message, denoted as $m_k{}^{(i)} = \{m_{c,k}{}^{(i)}, m_{p,k}{}^{(i)}\}, \forall i \in \{1, \ldots, Q_k\}$, resulting in a common message vector $\boldsymbol{m}_{c,k} = [m_{c,k}{}^{(1)}, \ldots, m_{c,k}{}^{(Q_k)}]^T$ and a private message vector $\boldsymbol{m}_{p,k} = [m_{p,k}{}^{(1)}, \ldots, m_{p,k}{}^{(Q_k)}]^T$ for user k. The common message vectors of all users $\{\boldsymbol{m}_{c,1}, \ldots, \boldsymbol{m}_{c,K}\}$ are combined into Q_c $\{Q_c \leq \min(M, N)\}$ common messages $\boldsymbol{m}_c \in \mathbb{C}^{Q_c \times 1}$, which are encoded into a common stream vector $\boldsymbol{s}_c = [s_c{}^{(1)}, \ldots, s_c{}^{(Q_c)}]^T$ to be decoded by all users. The private message vector $\boldsymbol{m}_{p,k}$ of user k is independently encoded into the private stream vector $\boldsymbol{s}_k = [s_k{}^{(1)}, \ldots, s_k{}^{(Q_c)}]^T$ to be decoded by user k only. The $K + 1$ data stream vectors $s_c, s_1, \ldots, s_K$ are linearly precoded by the precoders $P_c, P_1, \ldots, P_K$, respectively, and superposed at the transmitter to form the resulting transmit signal:

$$x = \boldsymbol{P}_c \boldsymbol{s}_c + \sum_{k=1}^{K} \boldsymbol{P}_k \boldsymbol{s}_k \tag{8.35}$$

The received signal at user k is

$$y = \boldsymbol{H}_k^H \boldsymbol{P}_c \boldsymbol{s}_c + \sum_{j=1}^{K} \boldsymbol{H}_k^H \boldsymbol{P}_j \boldsymbol{s}_j + n_k \tag{8.36}$$

User k employs SIC or joint decoding to decode two stream vectors $\boldsymbol{s}_c$ and $\boldsymbol{s}_k$. If SIC is used, only one-layer of SIC is required at each user. Assuming Gaussian signaling, the instantaneous rates of decoding the common and private stream vectors based on SIC are given by

$$\begin{aligned} R_{c,k} &= \log_2 \det(\mathbf{I} + \boldsymbol{P}_c^H \boldsymbol{H}_k (\boldsymbol{R}_{c,k})^{-1} \boldsymbol{H}_k^H \boldsymbol{P}_c) \\ R_k &= \log_2 \det(\mathbf{I} + \boldsymbol{P}_k^H \boldsymbol{H}_k (\boldsymbol{R}_k)^{-1} \boldsymbol{H}_k^H \boldsymbol{P}_k) \end{aligned} \tag{8.37}$$

where I is the identity matrix, and $R_{c,k}$ and R_k are noise plus interference covariance matrices defined as

$$\begin{aligned} \boldsymbol{R}_{c,k} &= \mathbf{I} + \sum_{j=1}^{K} \boldsymbol{H}_k^H \boldsymbol{P}_j \boldsymbol{P}_j^H \boldsymbol{H}_k \\ \boldsymbol{R}_k &= \mathbf{I} + \sum_{j=1, j\neq k}^{K} \boldsymbol{H}_k^H \boldsymbol{P}_j \boldsymbol{P}_j^H \boldsymbol{H}_k \end{aligned} \tag{8.38}$$

To ensure that the common stream vector is successfully decoded by all users, its achievable rate should be $R_c = \min_k\{R_{c,1}, \ldots, R_{c,K}\}$. Since $\boldsymbol{s}_c$ contains the sub-messages of K users, the sum of rates for all users equals $\sum_{k\in\mathcal{K}} C_k = R_c$, where each C_k denotes the portion of R_c allocated to user k for the transmission of $\boldsymbol{m}_{c,k}$. The overall achievable rate of each user k is $R_{k,\text{tot}} = C_k + R_k$.

8.4.3.1.2 Two-layer RS

In this subsection, the system model proposed for the MISO BC outlined in Ref. [40] is extended to encompass the MIMO BC scenario, where each node is equipped with multiple antennas. Within the framework of clustered RS (two-layer RS), users are organized into G distinct groups, denoted by $\mathcal{G} = \{1, \ldots, G\}$, based on the similarity of their channel covariance matrices.

Each group g contains a subset of users $\mathcal{K}_g$ such that $\cup_{g\in\mathcal{G}}\mathcal{K}_g = \mathcal{K}$. In contrast to K-user RS where each user message is split into two parts, in this clustered RS, each user message ${m_k}^{(i)}$, $i \in \{1, 2, \ldots, Q_k\}$ in $\boldsymbol{m}_k$

for user k (assuming user k is in group g) is split into three sub-messages as $m_k^{(i)} = \{m_{\mathcal{K},k}^{(i)}, m_{\mathcal{K}_g,k}^{(i)}, m_{k,k}^{(i)}\}$, resulting in three sub-message vectors, namely, an inter-group common message vector $\boldsymbol{m}_k^{\mathcal{K}} = \{m_{\mathcal{K},k}^{(1)}, \ldots, m_{\mathcal{K},k}^{(Q_k)}\}$, an inner-group common message vector $\boldsymbol{m}_k^{\mathcal{K}_g} = \{m_{\mathcal{K}_g,k}^{(1)}, \ldots, m_{\mathcal{K}_g,k}^{(Q_k)}\}$, and a private message vector $\boldsymbol{m}_k^{k} = \{m_{k,k}^{(1)}, \ldots, m_{k,k}^{(Q_k)}\}$. The inter-group common messages of all users $\{\boldsymbol{m}_1^{\mathcal{K}}, \ldots, \boldsymbol{m}_K^{\mathcal{K}}\}$ are combined into Q_c ($Q_c \leq \min(M, N)$) common messages $m_{\mathcal{K}} \in \mathbb{C}^{Q_c \times 1}$ and encoded into one common stream vector $\mathrm{s}_{\mathcal{K}} = [s_{\mathcal{K}}^{(1)}, \ldots, s_{\mathcal{K}}^{(Q_c)}]^T$, which is decoded by all users. For group g, $g \in \mathcal{G}$, the inner-group common message vector of users in the group g, $\{\boldsymbol{m}_k^{\mathcal{K}_g}, \forall k \in \mathcal{K}_g\}$ are combined into Q_{cg} $\{Q_{cg} \leq \min(M, N)\}$ common messages $m_{\mathcal{K}_g} \in \mathbb{C}^{Q_{cg} \times 1}$ and encoded into one common stream vector $\mathrm{s}_{\mathcal{K}_g} = [s_{\mathcal{K}_g}^{(1)}, \ldots, s_{\mathcal{K}_g}^{(Q_{cg})}]^T$ to be decoded by all users in group g. The private message vector $\boldsymbol{m}_k^k$ of user k is independently encoded into the private stream vector $\mathrm{s}_k = [s_k^{(1)}, \ldots, s_k^{(Q_k)}]^T$ to be decoded by user k only. The overall $K + G + 1$ data stream vectors $\mathrm{s}_{\mathcal{K}}, \mathrm{s}_{\mathcal{K}_1}, \ldots, \mathrm{s}_{\mathcal{K}_G}, \mathrm{s}_1, \ldots, \mathrm{s}_K$ are linearly precoded by the corresponding precoders $P_{\mathcal{K}}, P_{\mathcal{K}_1}, \ldots, P_{\mathcal{K}_G}, P_1, \ldots, P_K$. The resulting transmit signal is

$$x = \boldsymbol{P}_{\mathcal{K}} \boldsymbol{s}_{\mathcal{K}} + \sum_{g=1}^{G} \boldsymbol{P}_{\mathcal{K}_g} \boldsymbol{s}_{\mathcal{K}_g} + \sum_{k=1}^{K} \boldsymbol{P}_k \boldsymbol{s}_k \tag{8.39}$$

The received signal at user k is

$$y = \boldsymbol{H}_k^H \boldsymbol{P}_{\mathcal{K}} \boldsymbol{s}_{\mathcal{K}} + \sum_{g=1}^{G} \boldsymbol{H}_k^H \boldsymbol{P}_{\mathcal{K}_g} s_{\mathcal{K}_g} + \sum_{j=1}^{K} \boldsymbol{H}_k^H \boldsymbol{P}_j \boldsymbol{s}_j + n_k \tag{8.40}$$

In contrast to K-user RS where each user only decodes two streams, in the clustered RS, each user k ($k \in \mathcal{K}_g$) employs SICs or joint decoding to decode three stream vectors $\mathrm{s}_{\mathcal{K}}$, $\mathrm{s}_{\mathcal{K}_g}$, and s_k. Two-layer RS of SIC is required at each user if SIC is employed. Assuming Gaussian signaling, the instantaneous rates of decoding $\mathrm{s}_{\mathcal{K}}$, $\mathrm{s}_{\mathcal{K}_g}$, and s_k based on SIC are given by

$$R_{\mathcal{K},k} = \log_2 \det \left(\mathbf{I} + \boldsymbol{P}_{\mathcal{K}}^H \boldsymbol{H}_k (\boldsymbol{R}_{\mathcal{K},k})^{-1} \boldsymbol{H}_k^H \boldsymbol{P}_{\mathcal{K}} \right)$$

$$
\begin{aligned}
R_{\mathcal{K}_g,k} &= \log_2 \det\left(\mathbf{I} + \boldsymbol{P}_{\mathcal{K}_g}^H \boldsymbol{H}_k (\boldsymbol{R}_{\mathcal{K}_g,k})^{-1} \boldsymbol{H}_k^H \boldsymbol{P}_{\mathcal{K}_g}\right) \\
R_k &= \log_2 \det\left(\mathbf{I} + \boldsymbol{P}_k^H \boldsymbol{H}_k (\boldsymbol{R}_k)^{-1} \boldsymbol{H}_k^H \boldsymbol{P}_k\right)
\end{aligned} \tag{8.41}
$$

where the noise plus interference covariance matrices $R_{\mathcal{K},k}$, $R_{\mathcal{K}_g,k}$, and R_k are defined as

$$
\begin{aligned}
\boldsymbol{R}_{\mathcal{K},k} &= \mathbf{I} + \sum_{g=1}^{G} \boldsymbol{H}_k^H \boldsymbol{P}_{\mathcal{K}_g} \boldsymbol{P}_{\mathcal{K}_g}^H \boldsymbol{H}_k + \sum_{j=1}^{K} \boldsymbol{H}_k^H \boldsymbol{P}_j \boldsymbol{P}_j^H \boldsymbol{H}_k \\
\boldsymbol{R}_{\mathcal{K}_g,k} &= \mathbf{I} + \sum_{i=1, i\neq g}^{G} \boldsymbol{H}_k^H \boldsymbol{P}_{\mathcal{K}_i} \boldsymbol{P}_{\mathcal{K}_i}^H \boldsymbol{H}_k + \sum_{j=1}^{K} \boldsymbol{H}_k^H \boldsymbol{P}_j \boldsymbol{P}_j^H \boldsymbol{H}_k \\
\boldsymbol{R}_k &= \mathbf{I} + \sum_{i=1, i\neq g}^{G} \boldsymbol{H}_k^H \boldsymbol{P}_{\mathcal{K}_i} \boldsymbol{P}_{\mathcal{K}_i}^H \boldsymbol{H}_k + \sum_{j=1, j\neq k}^{K} \boldsymbol{H}_k^H \boldsymbol{P}_j \boldsymbol{P}_j^H \boldsymbol{H}_k
\end{aligned} \tag{8.42}
$$

To guarantee the successful decoding of the inter-group and inner-group common stream vectors, the achievable rates follow $R_{\mathcal{K}} = \min_k\{R_{\mathcal{K},k}, k \in \mathcal{K}\}$ and $R_{\mathcal{K}_g} = \min_k\{R_{\mathcal{K}_g,k}, k \in \mathcal{K}_g\}$. As $s_{\mathcal{K}}$ and $s_{\mathcal{K}_g}$ contain sub-messages of multiple users, we have $\sum_{k\in\mathcal{K}} C_k^{\mathcal{K}} = R_{\mathcal{K}}$ and $\sum_{k\in\mathcal{K}_g} C_k^{\mathcal{K}_g} = R_{\mathcal{K}_g}$, where $C_k^{\mathcal{K}}$ and $C_k^{\mathcal{K}_g}$, respectively, denote the portion of $R_{\mathcal{K}}$ and $R_{\mathcal{K}_g}$, allocated to user k for the transmission of $\boldsymbol{m}_k^{\mathcal{K}}$ and $\boldsymbol{m}_k^{\mathcal{K}_g}$. The overall achievable rate of user k is $R_{k,\text{tot}} = C_k^{\mathcal{K}} + C_k^{\mathcal{K}_g} + R_{\mathcal{K}}$. It is easy to observe that the clustered RS is a more general framework than K-user RS for MIMO BC.

In Fig. 8.11, a four-user two-group clustered RS example is illustrated with user 1 and user 2 in group 1 and user 3 and user 4 in group 2. The user sets $\mathcal{K} = \{1, 2, 3, 4\}$, $\mathcal{K}_1 = \{1, 2\}$, and $\mathcal{K}_2 = \{3, 4\}$ are simply denoted as 1234, 12, and 34, respectively. User 1 and user 2 are required to decode the inter-group common stream vector s_{1234} and the inner-group common stream vector s_{12}, respectively, while user 3 and user 4 are required to decode s_{1234} and s_{34}, respectively.

8.4.3.1.3 Generalized RS

Generalized RS proposed in Ref. [8] for MISO BC enhances SE by splitting and encoding multiple common streams, requiring decoding by different

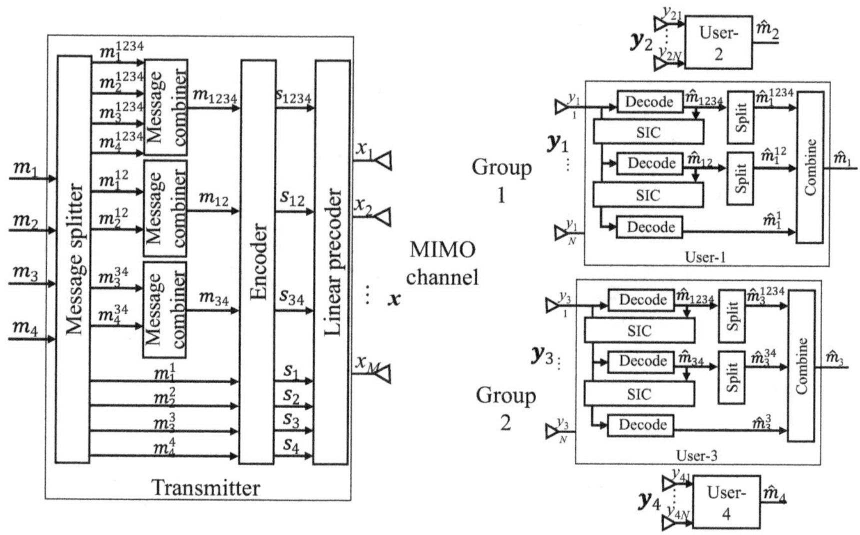

Fig. 8.11. Four-user downlink MIMO clustered RS.

subsets of users. It is a generalized transmission framework encompassing K-user RS, clustered RS, linearly precoded MU-MIMO, and MIMO NOMA as special cases.

The extension of generalized RS to MIMO BC is presented here. At the transmitter, each user message is split into 2^{K-1} parts, i.e., $m_k^{(i)} = \{M_{\mathcal{A}',k} | \mathcal{A}' \in \mathcal{K}, k \in \mathcal{A}'\}$. The message vector $\boldsymbol{m}_k$ is therefore split into 2^{K-1} sub-message vectors $\boldsymbol{m}_k^{\mathcal{A}'}\{\forall \mathcal{A}' \in \mathcal{K},\ k \in \mathcal{A}'\}$ with $\boldsymbol{m}_k^{\mathcal{A}'} = \{M_{\mathcal{A}',k}^{(1)}, \ldots, M_{\mathcal{A}',k}^{(Q_k)}\}$.

The sub-message vectors $\{\boldsymbol{m}_k^{\mathcal{A}} | k \in \mathcal{A}\}$ for all users in a given subset $\mathcal{A} \in \mathcal{K}$ are combined and encoded into one stream vector $\mathrm{s}_{\mathcal{A}}$ of dimension $Q_{c\mathcal{A}} \times 1$ which is only decoded by users in $\mathcal{A}$ and treated as noise by the remaining users. Expanding upon the stream-ordering concept introduced in the generalized RS for MISO BC [8], the stream vectors targeted for decoding by l number of users are designated as l-order stream vectors $\{\mathrm{s}_{\mathcal{A}'} | \mathcal{A}' \in \mathcal{K}, |\mathcal{A}'| = l\}$, which are then encapsulated into a larger dimensional vector $\mathrm{s}_l \in \mathbb{C}^{\sum_{\mathcal{A}' \in \mathcal{K}, |\mathcal{A}'|=l} Q_{c\mathcal{A}'} \times 1}$ and linearly precoded by the precoding matrix P_l composed by $\{P_{\mathcal{A}'} | \mathcal{A}' \in \mathcal{K}, |\mathcal{A}'| = l\}$. The resulting transmit signal is

$$x = \sum_{l=1}^{K} \boldsymbol{P}_l \boldsymbol{s}_l = \sum_{l=1}^{K} \sum_{\mathcal{A}' \subseteq \mathcal{K}, |\mathcal{A}'|=l} \boldsymbol{P}_{\mathcal{A}'} \boldsymbol{s}_{\mathcal{A}'} \tag{8.43}$$

At user k, $2^{K-1} - 1$ layers of SIC or joint decoding are employed to decode the required stream vectors. The set of l-order stream vectors required to be decoded at user k is denoted by

$$\mathcal{S}_{l,k} = \{\mathrm{s}_{\mathcal{A}'} | \mathcal{A}' \subseteq \mathcal{K}, |\mathcal{A}'| = l, k \in \mathcal{A}'\}.$$

Following the decoding order from the K-order stream vector down to the l-order stream vector, and assuming a certain decoding order $\pi_{l,k}$ to decode the l-order stream vectors in $\mathcal{S}_{l,k}$ (i.e., $s_{\pi_{l,k}(i)}$ is decoded before $s_{\pi_{l,k}(j)}$ if $i < j$), the rate of decoding the l-order stream vector at user k under Gaussian signaling and SIC decoding is obtained as

$$R_{\pi_{l,k}(i),k} = \log_2 \det\left(\mathbf{I} + \boldsymbol{P}^H_{\pi_{l,k}(i)} \boldsymbol{H}_k (\boldsymbol{R}_{\pi_{l,k}(i),k})^{-1} \boldsymbol{H}_k^H \boldsymbol{P}_{\pi_{i,k}(i)}\right) \tag{8.44}$$

where $R_{\pi_{l,k}(i),k}$ is given as

$$\begin{aligned}
\boldsymbol{R}_{\pi_{l,k}(i),k} = \mathbf{I} &+ \sum_{j>i} \boldsymbol{H}_k^H \boldsymbol{P}_{\pi_{l,k}(j)} \boldsymbol{P}^H_{\pi_{l,k}(j)} \boldsymbol{H}_k \\
&+ \sum_{l'=1}^{l-1} \sum_{j=1}^{|\mathcal{S}_{l',k}|} \boldsymbol{H}_k^H \boldsymbol{P}_{\pi_{l',k}(j)} \boldsymbol{P}^H_{\pi_{l,k}(j)} \boldsymbol{H}_k \\
&+ \sum_{\mathcal{A}' \subseteq \mathcal{K}, k \notin \mathcal{A}'} \boldsymbol{H}_k^H \boldsymbol{P}_{\mathcal{A}'} \boldsymbol{P}^H_{\mathcal{A}'} \boldsymbol{H}_k
\end{aligned} \tag{8.45}$$

To ensure successful decoding of each stream vector $\mathrm{s}_{\mathcal{A}}$, the achievable rates follow $R_{\mathcal{A}} = \min_k\{R_{\mathcal{A},k},\ k \in \mathcal{A}\}$. As $\mathrm{s}_{\mathcal{A}}$ contains sub-messages of users in $\mathcal{A}$, then $\sum_{k\in\mathcal{A}} C_k^{\mathcal{K}} = R_{\mathcal{A}}$, where $C_k^{\mathcal{A}}$ denotes the portion of $R_{\mathcal{A}}$ allocated to user k for the transmission of $\boldsymbol{m}_k^{\mathcal{A}}$. The overall achievable rate of user k is

$$R_{k,\mathrm{tot}} = \sum_{k\in\mathcal{A}'} C_k^{\mathcal{A}'} + R_{k,k}.$$

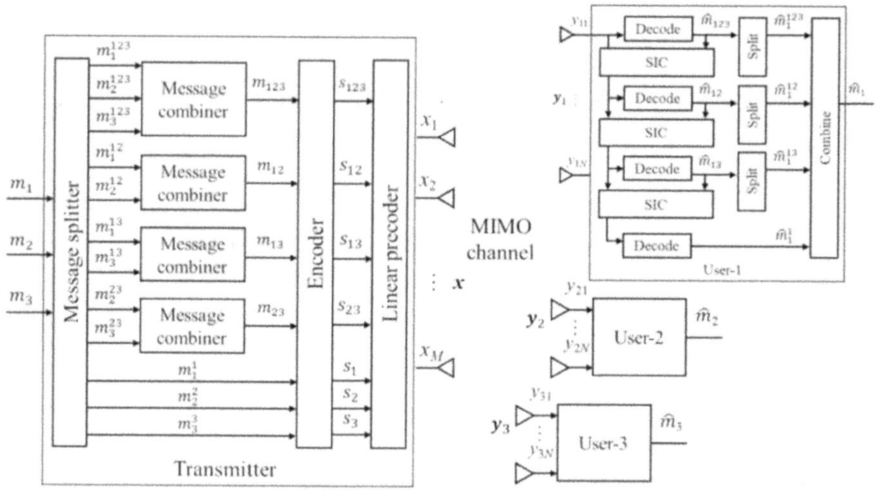

Fig. 8.12. Three-user downlink MIMO generalized RS.

A three-user generalized RS example is illustrated in Fig. 8.12. The user message vector of each user is split into four sub-message vectors. For user 1, $\boldsymbol{m}_1$ is split into $\{\boldsymbol{m}_1^{123}, \boldsymbol{m}_1^{12}, \boldsymbol{m}_1^{13}, \boldsymbol{m}_1^{1}\}$, where each is required to be decoded by different groups of users. $\{\boldsymbol{m}_1^{123}, \boldsymbol{m}_2^{123}, \boldsymbol{m}_3^{123}\}$ are combined into $\boldsymbol{m}_{123}$, which is then encoded into the three-order stream vector s_{123}. Similarly, $\{\boldsymbol{m}_1^{12}, \boldsymbol{m}_2^{12}\}$ are combined into $\boldsymbol{m}_{12}$ and encoded into the two-order stream vector s_{12}. Following this method, seven transmit streams are obtained. A three-layer SIC is adopted at each user to decode the intended four streams. In this example, user 1 decodes s_{123}, s_{12}, s_{13}, s_1 sequentially with $s_{\pi_{2,1}(1)} = s_{12}$ and $s_{\pi_{2,1}(2)} = s_{13}$.

8.4.4 *K-user uplink RSMA*

Uplink RSMA was initially introduced in Ref. [28] for SISO MAC and later extended to SIMO MAC in Ref. [30]. The primary benefit of uplink RSMA, as discovered in existing works, is that the capacity region of the Gaussian MAC can be achieved by uplink RSMA without time-sharing among users. In the following, a generic uplink RSMA transmission framework for MIMO MAC denoted as uplink MIMO RSMA is introduced.

Figure 8.13 illustrates the K-user uplink MIMO RSMA based on SC at the transmitters and SIC at the receivers [30]. A Q_k-dimensional message

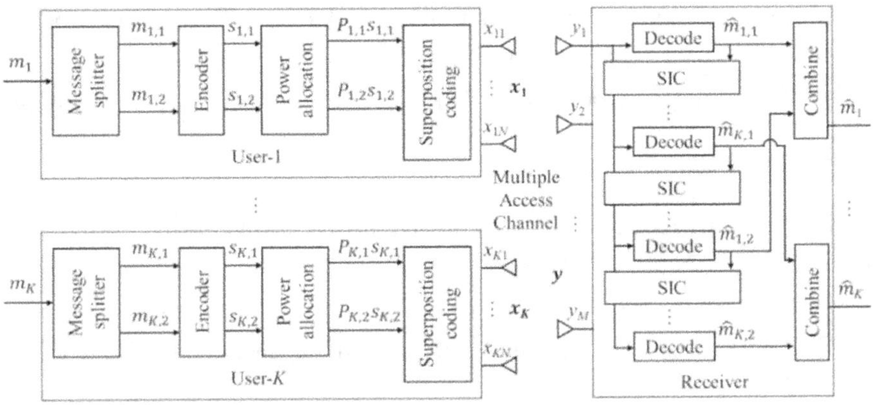

Fig. 8.13. K-user uplink MIMO RSMA.

vector $\boldsymbol{m}_k = [\boldsymbol{M}_k{}^{(1)}, \ldots, \boldsymbol{M}_k{}^{(Q_k)}]^T$ is transmitted from user k to the receiver, where $Q_k \leq \min(M, N)$.

At user k, each user message $\boldsymbol{M}_k{}^{(i)}$ is split into two sub-messages $\boldsymbol{M}_{k,1}{}^{(i)}$ and $\boldsymbol{M}_{k,2}{}^{(i)}$, resulting in two sub-message vectors $\boldsymbol{m}_{k,1}$ and $\boldsymbol{m}_{k,2}$. Note that splitting $\sum_{k=1}^{K} Q_k - 1$ messages is sufficient to achieve the capacity region without time-sharing in this setting. For illustrative simplicity, a more general model is considered where the user messages of all users are split.

At user k, the sub-message vectors $\boldsymbol{m}_{k,1}$ and $\boldsymbol{m}_{k,2}$ are independently encoded into $\boldsymbol{s}_{k,1}$ and $\boldsymbol{s}_{k,2}$, precoded by $P_{k,1}$ and $P_{k,2}$, and superposed at the transmitter. The resulting transmit signal is

$$x_k = \boldsymbol{P}_{k,1}\boldsymbol{s}_{k,1} + \boldsymbol{P}_{k,2}\boldsymbol{s}_{k,2}, \quad \forall k \in \mathcal{K} \tag{8.46}$$

The signals of all users are transmitted via MIMO MAC, and the resulting receive signal is

$$y = \sum_{k \in \mathcal{K}} \boldsymbol{H}_k x_k + n \tag{8.47}$$

where $\mathrm{H}_k \in \mathbb{C}^{M \times N}$ is the channel matrix between user k and the receiver, and $\mathrm{n} \sim \mathcal{CN}(0, \mathrm{I}_M)$ is the AWGN vector. The receiver then employs the receive filters $\mathbf{m}_{k,1}, \mathbf{m}_{k,2}, k \in \mathcal{K}$ to detect the stream vectors for all users. Denote the decoding order of the $2K$ received stream vectors $\{s_{k,i}k \in \{1, \ldots, K\}, i \in \{1, 2\}\}$ by π such that $\pi_{k,i} < \pi_{k',i'}$ if $s_{k,i}$ is decoded before $s_{k',i'}$. $2K - 1$ layers of SIC are needed to decode all stream vectors.

Assuming Gaussian signaling and infinite block length, the rate of decoding $s_{k,i}$ at the receiver is given as follows:

$$R_{k,i} = \log_2 \det(\mathbf{I} + \boldsymbol{P}_{k,i}^H \boldsymbol{H}_k^H \boldsymbol{M}_{k,i}^H (\boldsymbol{R}_{k,i})^{-1} \boldsymbol{M}_{k,i} \boldsymbol{H}_k \boldsymbol{P}_{k,i}) \tag{8.48}$$

where $\boldsymbol{R}_{k,i}$ is defined as

$$\boldsymbol{R}_{k,i} = \mathbf{I} + \sum_{\pi_{k',i'} > \pi_{k,i}} \boldsymbol{M}_{k,i} \boldsymbol{H}_{k'} \boldsymbol{P}_{k',i'} \boldsymbol{P}_{k',i'}^H \boldsymbol{H}_{k'}^H \boldsymbol{M}_{k,i}^H \tag{8.49}$$

The uplink MIMO RSMA model is generic, which encompasses uplink SISO RSMA from Ref. [28] and uplink SIMO RSMA from Ref. [30] as two sub-schemes.

8.5 Performance Issues

8.5.1 *System complexity*

8.5.1.1 *Encoder complexity*

RSMA introduces higher encoder complexity compared to other MA schemes due to message splitting, resulting in encoding multiple streams at the RSMA transmitter. Among linearly precoded RSMA schemes, one-layer RS exhibits the lowest encoding complexity, requiring only one additional stream encoded at the transmitter compared to SDMA and NOMA baselines, where each of the K messages is encoded into K streams in the K-user scenario. In the case of two-layer clustered RS, G additional common streams for G groups, corresponding to one group-specific common stream for each user group, are encoded at the transmitter besides the common stream for all users. Thus, the encoder complexity difference between one-layer RS/two-layer clustered RS and SDMA/NOMA remains constant regardless of the number of users K. Conversely, the encoder complexity difference between RS-common message decoding/generalized RS and SDMA/NOMA increases with K due to the larger number of sub-messages split from each message.

8.5.1.2 *Scheduler complexity*

SDMA offers the advantage of low encoder and receiver complexities. However, its scheduler complexity is relatively high because it requires pairing users with semi-orthogonal channels and similar channel strengths to achieve optimal performance. SDMA's benefits are most prominent when user channels exhibit sufficient orthogonality. Furthermore, effective user scheduling necessitates precise CSIT, leading to heightened signaling overhead and latency.

The SIC receivers utilized in NOMA and RSMA introduce additional complexity to the scheduler compared to OMA and SDMA. The utilization of SIC introduces two key considerations at the transmitter: the decoding order and user grouping. The interconnected nature of user grouping, decoding order, precoders, and user selection mandates joint optimization at the transmitter. Among strategies employing SIC, multi-antenna NOMA ($G > 1$), two-layer clustered RS, and RS-common message decoding entail the highest complexity in user grouping [8, 30].

In terms of decoding order, multi-antenna NOMA, generalized RS, and RS-common message decoding exhibit high decoding order complexity. In multi-antenna NOMA ($G = 1$), all users share the same decoding order to decode user messages, leading to $K!$ possible decoding orders for the K streams. It's worth noting that selecting users with aligned channels and significant channel strength differences, which is advantageous for multi-antenna NOMA ($G = 1$), increases scheduler complexity.

In generalized RS, the decoding order begins from the K-order streams downward to the 1-order streams, resulting in $\binom{K}{k}$ k-order streams for $2 \leq k \leq K-1$. This leads to $\binom{K}{k}!$ possible decoding orders, requiring the transmitter to select from $\prod_{k=2}^{K-1} \binom{K}{k}!$ possible decoding orders for users to decode all intended streams. For RS-common message, if it's assumed that the K common streams are decoded at all users, the transmitter must consider a total of $(K!)^K$ decoding orders without the need for user grouping. Alternatively, if user grouping is involved, the transmitter must jointly decide the decoding order and the group of common streams decoded at each user.

In summary, multi-antenna NOMA ($G > 1$) and RS-common message exhibit relatively high scheduler complexity due to the joint decision-making required for decoding order and user grouping. For multi-antenna NOMA ($G > 1$), users with semi-orthogonal channels need to be assigned to different user groups, and channels within the same group should be aligned. On the other hand, one-layer RS shows relatively low scheduler complexity compared to other MA schemes (including OMA, SDMA, NOMA, and all RSMA schemes) as it doesn't necessitate grouping or ordering and is less sensitive to user pairing. Consequently, one-layer RS doesn't require complex user scheduling and pairing. Generally, all RSMA schemes, including one-layer RS, two-layer clustered RS, generalized RS, and RS-common message decoding, are suitable for users with varying channel strength disparities and directions.

8.5.1.3 *Receiver complexity*

Among the multiple access schemes, OMA and SDMA demonstrate the lowest receiver complexity since they do not involve SIC. Conversely, MA schemes utilizing SIC, such as NOMA and RSMA variants, entail higher receiver complexities. Within SIC-based schemes, one-layer RS and two-layer clustered RS exhibit the least receiver complexity. In scenarios with K users, one-layer RS requires only one SIC operation per user, while two-layer clustered RS demands two, irrespective of K. Consequently, both schemes are less susceptible to error propagation [30].

Conversely, schemes such as NOMA (for fixed user groups G), generalized RS, and RS-common message decoding necessitate multiple SIC operations based on K, leading to increased transceiver complexity and error propagation risk. Specifically, multi-antenna NOMA ($G = 1$) [15] requires $K-1$ SIC layers per user, whereas multi-antenna NOMA ($G > 1$) demands $|\mathcal{K}_i|-1$ SIC layers per user from group i, where $|\mathcal{K}_i| \leq K$ denotes the group size for group i.

RS-common message decoding and generalized RS involve complex receivers, requiring K and $2^{K-1} - 1$ SIC layers, respectively. Although generalized RS offers superior flexibility in managing interference, it also entails higher receiver complexity. Efforts have been made to develop low-complexity versions of RS-common message decoding and generalized RS by controlling the number of common streams decoded per user and simplifying decoding order design [31, 32].

Additionally, it's worth noting that generalized RS serves two purposes: achieving optimal performance at the expense of higher transceiver complexity and identifying key common streams for performance enhancement, which could be deactivated to mitigate encoding and decoding complexities.

8.5.2 *System performance*

8.5.2.1 *User fairness*

In practice, the issue of user fairness arises due to disparate path losses experienced by users in different locations, leading to a situation where

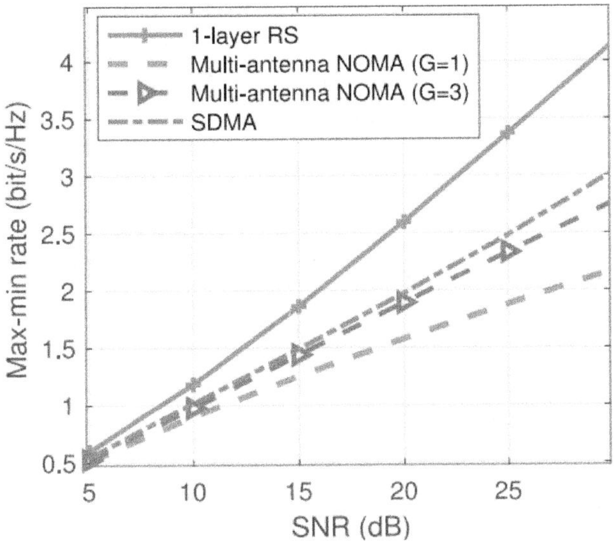

Fig. 8.14. Max–min rate performance for the MISO BC with imperfect CSIT $\alpha = 0.5$, $M = K = 6$.

users closer to the transmitter receive more resources, while users farther away receive fewer resources. To address this issue, the max–min rate utility function is preferred for precoder design, ensuring that a certain rate performance is guaranteed for all users.

Figure 8.14 depicts the max–min rate performance and the associated power allocation between the common and private streams for RSMA in the MISO BC under imperfect CSI conditions [30]. The transmitter is equipped with $M = 6$ antennas and serves $K = 6$ single-antenna users. Imperfect CSI is modeled with a CSI scale factor ($\alpha = 0.5$). Notably, one-layer RS exhibits superior max–min rate performance compared to multi-antenna NOMA and SDMA.

The scheduler complexity and receiver complexity of multi-antenna NOMA are notably higher compared to one-layer RS and SDMA. For multi-antenna NOMA (with $G = 3$), the transmitter must jointly optimize user grouping, decoding order, and precoders. In scenarios where $K = 6$, each user requires up to five layers of SIC for multi-antenna NOMA (with $G = 1$), although not all layers are always utilized. Despite its high complexity, multi-antenna NOMA demonstrates inferior performance

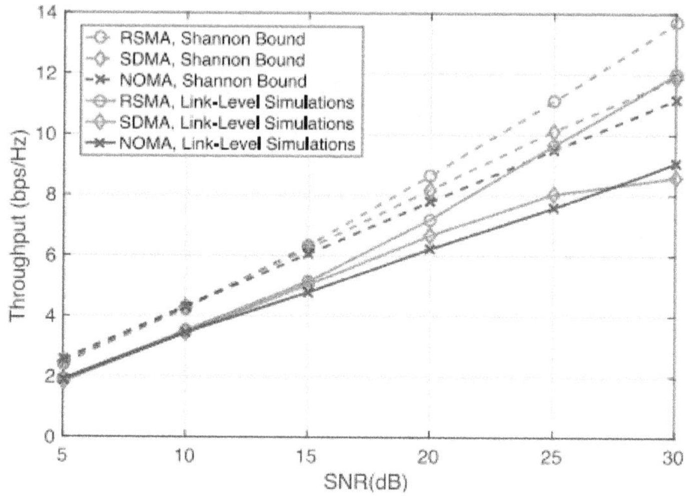

Fig. 8.15. Throughput vs. SNR for different MA schemes for the MISO BC.

compared to one-layer RS with a single layer of SIC at each user or multi-user linear precoding without SIC. Consequently, multi-antenna NOMA may inefficiently utilize SIC, as emphasized in Ref. [15].

8.5.2.2 *Throughput performance*

In this subsection, the throughput performance of RSMA is investigated in a two-user scenario compared with the Shannon bounds. The system throughput is defined as the total number of successfully recovered information bits at all users per channel use. In a link-level simulation platform, the PHY-layer architecture is specifically designed for RSMA, as demonstrated in previous studies [30, 33–35].

Figure 8.15 illustrates both the Shannon bounds and the achieved throughput levels of RSMA, SDMA, and NOMA for the MISO BC under conditions of imperfect CSI [30]. In this scenario, the transmitter is equipped with $M = 2$ antennas and serves $K = 2$ single-antenna users. Imperfect CSI is modeled with a scaling factor of 0.6.

The overall throughput performance achieved by the link-level simulation closely tracks the corresponding Shannon bounds, showing that RSMA provides a significant throughput gain compared to SDMA and NOMA. Notably, the throughput improvement of RSMA over SDMA exceeds expectations based on the Shannon bound. Additionally, it is observed that the throughput of SDMA saturates in the high SNR regime.

8.5.2.3 *Sum-rate performance*

To evaluate the performance gain of the RSMA scheme and for comparison purposes, the sum-rate maximization problems for uplink NOMA, FDMA, and TDMA schemes are presented [36–38].

8.5.2.3.1 System Model

In a single-cell uplink network with one BS serving a set of K users using RSMA, each user first transmits a superposition code of two messages to the BS. Subsequently, the BS utilizes an SIC technique to decode the messages of all users [28]. The transmitted message s_k of user k is expressed as

$$x_k = \sum_{j=1}^{2} \sqrt{p_{kj}} s_{kj}, \quad \forall k \in \mathcal{K} \tag{8.50}$$

where p_{kj} is the transmit power of message s_{kj} from user k and $\mathcal{K} \equiv \{1, 2, \ldots, K\}$.

The total received message y at the BS can be given by

$$y = \sum_{k=1}^{K} \sqrt{h_k} x_k + n = \sum_{k=1}^{K} \sum_{j=1}^{2} \sqrt{h_k p_{kj}} s_{kj} + n \tag{8.51}$$

where h_k is the channel gain between user k and the BS and n is the AWGN. Each user k has a maximum transmission power limit P_k such that $\sum_j p_{kj} \leq P_k$.

To decode all messages s_{kj} from the received message y, the BS will use SIC. The decoding order at the BS is denoted by a permutation π. The permutation π belongs to a set, Π, defined as the set of all possible decoding orders of all $2K$ messages from K users. The decoding order of message s_{kj} from user k is π_{kj}. The achievable rate of decoding message s_{kj} is

$$r_{kj} = B\log_2\left(1 + \frac{h_k p_{kj}}{\sum_{\{(l \in \mathcal{K}, m \in \mathcal{J}) | \pi_{lm} > \pi_{kj}\}} h_l p_{lm} + \sigma^2 B}\right) \tag{8.52}$$

where B is the bandwidth of the BS, and σ^2 is the power spectral density of the Gaussian noise. The set $\{(l \in \mathcal{K}, m \in \mathcal{J})|\pi_{lm} > \pi_{kj}\}$ in (8.52) represents the messages s_{lm} that are decoded after message s_{kj}.

Since the transmitted message of user k includes messages s_{k1} and s_{k2}, the achievable rate of user k is given by

$$r_k = \sum_{j=1}^{2} r_{kj} \tag{8.53}$$

The objective is to maximize the sum-rate $r = \sum_k r_k$ of all users with proportional rate constraints. Mathematically, the sum-rate maximization problem can be formally posed as follows:

$$\max_{\tau,\pi,p} \tau \tag{8.54}$$

$$s.t. \quad r_k = D_k\tau, \quad \forall k \in \mathcal{K} \tag{8.54a}$$

$$\sum_{j=1}^{2} p_{kj} \le P_k, \quad \forall k \in \mathcal{K} \tag{8.54b}$$

$$\pi \in \Pi, \; p_{kj} \ge 0, \quad \forall k \in \mathcal{K}, \; j \in \mathcal{J} \tag{8.54c}$$

where $\boldsymbol{p} = [p_{11}, p_{12}, p_{21}, p_{22}, \ldots, p_{K1}, p_{K2}]^T$, τ is the sum-rate of all users with $r = \sum_k r_k$ where r_k is defined in (8.53), and $\mathcal{J} = \{1, 2\}$. $D_1, \ldots, D_K$ is a set of predetermined non-negative values that are used to ensure proportional fairness among users. The fairness index is defined as

$$\frac{\left(\sum_{k=1}^{K} D_k\right)^2}{K \sum_{k=1}^{K} D_k^2}$$

with the maximum value of 1 to be the greatest fairness case, in which all users would achieve the same data rate [38]. With proper normalization, the relation is established as $\sum_{k=1}^{K} D_k = 1$.

Although it was stated in Ref. [28] that RSMA can reach the optimal rate region, no practical algorithm was proposed to compute the decoding order and power allocation. It is therefore necessary to quantify the uplink performance gains that RSMA can obtain compared to conventional multiple access schemes.

Due to the non-linear equality constraint (8.54a) and discrete variable π, problem (8.54) is a non-convex mixed integer problem. Hence, it is generally hard to solve problem (8.54).

The problem (8.54) can be equivalently transformed [28] to

$$\max_{\tau,\boldsymbol{r}} \tau \tag{8.55}$$

$$\text{s.t.} \quad r_k = D_k\tau, \quad \forall k \in \mathcal{K} \tag{8.55a}$$

$$\sum_{k\in\mathcal{K}'} r_k \le B\log_2\left(1 + \frac{\sum_{k\in\mathcal{K}'} h_k P_k}{\sigma^2 B}\right), \quad \forall \mathcal{K}' \subseteq \mathcal{K}\backslash\emptyset \tag{8.55b}$$

where $\boldsymbol{r} = [r_1 r_2, \ldots, r_K]^T$ and $\mathcal{K}' \subseteq \mathcal{K}\backslash\emptyset$ is the non-empty subset of K. In problem (8.55), the dimension of the variable is smaller than that in problem (8.54). Moreover, the discrete decoding order variable is replaced by a rate variable in problem (8.55). Regarding the optimal solution of problem (8.55), the following results [36] can be obtained.

Lemma 1. *For the optimal solution* (τ^*, r^*) *of problem* (8.55), *there exists at least one* $\mathcal{K}' \subseteq \mathcal{K}\backslash\emptyset$ *such that*

$$\sum_{k\in\mathcal{K}'} r_k^* = B\log_2\left(1 + \frac{\sum_{k\in\mathcal{K}'} h_k P_k}{\sigma^2 B}\right)$$

Proof. See Appendix A.8. □

Theorem 1. *The optimal solution of problem* (8.55) *is*

$$\tau^* = \min_{\mathcal{K}'\subseteq\mathcal{K}\backslash\,\emptyset} \frac{B\log_2\left(1 + \frac{\sum_{k\in\mathcal{K}'} h_k P_k}{\sigma^2 B}\right)}{\sum_{k\in\mathcal{K}'} D_k} \tag{8.56}$$

and

$$r_k{}^* = D_k \min_{\mathcal{K}'\subseteq\mathcal{K}\backslash\,\emptyset} \frac{B\log_2\left(1 + \frac{\sum_{k\in\mathcal{K}'} h_k P_k}{\sigma^2 B}\right)}{\sum_{k\in\mathcal{K}'} D_k} \tag{8.57}$$

Proof. According to Lemma 1, there exists at least one $\mathcal{K}' \subseteq \mathcal{K}\backslash\emptyset$ such that

$$\sum_{k\in\mathcal{K}'} r_k^* = B\log_2\left(1 + \frac{\sum_{k\in\mathcal{K}'} h_k P_k}{\sigma^2 B}\right) \tag{8.58}$$

To ensure the feasibility of (8.55b), the optimal τ^* is given by (8.56). Then, according to (8.55a), the optimal r_k^* is determined as in (8.57). □

8.5.2.3.2 Sum rate of NOMA

Without loss of generality, the channel gains with K users are sorted in order, i.e., $h_1 \geq h_2 \geq \cdots \geq h_K$. In NOMA, the BS first decodes the messages of users with high channel gains and then decodes the messages of users with low channel gains by subtracting the interference from the decoded strong user. The achievable rate of user k with NOMA is calculated as [39]

$$r_k^{\text{NOMA}} = B\log_2\left(1 + \frac{h_k q_k}{\sum_{j=k+1}^{K} h_j q_j + \sigma^2 B}\right) \tag{8.59}$$

where q_k is the transmit power of user k, σ^2 is the power spectral density of the Gaussian noise, and B is the bandwidth of the BS. The transmission power q_k has a maximum transmit power limit P_k, i.e., one can have $q_k \leq P_k$ $\forall k \in \mathcal{K}$.

Similar to (8.54), the sum-rate maximization problem for uplink NOMA can be given by

$$\max_{\tau,\boldsymbol{q}} \tau \tag{8.60}$$

$$\text{s.t.} \quad B\log_2\left(1 + \frac{h_k q_k}{\sum_{j=k+1}^{K} h_j q_j + \sigma^2 B}\right) = D_k\tau, \quad \forall k \in \mathcal{K} \tag{8.60a}$$

$$0 \leq q_k \leq P_k, \quad \forall k \in \mathcal{K} \tag{8.60b}$$

where $\boldsymbol{q} = [q_1, q_2, \ldots, q_K]^T$. To obtain the optimal solution of problem (8.36), the following theorem is provided.

Theorem 2. *The optimal solution of problem* (8.60) *is*

$$\tau^* = \min_{k\in\mathcal{K}} \tau_k \tag{8.61}$$

$$q_k^* = \frac{1}{h_k}\left(2^{\frac{D_k\tau^*}{B}}-1\right)\sum_{j=k+1}^{K} 2^{\frac{\sum_{l=k+1}^{j-1} D_l\tau^*}{B}}\left(2^{\frac{D_j\tau^*}{B}}-1\right)\sigma^2 B$$
$$+\frac{1}{h_k}\left(2^{\frac{D_k\tau^*}{B}}-1\right)\sigma^2 B,\quad \forall k\in\mathcal{K} \tag{8.62}$$

where τ_k *is the solution to*

$$P_k = \frac{1}{h_k}\left(2^{\frac{D_k\tau_k}{B}}-1\right)\sum_{j=k+1}^{K} 2^{\frac{\sum_{l=k+1}^{j-1} D_l\tau_k}{B}}\left(2^{\frac{D_j\tau_k}{B}}-1\right)\sigma^2 B$$
$$+\frac{1}{h_k}\left(2^{\frac{D_k\tau_k}{B}}-1\right)\sigma^2 B,\quad \forall k\in\mathcal{K} \tag{8.63}$$

Proof. See Appendix B.8. □

Since the right-hand side of (8.63) monotonically increases with τ_k, the solution of τ_k to (8.63) can be effectively obtained by the bisection method.

8.5.2.3.3 Sum rate of FDMA

In FDMA, each user will be allocated a fraction of the BS bandwidth of B. Let b_k denote the fraction of bandwidth allocated to user k. Then the data rate of user k is

$$r_k^{\text{FDMA}} = Bb_k\log_2\left(1+\frac{h_kP_k}{\sigma^2 Bb_k}\right) \tag{8.64}$$

Note that user k transmits with the maximum power in (8.64) since there is no inter-user interference and the large power leads to a high data rate. Due to limited bandwidth, one has $\sum_{k=1}^{K} b_k = 1$.

Similar to (8.54), the sum-rate maximization problem for uplink FDMA can be given by

$$\max_{\tau, \boldsymbol{b}} \tau \tag{8.65}$$

$$s.t. \quad Bb_k \log_2\left(1 + \frac{h_k P_k}{\sigma^2 B b_k}\right) = D_k \tau, \quad \forall k \in \mathcal{K} \tag{8.65a}$$

$$\sum_{k=1}^{K} b_k = 1 \tag{8.65b}$$

$$b_k \geq 0, \quad \forall k \in \mathcal{K} \tag{8.65c}$$

where $\boldsymbol{b} = [b_1, b_2, \ldots, b_K]^T$. Regarding the optimal solution of problem (8.65), the following theorem is given.

Theorem 3. *The optimal solution of problem* (8.65) *is* (τ^*, b^*), *where* τ^* *is the solution of*

$$-\sum_{k=1}^{K} \frac{(\ln 2) D_k h_k P_k \tau}{B h_k P_k W\left(-\frac{(\ln 2) D_k \tau \sigma^2}{h_k P_k} e^{-\frac{(\ln 2) D_k \sigma^2 \tau}{h_k P_k}}\right) + (\ln 2) D_k \sigma^2 B_\tau} = 1 \tag{8.66}$$

where $W(\cdot)$ *is the Lambert-W function and*

$$b_k^* = -\frac{(\ln 2) D_k h_k P_k \tau^*}{B h_k P_k W\left(-\frac{(\ln 2) D_k \sigma^2 \tau^*}{h_k P_k} e^{-\frac{(\ln 2) D_k \sigma^2 \tau^*}{h_k P_k}}\right) + (\ln 2) D_k \sigma^2 B_{\tau^*}} \tag{8.67}$$

Proof. See Appendix C.8. □

8.5.2.3.4 Sum rate of TDMA

In TDMA, each user will be assigned a fraction of time to use the whole B bandwidth. Let a_k be the fraction of time allocated to user k. The data rate of user k is

$$r_k^{\text{TDMA}} = B a_k \log_2\left(1 + \frac{h_k P_k}{\sigma^2 B}\right) \tag{8.68}$$

with $\sum_{k=1}^{K} a_k = 1$.

Similar to (8.54), the sum-rate maximization problem for uplink TDMA can be given by

$$\max_{\tau, \boldsymbol{a}} \tau \tag{8.69}$$

$$s.t. \quad Ba_k \log_2\left(1 + \frac{h_k P_k}{\sigma^2 B}\right) = D_k \tau, \quad \forall k \in \mathcal{K} \tag{8.69a}$$

$$\sum_{k=1}^{K} a_k = 1 \tag{8.69b}$$

$$a_k \geq 0, \quad \forall k \in \mathcal{K} \tag{8.69c}$$

where $\boldsymbol{a} = [a_1, a_2, \ldots, a_K]^T$. For problem (8.69), the optimal solution is given by the following theorem.

Theorem 4. *The optimal solution of problem* (8.69) *is*

$$\tau^* = \frac{1}{\sum_{k=1}^{K} \frac{D_k}{B\log_2(1+\frac{h_k P_k}{\sigma^2 B})}} \tag{8.70}$$

and

$$a_k^* = \frac{D_k \tau^*}{B\log_2(1 + \frac{h_k P_k}{\sigma^2 B})}, \quad \forall k \in \mathcal{K} \tag{8.71}$$

Proof. See Appendix D.8. □

8.5.2.3.5 Analysis and discussion

Let the optimal uplink sum rates denote for RSMA, NOMA, FDMA, and TDMA as τ^{RSMA}, τ^{NOMA}, τ^{FDMA}, and τ^{TDMA}, respectively. For the optimal sum rate with various uplink multiple access schemes, the following lemma [36] can be obtained.

Lemma 2. $\tau^{\mathrm{RSMA}} \geq \tau^{\mathrm{NOMA}}$ *and* $\tau^{\mathrm{RSMA}} \geq \tau^{\mathrm{FDMA}} \geq \tau^{\mathrm{TDMA}}$.

Lemma 2 can be easily demonstrated by the fact that for any feasible solution to a NOMA or an FDMA scheme, a feasible solution to RSMA with the same or better objective value can be constructed. Similarly, for any feasible solution to a TDMA scheme, a feasible solution to an FDMA scheme with the same or better objective value can be constructed. To illustrate *the sum-rate performance of different multiple access schemes, the special case with two users ($K = 2$) is considered.*

Based on Lemma 1, the rate region of RSMA with two users can be expressed by

$$\mathcal{R}_2^{\text{RSMA}} = \{(r_1, r_2)|0 \le r_1 \le R_1, 0 \le r_2 \le R_2, r_1 + r_2 \le R_{\max}\} \quad (8.72)$$

where

$$R_1 = B\log_2\left(1 + \frac{h_1 P_1}{\sigma^2 B}\right)$$
$$R_2 = B\log_2\left(1 + \frac{h_2 P_2}{\sigma^2 B}\right)$$
$$R_{\max} = B\log_2\left(1 + \frac{h_1 P_1 + h_2 P_2}{\sigma^2 B}\right) \quad (8.73)$$

Lemma 3. *For two-user NOMA/FDMA/TDMA, then*

$$\mathcal{R}_2^{\text{NOMA}} = \left\{ \begin{array}{c} (r_1, r_2)|0 \le r_1 \le R_1, 0 \le r_2 \le R_2, r_2 \\ + B\log_2(2^{\frac{r_1}{B}} - 1) \le B\log_2\left(\frac{h_1 P_1}{\sigma^2 B}\right) \end{array} \right\} \quad (8.74)$$

and

$$\mathcal{R}_2^{\text{FDMA}} = \{(r_1, r_2)|0 \le r_1 \le R_1, 0 \le r_2 \le R_2, f_1(r_1) + f_2(r_2) \le 1\} \quad (8.75)$$

where

$$f_k(r_k) = -\frac{(\ln 2) h_k P_k r_k}{B h_k P_k W\left(-\frac{(\ln 2)\sigma^2 r_k}{h_k P_k} e^{-\frac{(\ln 2)\sigma^2 r_k}{h_k P_k}}\right) + (\ln 2)\sigma^2 B r_k}, \quad k = 1, 2 \quad (8.76)$$

and

$$\mathcal{R}_2^{\text{TDMA}} = \left\{ (r_1, r_2) \middle| 0 \le r_1 \le R_1, 0 \le r_2 \le R_2, \frac{r_1}{R_1} + \frac{r_2}{R_2} \le 1 \right\}. \quad (8.77)$$

Proof. See Appendix E.8. □

Based on Lemma 3, an example of the rate region for multiple access schemes is provided in Fig. 8.16. From the figure, it is observed that RSMA has the largest rate region, while TDMA has the smallest rate region.

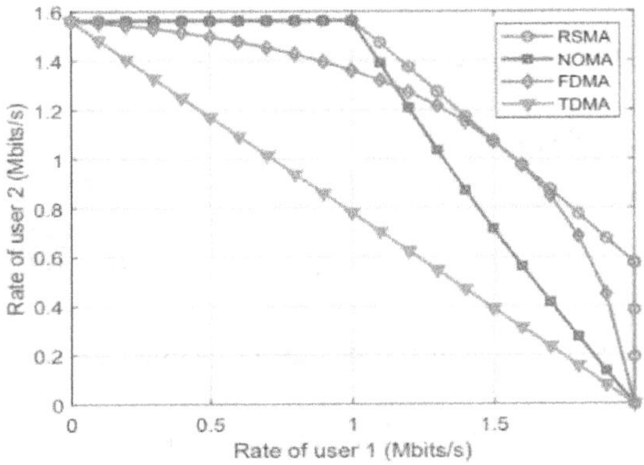

Fig. 8.16. Rate region for RSMA, NOMA, FDMA, and TDMA with $K = 2$ users, $h_1 = 9.45 \times 10^{-9}$, $h_2 = 6.17 \times 10^{-9}$, $P_1 = P_2 = 1$ dBm, $\sigma^2 = -174$ dBm/Hz, and $B = 1$ MHz.

According to the proportional rate constraints (8.54a), the relation of $r_1/r_2 = D_1/D_2$ represents a linear function through the original point in the rate coordinates (r_1, r_2). Thus, the optimal rate point can be obtained by finding the intersection between the rate region and the line $r_2 = r_1 D_1/D_2$. Thus, it is shown [36] from Fig. 8.16 that RSMA has the best sum-rate performance and TDMA has the worst sum-rate performance, which verifies the theoretical findings in Lemma 3.

In this section, the power optimization in an uplink RSMA system is investigated. The problem is formulated as a sum-rate maximization problem, which is transformed into an equivalent problem with only rate splitting variables, leading to a closed-form optimal solution. Given the optimal rate requirement of each user, the optimal transmit power of each user is determined under the specified decoding order. Simulation results demonstrate that RSMA achieves a higher sum-rate compared to NOMA, FDMA, and TDMA.

8.6 Properties of RSMA

8.6.1 *Advantages of RSMA*

The preceding analysis compares RSMA with existing multiple access (MA) schemes, including SDMA, NOMA, OMA, and multicasting, with respect to their underlying frameworks, complexities, and system performances. The results of this comparison underscore several critical advantages of RSMA, which are outlined as follows [29, 30]:

- **Universality:** RSMA serves as a universal MA scheme encompassing SDMA, NOMA, OMA, and multicasting as sub-schemes. Essentially, RSMA represents a superset of existing MA schemes, consistently achieving equal or superior performance compared to SDMA, NOMA, OMA, and multicasting. Consequently, there appears to be no necessity to explore additional MA schemes but rather adopt a unified RSMA framework.
- **Flexibility:** RSMA is a versatile and adaptive strategy for managing multi-user interference in wireless networks. Unlike SDMA, which addresses interference at the transmitter, or NOMA, which handles interference at the receiver, RSMA employs a combination of both approaches by strategically splitting user messages into common and private streams. By adjusting the allocation of resources to the common stream, RSMA allows for flexible interference cancellation levels at both ends of the communication link. This approach provides a balance between fully treating interference as noise (SDMA) and fully decoding interference (NOMA). RSMA optimizes power distribution between common and private streams, transitioning between SDMA and NOMA based on network load and user channel characteristics. In scenarios where user channels lack orthogonality and alignment, RSMA's use of common streams enhances interference management, making it superior to other multiple access schemes.
- **Robustness:** RSMA's robustness is rooted in its application in multi-antenna networks, driven by its optimality in the multi-antenna BC with imperfect CSIT. This makes RSMA particularly adept at managing multi-user interference arising from imperfect CSIT and renders it resilient to uncertainties stemming from various impairments, such as quantized feedback [40], pilot contamination [41], channel estimation errors [42], and user mobility [34]. Among these, user mobility emerges as a significant source of interference in next-generation networks and is anticipated to escalate in severity and frequency within 6G systems. Notably, previous studies have highlighted RSMA's ability to address the heightened mobility envisioned in 6G scenarios. Unlike SDMA, which saturates rapidly in sum-rate performance with increasing SNR and experiences substantial rate drops with escalating user speed, RSMA ensures robust multi-user connectivity and delivers notable sum-rate and throughput enhancements under conditions of user mobility. This stands in stark

contrast to other MA schemes (OMA, SDMA, and NOMA), which are primarily tailored for perfect CSIT scenarios and exhibit susceptibility to imperfect CSIT conditions [43].

- **Enhanced SE:** RSMA consistently offers higher SE compared to other MA schemes across scenarios with both perfect and imperfect CSI at the transmitter. In scenarios with perfect CSI, the SE of linearly precoded RSMA surpasses or matches that of schemes like MU-LP and NOMA, approaching the capacity region achieved by DPC. Even in scenarios with imperfect CSIT, linearly precoded RSMA can achieve a larger rate region than DPC, showcasing its adaptability to imperfect channel knowledge. Additionally, the integration of the benefits of RS and DPC in non-linearly precoded DPCRS further extends the rate region of the MISO BC with imperfect CSIT, surpassing both DPC and linearly precoded RS. RSMA maximizes the utilization of available CSIT and spatial dimensions in multi-antenna BC scenarios, achieving optimal sum-DoF in both perfect and imperfect CSIT conditions [30].
- **Enhanced EE:** RSMA's advantages extend beyond SE to include EE and their trade-off considerations. As RSMA encompasses OMA, SDMA, NOMA, and multicasting, its performance improvement is not limited to SE but also encompasses EE and the balance between them. Previous studies [44, 45] have demonstrated that RSMA outperforms existing MA schemes in terms of EE across various applications, accommodating a wide range of network loads and user deployments [46, 47].
- **Improved QoS and user fairness:** RSMA not only maximizes the utilization of sum-DoF but also enhances the MU-MIMO DoF in multi-antenna BC with both perfect and imperfect CSIT [30]. In scenarios with finite SNR, RSMA exhibits significant improvements in max−min rates compared to other MA schemes, as evidenced in Fig. 8.14 and various recent studies [30, 48–51]. Furthermore, RSMA's performance benefits become more pronounced when user rates are constrained by QoS requirements or when greater emphasis is placed on users with weaker channel conditions [8].
- **Reduced complexity:** RSMA offers a simpler transceiver architecture compared to many existing MA schemes, while linearly precoded RSMA achieves a larger rate region than DPC. It is shown [30] that the gains in sum/multi-user multi-input multi-output-DoF and the sum rate achieved by one-layer RS over multi-antenna NOMA are accompanied by lower scheduling and receiver complexities. Unlike multi-antenna NOMA, which necessitates joint user grouping and ordering at the transmitter

and multiple SIC layers at each user, one-layer RS doesn't require user ordering and grouping at the transmitter and only involves one SIC layer at each user. Furthermore, one-layer RS is less affected by disparate channel conditions, simplifying the scheduler further. In addition to reducing the scheduler and receiver complexities, RSMA also has the capability to decrease CSI feedback overhead in the presence of quantized feedback [52, 53].

- **Coverage extension:** Research studies such as Ref. [54, 55] have demonstrated that cooperative rate-splitting, which integrates cooperative user relaying (CUR) with one-layer RS, enhances the max−min rate significantly, particularly when serving users with significant differences in channel strengths jointly. Consequently, RSMA augmented with coverage extension methodologies such as unmanned aerial vehicles (UAVs), intelligent reconfigurable surfaces (IRS), or relay stations can further elevate user rates at the cell periphery, providing substantial coverage extension advantages.
- **Low latency:** Ultra-reliable low-latency communication (URLLC) stands as a key application scenario in next-generation wireless networks (NR) and 6G. One avenue to achieve low-latency communication involves reducing the size of transmitted packets. RSMA has demonstrated throughput improvements over SDMA and NOMA with finite-length polar codes in Ref. [33]. Moreover, it has been shown to achieve the same transmission rate as SDMA and NOMA but with shorter block lengths, consequently leading to lower latency, as evidenced in Ref. [56]. Therefore, RSMA emerges as a promising enabling technology to mitigate latency for URLLC services in 6G.

8.6.2 *Disadvantages of RSMA*

Although RSMA offers numerous advantages, it also introduces certain inherent disadvantages, which are outlined below [29, 30]:

- **Receiver complexity due to SIC:** RSMA necessitates at least one layer of SIC at each receiver. While schemes such as one-layer RS, two-layer clustered RS, 1-DPCRS, and Tomlinson–Harashima precoding RS [65] maintain a constant number of SIC layers per user regardless of the number of users K, generalized RS and M-DPCRS require an exponential increase in SIC layers with K. This exponential growth in SIC layers significantly escalates receiver complexity.

- **Increased encoding complexity:** Unlike traditional MA schemes such as SDMA, NOMA, and OMA, where the number of encoded data streams matches the number of users, RSMA entails encoding additional data streams because of message splitting and recombination at the transmitter. Consequently, the encoding complexity of RSMA surpasses that of conventional MA schemes.
- **Increased signaling burden:** Since RSMA involves splitting and combining user messages at the transmitter, each receiver must understand how to decode and combine the received messages to retrieve the intended information from both the common and private streams. Consequently, RSMA may impose a higher downlink signaling burden on transmission networks compared to other MA schemes, as it requires coordination between the transmitters and receivers to manage message splitting and combining effectively.
- **Increased optimization burden:** RSMA imposes a higher optimization burden compared to conventional SDMA schemes. Precoders for both the common and private streams must be jointly optimized along with common rate allocation. Although this enlarges the optimization space and offers more DoF for precoder design, it results in a greater optimization burden. Efforts to mitigate this burden include the design of low-complexity precoders. However, optimizing the power allocation between the common and private streams is critical for RSMA performance. The fraction of transmit power allocated to the private streams needs to be optimized for various operational scenarios to fully realize the potential performance gains of RSMA. Despite these challenges, the significant advantages of RSMA outweigh its drawbacks. As research on RSMA is still in its early stages, these challenges warrant further investigation and exploration in the future.

8.7 Enabling Applications and Challenges

8.7.1 *Enabling technologies*

- **Massive MIMO:** Massive MIMO is pivotal for achieving high SE and EE in 5G new radio (NR) systems. However, its effectiveness relies heavily on accurate CSIT, which poses challenges, particularly in systems with a large number of transmit antennas. FDD massive MIMO systems face issues related to feedback overhead, while TDD systems encounter challenges from channel estimation errors and pilot contamination. RSMA offers a solution by enhancing resilience to uncertainties in CSIT, presenting a promising approach for overcoming obstacles in massive MIMO deployments. Pioneering studies, exemplified by Ref. [40], explore integrating RSMA into FDD massive MIMO setups, introducing innovative two-tier precoding schemes to mitigate CSI feedback overhead. Similarly, investigations such as those by Ref. [57] examine RSMA's performance in TDD massive MIMO under realistic conditions marked by hardware impairments impacting channel estimation. Furthermore, RSMA effectively addresses self-interference challenges in massive MIMO relay systems [58]. Future research trajectories involve exploring RSMA's applicability in cell-free massive MIMO scenarios [41], where non-orthogonal uplink pilot sequences present unique challenges mitigated by RSMA techniques. Additionally, investigations into the implications of low-resolution analog-to-digital converters (ADCs) and digital-to-analog converters (DACs) on RSMA performance and power consumption optimization represent vital areas for further exploration.

- **Millimeter-wave (mmWave):** mmWave communication emerges as a solution to address spectrum scarcity, despite encountering challenges, such as high propagation loss [59]. To mitigate this issue, proposals advocate for the use of massive antenna arrays [60]. However, due to the impracticality of fully digital arrays, there is a preference for hybrid analog/digital processing. Conventional feedback mechanisms for mmWave massive MIMO systems are complex [53]. Recent studies have drawn inspiration from the advantages of RSMA and integrated RSMA with mmWave technology, aiming to reduce complexity while maintaining comparable performance levels [53]. These efforts include exploring joint optimization of analog and digital precoders, showcasing the potential for achieving EE in UAV networks compared to NOMA [53]. Future research directions include endeavors to reduce feedback overhead through techniques such as compressive sensing or deep learning, as well as leveraging mmWave channel sparsity in conjunction with RSMA [53].
- **Multigroup multicasting:** The integration of RSMA into multigroup multicasting networks has garnered significant research attention,

driven by its superior interference management capabilities [61–64]. Initial investigations have illustrated RSMA's efficacy in achieving notable enhancements in max–min fairness and EE across diverse deployment scenarios [61–65]. Studies focusing on PHY layer design and link-level simulations have shed light on RSMA's advantages, particularly in cellular and satellite communication contexts [64]. However, challenges persist in addressing worst-case rate limitations and channel fluctuations in terrestrial environments. Additionally, the integration of terrestrial-satellite networks presents promising avenues for future research, leveraging RSMA's benefits across both domains [66].

- **Non-orthogonal unicast and multicast:** Research on non-orthogonal unicast and multicast (NOUM) has gained significant attention recently, extending from OMA of broadcast/multicast signals to OMA of multiple services [20, 67, 68]. RSMA presents opportunities for improvement over conventional methods in multi-antenna NOUM. It splits unicast messages into common and private parts at the transmitter, encoding them with multicast messages into a super-common stream for all users [47]. This approach optimally exploits SIC, resulting in SE and EE gains without increased receiver complexity. Additionally, a DPC RS-aided NOUM enlarges the rate region compared to conventional methods [69]. Challenges remain in extending these approaches to scenarios with multiple multicast groups, necessitating novel transceiver designs to address three-dimensional interference [47].
- **Multi-cell MIMO including coordinated multi-point:** Research on RSMA in multi-cell MIMO networks, including coordinated multi-point (CoMP), cloud-radio access network (C-RAN), and fog-radio access network (F-RAN), has demonstrated promising outcomes [70, 71]. In coordinated multi-cell networks, RSMA enhances both SE and EE by effectively managing inter-cell interference, employing strategies, such as treating it as noise or partially decoding it [72–74]. Cooperative multi-cell setups with shared CSI and data via fronthaul links have been explored, leading to simplified receiver designs [82]. RSMA in C-RAN addresses fronthaul capacity limitations under both perfect and imperfect CSI scenarios [30, 70, 71]. Furthermore, in F-RAN, RSMA-based edge caching outperforms techniques like MU-LP and NOMA, offering the potential for improved user performance and streamlined transceiver designs [46, 75, 76]. Future research may focus on optimizing content placement and delivery to maximize system performance

in RSMA-aided F-RAN, considering constraints such as limited cache sizes [46, 75, 76].

- **Simultaneous wireless information and power transfer (SWIPT):** Research on SWIPT has gained momentum due to its integration of wireless information transfer (WIT) and wireless power transfer (WPT) [77]. Conventional SWIPT networks encounter challenges, particularly when energy receivers and information receivers are spatially separated, necessitating dedicated energy-carrying signals [78]. RSMA has emerged as a promising solution for multi-antenna SWIPT networks, offering improved interference management among information receivers and ensuring reliable harvested power at energy receivers [78–80]. However, challenges persist, including the limited transmission range to energy receivers and the imperative of ensuring information security. RSMA's capability to encode information into common and private streams presents potential solutions to these challenges [78–80]. Further exploration is warranted to evaluate RSMA's performance in SWIPT scenarios and its contributions to enhancing security [78–80].
- **Cooperative user relaying (CUR):** CUR is a technique enhancing system capacity, transmission reliability, and coverage without extra antennas at transceivers. Users overhear others' information, forwarding it to intended users, improving spatial diversity and signal strength [81, 82]. Cooperative NOMA integrates easily with CUR. However, NOMA's full decoding limitation hinders CUR [54]. In a two-user MISO BC, CRS was proposed, extending to K-user cases [54, 83]. CRS involves two phases: conventional BC serving in the first phase and selected relaying user(s) forwarding decoded common streams in the second, enhancing common stream rates. CRS, a general strategy, outperforms cooperative NOMA, one-layer RS, and MU-LP even at low SNR [54, 83], enhancing PHY layer security [84]. Future research on CRS should focus on imperfect CSIT and sophisticated relaying user selection algorithms, including joint optimization of precoding and user selection.
- **Wireless caching:** Wireless caching presents promising solutions for alleviating backhaul traffic and tackling spectrum scarcity in forthcoming networks [85]. In cache-aided transmission, content prefetching during placement and subsequent delivery phases is integral [85]. The interference management capabilities of RSMA have spurred interest in its deployment within cache-aided networks [76, 85–87]. RSMA-aided caching involves partitioning user messages into cached and uncached

segments, leveraging single-layer RS during delivery to enhance DoF and reduce CSIT requirements [76, 85–87]. Future research endeavors should concentrate on optimizing caching methodologies to SE and EE while mitigating transmission delays, particularly in scenarios constrained by fronthaul limitations [76, 85–87].

- **UAV-aided communication:** UAVs play an increasingly vital role in both 5G NR and envisioned 6G networks, offering extended coverage and flexible, cost-effective mobility [88–90]. RSMA, when applied in UAV-aided communications, has exhibited promising outcomes by enhancing sum-rate performance and EE compared to conventional techniques, such as SDMA and NOMA [88–90]. Nevertheless, challenges persist, particularly in UAV deployment and trajectory design, attributed to imperfect channel tracking and high mobility, resulting in co-channel interference [88–90]. While RSMA's resilience to channel inaccuracies presents a potential solution, further investigation is required to optimize trajectory design, evaluate the impact of UAV mobility on performance, and delineate RSMA's role in addressing UAV lifetime limitations [88–90].
- **Physical-layer security (PLS):** Privacy and security are of utmost importance in wireless communications, with the focus increasingly turning to PLS. RSMA emerges as a promising avenue for enhancing security by thwarting eavesdroppers' attempts to decode data while ensuring successful decoding at intended receivers [51, 84]. Initial studies highlight RSMA's potential in achieving secrecy rate improvements over conventional techniques like MU-LP and NOMA [51, 84]. Additionally, RSMA facilitates joint communication and jamming strategies, resulting in higher sum rates for legitimate users compared to SDMA [91]. However, challenges persist, particularly in addressing internal eavesdropping by authorized users and countering pilot contamination attacks in TDD massive MIMO scenarios [51, 84, 91]. Future research endeavors should concentrate on assessing RSMA's effectiveness in these scenarios to ensure robust secrecy performance.
- **Satellite communications:** Satellite communications play a crucial role in extending coverage, especially where terrestrial infrastructure is challenging to deploy. The evolution of satellite networks toward multibeam architectures introduces inter-beam interference, akin to cellular multigroup multicasting. Inspired by the success of RSMA in cellular networks, recent studies have explored its application in multibeam satellite communications [62]. Initial investigations demonstrate

the effectiveness of RSMA in managing inter-beam interference and improving rate regions compared to traditional strategies [62]. Additionally, RSMA exhibits promising gains over MU-LP in scenarios with imperfect CSIT and multiple beams [62]. Furthermore, RSMA proves robust in handling inter-gateway interference and achieving throughput gains in satellite systems with multiple gateways [128,140]. Integrated terrestrial-satellite networks represent a promising avenue for seamless connectivity, where RSMA offers significant performance enhancements in terms of sum rate, secrecy rate, and EE [90, 92–94]. However, current research on RSMA in integrated terrestrial-satellite networks primarily focuses on single-beam scenarios without inter-beam interference. Moreover, the full potential of satellite cooperation with ground-based BSs remains largely unexplored. Future work should address these limitations by investigating RSMA strategies for scenarios with inter-beam interference and maximizing the collaborative capabilities of satellites and ground-based infrastructure to efficiently serve users with poor terrestrial channels.

- **Integrated radar/sensing and communication (ISAC):** ISAC systems integrate radar sensing and communication functionalities, sharing the spectrum among different RF systems. Recent studies propose an RSMA-aided distributed framework for radar communication (DFRC), aiming to improve the trade-offs between communication rates and radar performance [95–97]. RSMA addresses challenges such as imperfect channel information and hardware constraints, resulting in significant sum-rate gains over traditional methods. However, security remains a concern in ISAC systems, particularly in military applications. Future research should investigate whether RSMA-aided DFRC can enhance PLS and explore the trade-off between radar detection performance and secrecy rates. Addressing these challenges will facilitate the deployment of ISAC systems in real-world scenarios.
- **Cognitive radio (CR):** CR presents a solution to address radio resource scarcity by enabling secondary users (SUs) to opportunistically access frequency bands allocated to licensed primary users (PUs), provided interference to PUs is mitigated. RSMA's robust interference management capabilities make it well suited for MIMO CR networks. In Ref. [80], RSMA was shown to minimize transmit power while satisfying SU data rate and PU interference constraints, resulting in reduced power consumption compared to SDMA. Additionally, Ref. [98] underscores the advantages of RSMA for joint communications and jamming in MIMO CR networks with imperfect CSIT. However, addressing security

concerns arising from spectrum sharing between SUs and PUs remains challenging, particularly concerning the secrecy performance of RSMA in this context.

- **Massive machine-type communication (mMTC):** mMTC plays a crucial role in facilitating IoT applications in 5G and beyond, offering solutions for various IoT deployments in smart cities. Featured in 3GPP Release 13, mMTC aims to establish connectivity among a massive number of IoT devices with low complexity and power consumption. However, serving numerous devices while minimizing control signaling and receiver complexity poses significant challenges. RSMA has emerged as a promising solution, demonstrating higher SE, reduced receiver complexity, and enhanced robustness against CSIT inaccuracies and network loads [99]. Yet, its applicability to address the hybrid requirements of IoT users in 6G, such as enhancing throughput while reducing transmission delay, remains largely unexplored. This hybrid service, known as enhanced mobile broadband (eMBB)-URLLC-mMTC, is identified as a core service in 6G, warranting further investigation into the suitability of RSMA as a multiple access technique.
- **Visible light communication (VLC):** VLC, utilizing the visible light spectrum for communication, offers high-speed connectivity, particularly in indoor environments, complementing RF-based mobile networks. However, multi-user interference remains a challenge in MU-MIMO VLC, where multiple light-emitting diodes (LEDs) serve multiple users simultaneously. RSMA, known for its strong interference management in RF communications and the high operating SNR of VLC, shows promise in MU-MIMO VLC. Studies in Ref. [100, 101] demonstrate RSMA's clear sum-rate gains over conventional schemes, even with highly correlated user channels. Practical limitations like imperfect CSIT require further exploration for RSMA-aided VLC, as discussed in Ref. [101].
- **Intelligent reconfigurable surface (IRS)-aided communications:** IRS-aided communications is a promising 6G technology that utilizes programmable meta-atoms to control signal reflection and enhance wireless networks. RSMA's effectiveness in multi-antenna systems has prompted research into its integration with IRS for improved communication. Initial studies demonstrate RSMA's superior EE in MISO BCs, with multiple IRSs aiding downlink transmission. Further investigations explore hybrid RS and IRS-aided strategies, showing enhanced performance. Challenges remain in acquiring CSI at the IRS, and integrating RSMA with IRS warrants further study for optimizing system performance [16, 102–105].

- **Other open issues:** While RSMA offers promising solutions in various areas of 6G technology, there are additional research directions where its potential can be further explored:
  - **Space–air–ground integrated networks (SAGIN):** SAGIN aims to integrate space, air, and ground communications seamlessly to balance network load and mitigate interference [106]. RSMA's advanced interference management capabilities position it as a suitable candidate to address interference challenges within SAGIN networks effectively.
  - **Vehicle-to-everything (V2X) communications:** V2X communication, including V2N, V2I, V2P, and V2V communications, is critical for intelligent transportation systems [107, 108]. RSMA's resilience against interference caused by user mobility and feedback delays makes it a promising technology for enhancing V2X communication reliability and efficiency.
  - **3D eMBB: eMBB, URLLC, massive mMTC services:** With 6G expected to deliver three-dimensional (3D) services to cater to diverse use cases [109], RSMA stands out as a unified solution capable of optimizing performance across key metrics, such as data rate, mobility, latency, reliability, and EE. These areas present numerous open research topics where RSMA can contribute significantly.

8.7.2 *Challenges and future research*

RSMA's benefits in the context of the multi-antenna BC have sparked significant interest in integrating it with various 5G/6G communication technologies. These include massive MIMO, IRS, VLC, UAVs, joint communication and sensing, and satellite communication networks. Despite this growing interest, RSMA research is still in its nascent stages, with several open challenges across different applications. These challenges span PHY-layer design, cross-layer design, and performance analysis [30, 121].

- **MIMO BC:** Previous studies primarily explored RSMA in the MISO BC with single-antenna receivers. However, extending RSMA to scenarios with multi-antenna receivers remains relatively unexplored. It's crucial to revisit considerations, such as the number of common streams and transceiver design to adapt RSMA for the MIMO BC [110].

  Significant advancements in information theory have shown that RSMA approaches the capacity region in two-user MIMO BCs with perfect CSIT, indicating its versatility. Efforts in communication theory aim to optimize precoders for various MIMO BC scenarios, even with imperfect CSIT, leading to improvements in ergodic sum rate and DoF [111–113].

While RSMA research historically focused on single-antenna receivers in the MIMO BC, recent efforts have expanded its scope. Challenges remain in maximizing RSMA's performance and addressing its limitations, particularly in scenarios with imperfect CSIT. Further research is needed to fully explore RSMA's potential across diverse MIMO BC setups and optimize its practical implementation [110].

- **MIMO interference channel (IC):** Understanding the entire capacity region and the optimal strategy for the MIMO IC remains a challenge, especially in scenarios with imperfect CSIT. While the HK scheme [19] has shown capacity achievement in the SISO IC, its extension to the MIMO IC with multiple cells and a constant gap is established. RSMA demonstrates promising results in achieving the best-known DoF region for the K-cell MISO IC with imperfect CSIT, and extensions to the MIMO IC with various antenna configurations are being explored. Coordinated beamforming for RSMA in the multi-cell MISO IC with perfect CSIT shows superior trade-offs between SE and EE compared to SDMA and NOMA. However, fully realizing the potential of RSMA in unlocking the full DoF and capacity region of the MIMO IC requires further exploration, particularly in PHY-layer design and addressing link-level simulations. Integrating RSMA with interference alignment, while considering CSIT uncertainties, could enhance transmission SE. Tackling open challenges in transceiver design and cross-layer resource allocation for diverse objectives is crucial for advancing the applicability of RSMA in the MIMO IC [74, 114, 115].
- **Alternative receiver design:** While SIC is essential in RSMA for interference management, its vulnerability to error propagation poses challenges, especially in scenarios with imperfect CSI at the receiver. Despite RSMA's robust performance with SIC under perfect CSIR, degradation occurs in the presence of imperfections, even with correctly decoded common messages. Recent studies propose deep learning-based receiver algorithms to mitigate this issue, aiming to improve symbol detection pre- and post-interference cancellation and enhance error rate performance [33–35]. The results indicate promising mitigation of interference cancellation effects and enhanced error rate performance under imperfect CSIR conditions, highlighting the importance of further investigating RSMA's error rate performance under imperfect CSIR to enhance its robustness in real-world scenarios.

- **Hybrid automatic repeat request (HARQ) design:** The integration of HARQ with RSMA remains largely unexplored in current literature. HARQ significantly enhances packet-based transmission efficiency by combining MAC-layer ACK-NACK feedback with channel coding to facilitate efficient re-transmission schemes for recovering lost packets in wireless networks. While HARQ is a well-established mechanism in modern wireless standards, its application to RSMA has yet to be thoroughly investigated. Unlike in SDMA, where HARQ involves encoding packets separately for different users, its implementation in RSMA is more intricate due to message splitting and combining operations, as well as potential scenarios of incorrect decoding of common and/or private messages at each user. Thus, further research is needed to explore optimal encoding strategies and re-transmission schemes within the RSMA framework when combined with HARQ [30].
- **Waveform design:** Waveform design is a critical yet often overlooked aspect in RSMA research. While resource allocation and precoder optimization have been studied for multi-carrier waveforms in some works [64, 116–118], issues related to waveform structure imperfections remain unaddressed. In practical scenarios, waveforms may suffer from impairments like intercarrier interference due to Doppler shift, especially in OFDM systems. RSMA holds promise in bolstering waveform robustness by mitigating these vulnerabilities and providing flexibility in waveform design, ultimately enhancing SE. Further research is necessary to explore how RSMA can effectively tackle waveform imperfections and optimize waveform design for greater SE.

8.7.3 *Standardization and implementation*

Standardizing and implementing RSMA is crucial for its integration into wireless networks [30]. Key areas requiring attention include the design of a compatible HARQ scheme, which is essential for efficient packet-based transmission. Additionally, downlink and uplink signaling must be enhanced to ensure the synchronization of message-splitting ratios between transmitters and receivers. While the PHY-layer design and the performance of RSMA have been explored, there's a need for system-level evaluations considering higher-layer designs, such as HARQ, user scheduling, and QoS provisioning.

Previous 3GPP releases have already tackled aspects relevant to RSMA, offering foundational insights for its inclusion in future releases with minimal additional mechanisms. For instance, MU-MIMO and CoMP mechanisms provide valuable precedents for multi-user transmission, while MBS addresses broadcast/multicast transmission, which can be adapted to support RSMA. Moreover, SIC mechanisms considered in previous releases for network-assisted interference cancellation and suppression can readily be adopted for RSMA without significant additional standardization and implementation burden. These efforts pave the way for RSMA integration into future wireless standards, facilitating its deployment and adoption in real-world networks [106, 107, 109, 119, 120].

8.8 Summary

This chapter provides an in-depth overview of RSMA, which diverges from the OMA vs. NOMA debate by emphasizing interference management in multiple access techniques. RSMA employs the RS principle to decode interference partially, leading to improved SE, EE, and computation efficiency compared to traditional methods. It serves as a versatile framework for designing non-orthogonal transmission, MA, and interference management strategies in future wireless networks, bridging the gap between fully decoding interference and treating it as noise. RSMA encompasses four existing MA schemes and offers enhancements in SE, EE, coverage, user fairness, reliability, and QoS under various network loads and channel conditions. Despite its potential, RSMA faces challenges in standardization but presents opportunities for revolutionizing PHY and MAC layers. The chapter surveys state-of-the-art research, discusses RSMA's architecture, applications, performance, and complexity compared to existing MA schemes, and outlines future research challenges for RSMA-enabled wireless communication systems in beyond 5G networks.

Appendix A.8. Proof of Lemma 1

Assume that for the optimal solution $(r^*, \boldsymbol{r}^*)$ of problem (8.55), one can have

$$\sum_{k\in\mathcal{K}'} r_k^* < B\log_2\left(1+\frac{\sum_{k\in\mathcal{K}'} h_k P_k}{\sigma^2 B}\right), \quad \forall\mathcal{K}' \subseteq \mathcal{K}\backslash\emptyset \tag{A.1}$$

In this case, it is possible to construct a new rate solution $r' = [r'_1, r'_2, \ldots, r'_K]^T$ with $r'_k = \epsilon r^*_k$ and

$$\epsilon = \min_{\mathcal{K}' \subseteq \mathcal{K} \setminus \emptyset} \frac{B \log_2 \left(1 + \frac{\sum_{k \in \mathcal{K}'} h_k P_k}{\sigma^2 B}\right)}{\sum_{k \in \mathcal{K}'} r^*_k} > 1 \tag{A.2}$$

According to (A.2), it can be shown that

$$\sum_{k \in \mathcal{K}'} r'_k \leq B \log_2 \left(1 + \frac{\sum_{k \in \mathcal{K}'} h_k P_k}{\sigma^2 B}\right), \quad \forall \mathcal{K}' \subseteq \mathcal{K} \setminus \emptyset \tag{A.3}$$

which ensures that r' satisfies constraints (8.55b).

Based on (8.54a), one can have $\tau^* = \frac{r^*_k}{D_k}, \forall k \in \mathcal{K}$. Define τ' as

$$\tau' = \frac{r'_k}{D_k} = \frac{\epsilon r^*_k}{D_k} > \tau^* \tag{A.4}$$

According to (A.3) and (A.4), it is evident that the new solution (τ', r') is feasible and the objective value (8.55) of the new solution is better than that of solution (τ^*, r^*), which contradicts the fact that (τ^*, r^*) is the optimal solution. Lemma 1 is proved.

Appendix B.8. Proof of Theorem 2

Denote

$$z_k = \sum_{j=k}^{K} h_j q_j, \quad \forall k \in \mathcal{K} \tag{B.1}$$

According to (8.60a), we can obtain

$$B \log_2 \left(\frac{z_k + \sigma^2 B}{z_{k+1} + \sigma^2 B} \right) = D_k \tau \tag{B.2}$$

Based on (B.2), we have

$$z_k = 2^{\frac{D_k \tau}{B}} z_{k+1} + \left(2^{\frac{D_k \tau}{B}} - 1\right) \sigma^2 B \tag{B.3}$$

Using the recursive formulation (B.3) and $z_{K+1} = \sum_{k=K+1}^{K} q_k = 0$, we can calculate

$$z_k = \sum_{j=k}^{K} 2^{\frac{\sum_{l=k}^{j-1} D_l \tau}{B}} \left(2^{\frac{D_j \tau}{B}} - 1\right) \sigma^2 B \tag{B.4}$$

where we set $\sum_{l=k}^{k-1} D_l = 0$. Based on (B.1), we have

$$q_k = \frac{z_k - z_{k+1}}{h_k}, \quad \forall k \in \mathcal{K} \tag{B.5}$$

Combining (B.4) and (B.5) yields

$$q_k = \frac{1}{h_k}\left(2^{\frac{D_k \tau}{B}} - 1\right) \sum_{j=k+1}^{K} 2^{\frac{\sum_{l=k+1}^{j-1} D_l \tau}{B}} \left(2^{\frac{D_j \tau}{B}} - 1\right) \sigma^2 B + \frac{1}{h_k}\left(2^{\frac{D_k \tau}{B}} - 1\right) \sigma^2 B, \quad \forall k \in \mathcal{K} \tag{B.6}$$

which monotonically increases with τ. Considering the maximum uplink transmission power constraints (8.60b), one can obtain that

$$\tau \le \tau_k, \quad \forall k \in \mathcal{K} \tag{B.7}$$

where τ_k is the solution to (8.63). To maximize sum rate τ, the optimal τ^* is given by (8.61). Consequently, the optimal transmission power is provided by (8.62).

Appendix C.8. Proof of Theorem 3

Solving (8.41a) yields

$$b_k = -\frac{(\ln 2) D_k h_k P_k \tau}{B h_k P_k W\left(-\frac{(\ln 2) D_k \sigma^2 \tau}{h_k P_k} e^{-\frac{(\ln 2) D_k \sigma^2 \tau}{h_k P_k}}\right) + (\ln 2) D_k \sigma^2 B \tau} \tag{C.1}$$

Substituting (C.1) to constraint (8.65b), the optimal τ^* is calculated as (8.66), and the optimal bandwidth allocation is accordingly given by (8.67).

Appendix D.8. Proof of Theorem 4

Solving (8.69a) yields

$$a_k = \frac{D_k \tau}{B\log_2(1 + \frac{h_k P_k}{\sigma^2 B})} \tag{D.1}$$

Combining (D.1) and constraint (8.69b), the optimal sum rate is given by (8.70). Then, the optimal time sharing is given by (8.71).

Appendix E.8. Proof of Lemma 3

For two-user NOMA, according to (8.59), one can obtain

$$r_1 = B\log_2\left(1 + \frac{h_1 q_1}{h_2 q_2 + \sigma^2 B}\right), \quad r_2 = B\log_2\left(1 + \frac{h_2 q_2}{\sigma^2 B}\right)$$
$$0 \le q_1 \le P_1, \quad 0 \le q_2 \le P_2 \tag{E.1}$$

Based on (E.1), the following is derived:

$$r_2 = B\log_2\left(1 + \frac{h_2 q_2}{\sigma^2 B}\right) \le B\log_2\left(1 + \frac{h_2 P_2}{\sigma^2 B}\right) = R_2 \tag{E.2}$$

and

$$h_2 q_2 + \sigma^2 B = \frac{h_1 q_1}{2^{\frac{r_1}{B}} - 1} \tag{E.3}$$

Substituting (E.3) into (E.2) yields

$$\begin{aligned} r_2 &= B\log_2\left(1 + \frac{h_2 q_2}{\sigma^2 B}\right) = B\log_2\left(\frac{h_1 q_1}{(2^{\frac{r_1}{B}} - 1)\sigma^2 B}\right) \\ &= B\log_2\left(\frac{h_1 q_1}{\sigma^2 B}\right) - B\log_2\left(2^{\frac{r_1}{B}} - 1\right) \le B\log_2\left(\frac{h_1 P_1}{\sigma^2 B}\right) \\ &\quad - B\log_2\left(2^{\frac{r_1}{B}} - 1\right) \end{aligned} \tag{E.4}$$

where the inequality follows from the fact that $0 \le q_1 \le P_1$. Combining (E.1) and (E.4), the two-user NOMA rate region is given in (8.73).

For two-user FDMA, according to (8.64), the following is derived:

$$r_1 = Bb_1\log_2\left(1+\frac{h_1P_1}{\sigma^2Bb_1}\right)$$

$$r_2 = Bb_1\log_2\left(1+\frac{h_2P_2}{\sigma^2Bb_2}\right)$$

$$b_1+b_2=1,\quad b_1,b_2\geq 0 \tag{E.5}$$

Based on (E.5), we have

$$b_k = f_k(r_k),\quad k=1,2 \tag{E.6}$$

where $f_k(r_k)$ is defined in (8.76). Combining (E.5) and (E.6), the two-user FDMA rate region is given in (8.74).

For two-user TDMA, according to (8.68), one can calculate

$$r_1 = a_1R_1,\quad r_2=a_2R_2,\quad a_1+a_2=1,\quad a_1,\quad a_2\geq 0 \tag{E.7}$$

Based on (E.7), two-user TDMA rate is given as (8.77).

References

[1] H. Tataria, M. Shafi, A. F. Molisch, M. Dohler, H. Sjöland, and F. Tufvesson, 6G wireless systems: Vision, requirements, challenges, insights, and opportunities, *Proc. IEEE*, 109(7), 1166–1199, Jul 2021.

[2] D. Vargas and Y. J. D. Kim, Two-layered superposition of broadcast/multicast and unicast signals in multiuser OFDMA systems, *IEEE Trans. Wireless Commun.*, 19(2), 979–994, Feb 2020.

[3] A. Goldsmith, S. A. Jafar, N. Jindal, and S. Vishwanath, Capacity limits of MIMO channels, *IEEE J. Sel. Areas Commun.*, 21(5), 684–702, Jun 2003.

[4] B. Clerckx, H. Joudeh, C. Hao, M. Dai, and B. Rassouli, Rate splitting for MIMO wireless networks: A promising PHY-layer strategy for LTE evolution, *IEEE Commun. Mag.*, 54(5), 98–105, May 2016.

[5] Z. Ding *et al.*, Application of non-orthogonal multiple access in LTE and 5G networks, *IEEE Commun. Mag.*, 55(2), 185–191, Feb 2017.

[6] S. M. Riazul Islam, Nurilla Avazov, Octavia A. Dobre and Kyung-Sup Kwak, Power-domain non-orthogonal multiple access (NOMA) in 5G systems: Potentials and challenges. *IEEE Commun. Surv. Tutor.*, 19(2), 721–742, 2017

[7] W. Shin, M. Vaezi, B. Lee, D. J. Love, J. Lee, and H. V. Poor, Non-orthogonal multiple access in multi-cell networks: Theory, performance, and practical challenges, *IEEE Commun. Mag.*, 55(10), 176–183, Oct 2017.

[8] Y. Mao, B. Clerckx, and V. O. K. Li, Rate-splitting multiple access for downlink communication systems: Bridging, generalizing, and outperforming SDMA and

NOMA, *EURASIP Journal on Wireless Communications and Networking*, 2018(1), pp. 1–54, 2018.

[9] B. Clerckx, Y. Mao, R. Schober, and H. V. Poor, Rate-splitting unifying SDMA, OMA, NOMA, and multicasting in MISO broadcast channel: A simple two-user rate analysis, *IEEE Wireless Commun. Lett.*, 9(3), 349–353, Mar 2020.

[10] N. Jindal, MIMO broadcast channels with finite-rate feedback, *IEEE Trans. Inf. Theory*, 52(11), 5045–5060, Nov 2006.

[11] Y. Saito, Y. Kishiyama, A. Benjebbour, T. Nakamura, A. Li, and K. Higuchi, Non-orthogonal multiple access (NOMA) for cellular future radio access, in *Proc. IEEE 77th Veh. Tech. Conf. (VTC Spring)*, 1–5, 2013.

[12] Y. Liu *et al.*, Evolution of NOMA toward next generation multiple access (NGMA) for 6G, *IEEE J. Sel. Areas Commun.*, 40(4), 1037–1071, Apr 2022.

[13] Z. Ding, F. Adachi, and H. V. Poor, The application of MIMO to non-orthogonal multiple access, *IEEE Trans. Wireless Commun.*, 15(1), 537–552, Jan 2016.

[14] T. Han, J. Gong, X. Liu, S. M. R. Islam, Q. Li, Z. Bai, K. S. Kwak, On downlink NOMA in heterogeneous networks with non-uniform small cell deployment, *IEEE Access*, 6, 31099–31109, Jun 2018.

[15] B. Clerckx *et al.*, Is NOMA efficient in multi-antenna networks? A critical look at next generation multiple access techniques, *IEEE Open J. Commun. Soc.*, 2, 1310–1343, 2021.

[16] H. You, Z. Bai, H. Liu, T. A. Tsiftsis and K.S. Kwak, Rate-splitting for intelligent reflecting surface-assisted CR-NOMA systems, *J. of Commun. and Netw.*, 25(1), 15–24, Feb 2023.

[17] A. Carleial, A case where interference does not reduce capacity, *IEEE Trans. Inf. Theory*, 21(5), 569–570, 1975.

[18] A. Carleial, Interference channels, *IEEE Trans. Inf. Theory*, 24(1), 60–70, Jan 1978.

[19] T. Han and K. Kobayashi, A new achievable rate region for the interference channel, *IEEE Trans. Inf. Theory*, 27(1), 49– 60, Jan 1981.

[20] M. Costa, Writing on dirty paper, *IEEE Trans. Inf. Theory*, 29(3), 439–441, May 1983.

[21] H. Weingarten, Y. Steinberg, and S. S. Shamai, The capacity region of the Gaussian multiple-input multiple-output broadcast channel, *IEEE Trans. Inf. Theory*, 52(9), 3936–3964, Sep 2006.

[22] H. Joudeh and B. Clerckx, Sum-rate maximization for linearly precoded downlink multiuser MISO systems with partial CSIT: A rate splitting approach, *IEEE Trans. Commun.*, 64(11), 4847–4861, Nov 2016.

[23] E. Piovano and B. Clerckx, Optimal DoF region of the K-user MISO BC with partial CSIT, *IEEE Commun. Lett.*, 21(11), 2368–2371, Nov 2017.

[24] Y. Mao and B. Clerckx, Beyond dirty paper coding for multi-antenna broadcast channel with partial CSIT: A rate-splitting approach, *IEEE Trans. Commun.*, 68(11), 6775–6791, Nov 2020.

[25] A. Mishra, Y. Mao, O. Dizdar, and B. Clerckx, Rate-splitting multiple access for downlink multiuser MIMO: Precoder optimization and PHY-layer design, *IEEE Trans. Commun.*, (early access) 2021.

[26] A. Gholami Davoodi and S. Jafar, Degrees of freedom region of the (m, n1, n2) MIMO broadcast channel with partial CSIT: An application of sum-set inequalities based on aligned image sets, *IEEE Trans. Inf. Theory*, 66(10), 6256–6279, Oct 2020.

[27] C. Hao, B. Rassouli, and B. Clerckx, Achievable DoF regions of MIMO networks with imperfect CSIT, *IEEE Trans. Inf. Theory*, 63(10), 6587–6606, Oct 2017.

[28] B. Rimoldi and R. Urbanke, A rate-splitting approach to the Gaussian multiple-access channel, *IEEE Trans. Inf. Theory*, 42(2), 364–375, Mar 1996.

[29] B. Clerckx, Y. Mao, E. A. Jorswieck, J. Yuan, D. J. Love, E. Erkip and D. Niyato, A primer on rate-splitting multiple access: Tutorial, myths, and frequently asked questions, *IEEE J. on Sel. Areas in Commun.*, 41(5), 1265–1308, May 2023.

[30] Y. Mao, O. Dizdar, B. Clerckx, R. Schober, P. Popovski, and H. V. Poor, Rate-splitting multiple access: Fundamentals, survey, and future research trends, *IEEE Commun. Surv. Tutor.*, 24(4), 2073–2126, 2022.

[31] Z. Li, C. Ye, Y. Cui, S. Yang, and S. Shamai, Rate splitting for multi-antenna downlink: Precoder design and practical implementation, *IEEE J. Sel. Areas Commun.*, 38(8), 1910–1924, Aug 2020.

[32] A. Alameer Ahmad, H. Dahrouj, A. Chaaban, A. Sezgin, and M. Alouini, Interference mitigation via rate-splitting and common message decoding in cloud radio access networks, *IEEE Access*, 7, 80350–80365, 2019.

[33] O. Dizdar, Y. Mao, W. Han, and B. Clerckx, Rate-splitting multiple access for Downlink multi-antenna communications: Physical layer design and link-level simulations, in *Proc. IEEE Annu. Symp. Pers. Indoor Mobile Radio Commun. (PIMRC)*, 1–6, 2020.

[34] O. Dizdar, Y. Mao, and B. Clerckx, Rate-splitting multiple access to mitigate the curse of mobility in (massive) MIMO networks, *IEEE Trans. Commun.*, 69(10), 6765–6780, Oct 2021.

[35] A. Mishra, Y. Mao, O. Dizdar, and B. Clerckx, Rate-splitting multiple access for downlink multiuser MIMO: Precoder optimization and PHY-layer design, *IEEE Trans. Commun.*, 70(2), 874–890, Feb 2022.

[36] Z. Yang, M. Chen, W. Saad, W. Xu, and M. Shikh-Bahaei, Sum-rate maximization of uplink rate splitting multiple access (RSMA) communication, *IEEE Trans. Mob. Comput.*, 21(7), 2596–2609, Jul 2022.

[37] M. Vaezi, Z. Ding, and H. V. Poor, *Multiple Access Techniques for 5G Wireless Networks and Beyond*, Springer, 2018.

[38] Z. Shen, J. G. Andrews, and B. L. Evans, Adaptive resource allocation in multiuser OFDM systems with proportional rate constraints, *IEEE Trans. Wireless Commun.*, 4(6), 2726–2737, Nov 2005.

[39] Y. Saito, Y. Kishiyama, A. Benjebbour, T. Nakamura, A. Li, and K. Higuchi, Non-orthogonal multiple access (NOMA) for cellular future radio access, in *Proc. IEEE Veh. Technol. Conf.* Dresden, German, 1–5 Jun 2013; e-splitting approach, *IEEE Trans. Commun.*, 63(9), 3232–3246, Sep 2015.

[40] M. Dai, B. Clerckx, D. Gesbert, and G. Caire, A rate splitting strategy for massive MIMO with imperfect CSIT, *IEEE Trans. Wireless Commun.*, 15(7), 4611–4624, Jul 2016.

[41] A. Mishra, Y. Mao, L. Sanguinetti, and B. Clerckx, Rate-splitting assisted massive machine-type communications in cell-free massive MIMO, *IEEE Commun. Lett.*, 26(6), 1358–1362, Jun 2022.

[42] H. Joudeh and B. Clerckx, Sum-rate maximization for linearly precoded Downlink multiuser MISO systems with partial CSIT: A rate-splitting approach, *IEEE Trans. Commun.*, 64(11), 4847–4861, Nov 2016.

[43] Y. Mao and B. Clerckx, Beyond dirty paper coding for multi-antenna broadcast channel with partial CSIT: A rate-splitting approach, *IEEE Trans. Commun.*, 68(11), 6775–6791, Nov 2020.

[44] G. Zhou, Y. Mao, and B. Clerckx, Rate-splitting multiple access for multi-antenna downlink communication systems: Spectral and energy efficiency tradeoff, *IEEE Trans. Wireless Commun.*, 21(7), 4816–4828, Jul 2022.

[45] J. Zhang, J. Zhang, Y. Zhou, H. Ji, J. Sun, and N. Al-Dhahir, Energy and spectral efficiency tradeoff via rate splitting and common beam-forming coordination in multicell networks, *IEEE Trans. Commun.*, 68(12), 7719–7731, Dec 2020.

[46] M. Z. Hassan, M. J. Hossain, J. Cheng, and V. C. M. Leung, Energy-spectrum efficient content distribution in fog-RAN using rate-splitting, common message decoding, and 3D-resource matching, *IEEE Trans. Wireless Commun.*, 20(8), 4929–4946, Aug 2021

[47] Y. Mao, B. Clerckx, and V. O. K. Li, Rate-splitting for multi-antenna non-orthogonal unicast and multicast transmission: Spectral and energy efficiency analysis, *IEEE Trans. Commun.*, 67(12), 8754–8770, Dec 2019.

[48] H. Joudeh and B. Clerckx, Robust transmission in downlink multiuser MISO systems: A rate-splitting approach, *IEEE Trans. Sig. Process.*, 64(23), 6227–6242, Dec 2016.

[49] Y. Mao, B. Clerckx, J. Zhang, V. O. K. Li, and M. A. Arafah, Max-min fairness of K-user cooperative rate-splitting in MISO broadcast channel with user relaying, *IEEE Trans. Wireless Commun.*, 19(10), 6362–6376, Oct 2020.

[50] H. Chen, D. Mi, B. Clerckx, Z. Chu, J. Shi, and P. Xiao, Joint power and subcarrier allocation optimization for multigroup multicast systems with rate splitting, *IEEE Trans. Veh. Technol.*, 69(2), 2306–2310, Feb 2020.

[51] H. Fu, S. Feng, W. Tang, and D. W. K. Ng, Robust secure beamforming design for two-user downlink MISO rate-splitting systems, *IEEE Trans. Wireless Commun.*, 19(12), 8351–8365, Dec 2020.

[52] C. Hao, Y. Wu, and B. Clerckx, Rate analysis of two-receiver MISO broadcast channel with finite rate feedback: A rate-splitting approach, *IEEE Trans. Commun.*, 63(9), 3232–3246, Sep 2015.

[53] M. Dai and B. Clerckx, Multiuser millimeter wave beamforming strategies with quantized and statistical CSIT, *IEEE Trans. Wireless Commun.*, 16(11), 7025–7038, Nov 2017.

[54] J. Zhang, B. Clerckx, J. Ge, and Y. Mao, Cooperative rate splitting for MISO broadcast channel with user relaying, and performance benefits over cooperative NOMA, *IEEE Sig. Process. Lett.*, 26(11), 1678–1682, Nov 2019.

[55] Y. Mao, B. Clerckx, J. Zhang, V. O. K. Li, and M. A. Arafah, Max-min fairness of K-user cooperative rate-splitting in MISO broadcast channel with user relaying, *IEEE Trans. Wireless Commun.*, 19(10), 6362–6376, Oct 2020.

[56] Y. Xu, Y. Mao, O. Dizdar, and B. Clerckx, Rate-splitting multiple access with finite blocklength for short-packet and low-latency Downlink communications, *IEEE Trans. Veh. Tech.*, 71(11), 12333–12337, Nov 2022.

[57] A. Papazafeiropoulos, B. Clerckx, and T. Ratnarajah, Rate-splitting to mitigate residual transceiver hardware impairments in massive MIMO systems, *IEEE Trans. Veh. Tech.*, 66(9), 8196–8211, Sep 2017.

[58] A. Papazafeiropoulos and T. Ratnarajah, Rate-splitting robustness in multi-pair massive MIMO relay systems, *IEEE Trans. Wireless Commun.*, 17(8), 5623–5636, Aug 2018.

[59] A. Alkhateeb, G. Leus, and R. W. Heath, Limited feedback hybrid precoding for multi-user millimeter wave systems, *IEEE Trans. Wireless Commun.*, 14 (11), 6481–6494, Nov 2015.

[60] O. E. Ayach, S. Rajagopal, S. Abu-Surra, Z. Pi, and R. W. Heath, Spatially sparse precoding in millimeter wave MIMO systems, *IEEE Trans. Wireless Commun.*, 13(3), 1499–1513, Mar 2014.

[61] H. Joudeh and B. Clerckx, Rate-splitting for max−min fair multigroup multicast beamforming in overloaded systems, *IEEE Trans. Wireless Commun.*, 16(11), 7276–7289, Nov 2017.

[62] L. Yin and B. Clerckx, Rate-splitting multiple access for Multigroup multicast and multibeam satellite systems, *IEEE Trans. Commun.*, 69(2), 976–990, Feb 2021.

[63] A. Z. Yalcin, M. Yuksel, and B. Clerckx, Rate splitting for multi-group Multicasting with a common message, *IEEE Trans. Veh. Tech.*, 69(10), 12281–12285, Oct 2020.

[64] H. Chen *et al.*, Rate-splitting for multicarrier multigroup multicast: Precoder design and error performance, *IEEE Trans. Broadcast.*, 67(3), 619–630, Sep 2021.

[65] A. R. Flores, R. C. De Lamare, and B. Clerckx, Tomlinson–Harashima precoded rate-splitting with stream combiners for MU-MIMO systems, *IEEE Trans. Commun.*, 69(6), pp. 3833–3845, Jun. 2021

[66] X. Zhu, C. Jiang, L. Yin, L. Kuang, N. Ge, and J. Lu, Cooperative multigroup multicast transmission in integrated terrestrial-satellite networks, *IEEE J. Sel. Areas Commun.*, 36(5), 981–992, May 2018.

[67] L. Fay, L. Michael, D. Gómez-Barquero, N. Ammar, and M. W. Caldwell, An overview of the ATSC 3.0 physical layer specification, *IEEE Trans. Broadcast.*, 62(1), 159–171, Mar 2016.

[68] S. M. Riazul Islam, Ming Zeng, Octavia A. Dobre, Kyung-Sup Kwak, Resource allocation for downlink NOMA systems: Key techniques and open issues, *IEEE Wireless Commun.*, 25(2), 40–47, Apr 2018.

[69] Y. Mao and B. Clerckx, Dirty paper coded rate-splitting for non-orthogonal unicast and multicast transmission with partial CSIT, in *Proc. 54th Asilomar Conf. Sigs. Syst. Comput.*, 990–995, 2020.

[70] A. A. Ahmad, Y. Mao, A. Sezgin, and B. Clerckx, Rate splitting multiple access in C-RAN: A scalable and robust design, *IEEE Trans. Commun.*, 69(9), 5727–5743, Sep 2021.

[71] D. Yu, J. Kim, and S. Park, An efficient rate-splitting multiple access scheme for the downlink of C-RAN systems, *IEEE Wireless Commun. Lett.*, 8(6), 1555–1558, Dec 2019.

[72] R. H. Etkin, D. N. C. Tse, and H. Wang, Gaussian interference channel capacity to within one bit, *IEEE Trans. Inf. Theory*, 54(12), 5534–5562, Dec 2008.

[73] A. Carleial, Interference channels, *IEEE Trans. Inf. Theory*, 24(1), 60–70, Jan 1978.

[74] C. Hao and B. Clerckx, MISO networks with imperfect CSIT: A topological rate-splitting approach, *IEEE Trans. Commun.*, 65(5), 2164–2179, May 2017.

[75] M. Z. Hassan, M. J. Hossain, J. Cheng, and V. C. M. Leung, Device-clustering and rate-splitting enabled device-to-device cooperation framework in fog radio access network, *IEEE Trans. Green Commun. Netw.*, 5(3), 1482–1501, Sep 2021.

[76] R.-J. Reifert, A. A. Ahmad, Y. Mao, A. Sezgin, and B. Clerckx, Rate-splitting multiple access in cache-aided cloud-radio access networks, *Front. Commun. Netw.*, 2, Sep 2021, Art. no. 716620.

[77] J. Xu, L. Liu, and R. Zhang, Multiuser MISO beamforming for simultaneous wireless information and power transfer, *IEEE Trans. Sig. Process.*, 62(18), 4798–4810, Sep 2014.

[78] Y. Mao, B. Clerckx, and V. O. K. Li, Rate-splitting for multi-user multi-antenna wireless information and power transfer, in *Proc. IEEE Int. Workshop Signal Process. Adv. Wireless Commun. (SPAWC)*, pp. 1–5, 2019.

[79] X. Su, L. Li, H. Yin, and P. Zhang, Robust power- and rate- splitting-based transceiver design in K-user MISO SWIPT interference channel under imperfect CSIT, *IEEE Commun. Lett.*, 23(3), 514–517, Mar 2019.

[80] M. R. C. Acosta, C. E. G. Moreta, and I. Koo, Joint power allocation and power splitting for MISO-RSMA cognitive radio systems with SWIPT and information decoder users, *IEEE Syst. J.*, 15(4), 5289–5300, Dec 2021.

[81] A. Ghosh, D. Wolter, J. Andrews, and R. Chen, Broadband wireless access with WiMax/802.16: Current performance benchmarks and future potential, *IEEE Commun. Mag.*, 43(2), 129–136, Feb 2005.

[82] M. Lee, J. Zheng, Y.-B. Ko, and D. Shrestha, Emerging standards for wireless mesh technology, *IEEE Wireless Commun.*, 13(2), 56–63, Apr 2006.

[83] Y. Mao, B. Clerckx, J. Zhang, V. O. K. Li, and M. A. Arafah, Max-min fairness of K-user cooperative rate-splitting in MISO broadcast channel with user relaying, *IEEE Trans. Wireless Commun.*, 19(10), 6362–6376, Oct 2020.

[84] P. Li, M. Chen, Y. Mao, Z. Yang, B. Clerckx, and M. Shikh-Bahaei, Cooperative rate-splitting for secrecy sum-rate enhancement in multi-antenna broadcast channels, in *Proc. IEEE Annu. Symp. Pers. Indoor Mobile Radio Commun. (PIMRC)*, 1–6, 2020.

[85] N. Naderializadeh, M. A. Maddah-Ali, and A. S. Avestimehr, Fundamental limits of cache-aided interference management, *IEEE Trans. Inf. Theory*, 63(5), 3092–3107, May 2017.

[86] E. Demarchou, C. Psomas, and I. Krikidis, Rate splitting with wireless edge caching: A system-level-based co-design, *IEEE Trans. Commun.*, 70(1), 664–679, Jan 2022.

[87] J. Zhang and P. Elia, Fundamental limits of cache-aided wireless BC: Interplay of coded-caching and CSIT feedback, *IEEE Trans. Inf. Theory*, 63(5), 3142–3160, May 2017.

[88] W. Jaafar, S. Naser, S. Muhaidat, P. C. Sofotasios, and H. Yanikomeroglu, Multiple access in aerial networks: From orthogonal and non-orthogonal to rate-splitting, *IEEE Open J. Veh. Technol.*, 1, 372–392, Oct 2020.

[89] W. Jaafar, S. Naser, S. Muhaidat, P. C. Sofotasios, and H. Yanikomeroglu, On the downlink performance of RSMA-based UAV communications, *IEEE Trans. Veh. Technol.*, 69(12), 16258–16263, Dec 2020.

[90] Z. Lin, M. Lin, B. Champagne, W.-P. Zhu, and N. Al-Dhahir, Secure and energy efficient transmission for RSMA-based cognitive satellite-terrestrial networks, *IEEE Wireless Commun. Lett.*, 10(2), 251–255, Feb 2021.

[91] O. Dizdar and B. Clerckx, Rate splitting multiple access for multi-antenna multi-carrier joint communications and jamming, in *Proc. Int. Conf. Sens. Sig. Process. Defence (SSPD) Sens. Decis.*, 1–5, 2021.

[92] Z. W. Si, L. Yin, and B. Clerckx, Rate-splitting multiple access for multigateway multibeam satellite systems with feeder link interference, *IEEE Trans. Commun.*, 70(3), 2147–2162, Mar 2022.

[93] Z. Lin, M. Lin, T. de Cola, J.-B. Wang, W.-P. Zhu, and J. Cheng, Supporting IoT with rate-splitting multiple access in satellite and aerial integrated networks, *IEEE Internet Things J.*, 8(14), 11123–11134, Jul 2021.

[94] L. Yin and B. Clerckx, Rate-splitting multiple access for satellite-terrestrial integrated networks: Benefits of coordination and cooperation, *IEEE Trans. on Wireless Commun.* 22(1), 317–332, Jan 2023.

[95] C. Xu, B. Clerckx, S. Chen, Y. Mao, and J. Zhang, Rate-splitting multiple access for multi-antenna joint radar and communications, *IEEE J. Sel. Topics Sig. Process.*, 15(6), 1332–1347, Nov 2021.

[96] O. Dizdar, A. Kaushik, B. Clerckx, and C. Masouros, Energy efficient dual-functional radar-communication: Rate-splitting multiple access, low-resolution DACs, and RF chain selection, *IEEE Open J. Commun. Soc.*, 3, 986–1006, 2022.

[97] R. Cerna-Loli, O. Dizdar, and B. Clerckx, A rate-splitting strategy to enable joint radar sensing and communication with partial CSIT, in *Proc. IEEE 22nd Int. Conf. Signal Process. Adv Wireless Commun. (SPAWC) Workshop*, 491–495, 2021.

[98] O. Dizdar and B. Clerckx, Rate-splitting multiple access for communications and jamming in multi-antenna multi-carrier cognitive radio systems, *IEEE Trans. Inf. Forensics Security*, 17, pp. 628–643, 2022.

[99] E. Piovano, H. Joudeh, and B. Clerckx, Overloaded multiuser MISO transmission with imperfect CSIT, in *Proc. 50th Asilomar Conf. Signals, Syst. Comput.*, 2016, pp. 34–38.

[100] S. Tao, H. Yu, Q. Li, Y. Tang, and D. Zhang, One-layer rate-splitting multiple access with benefits over power-domain NOMA in indoor multi-cell visible light communication networks, in *Proc. IEEE Int. Conf. Commun. (ICC) Workshop*, 1–7, 2020.

[101] S. Naser *et al.*, Rate-splitting multiple access: Unifying NOMA and SDMA in MISO VLC channels, *IEEE Open J. Veh. Technol.*, 1, 393–413, Oct 2020.

[102] Z. Yang, J. Shi, Z. Li, M. Chen, W. Xu, and M. Shikh-Bahaei, Energy efficient rate splitting multiple access (RSMA) with reconfigurable intelligent surface, in *Proc. IEEE Int. Conf. Commun. (ICC) Workshop*, pp. 1–6, 2020.

[103] A. Bansal, K. Singh, and C.-P. Li, Analysis of hierarchical rate splitting for intelligent reflecting surfaces-aided downlink multiuser MISO communications, *IEEE Open J. Commun. Soc.*, 2, 785–798, 2021.

[104] A. Bansal, K. Singh, B. Clerckx, C.-P. Li, and M.-S. Alouini, Rate-splitting multiple access for intelligent reflecting surface aided multi-user communications, *IEEE Trans. Veh. Technol.*, 70(9), 9217–9229, Sep 2021.

[105] Y. Gao, C. Yong, Z. Xiong, J. Zhao, Y. Xiao, and D. Niyato, Reflection resource management for intelligent reflecting surface aided wireless networks, *IEEE Trans. Commun.*, 69(10), 6971–6986, Oct 2021.

[106] Evolved universal terrestrial radio access (E-UTRA); LTE physical layer; general description (release 8), 3GPP, Sophia Antipolis, France, Rep. TS 36.201, Mar. 2009.

[107] Coordinated multi-point operation for LTE physical layer aspects (release 11), 3GPP, Sophia Antipolis, France, Rep. TR 36.819, Aug 2016.

[108] E. Garro *et al.*, 5G mixed mode: NR multicast-broadcast services, *IEEE Trans. Broadcast.*, 66(2), 390–403, Jun 2020.

[109] Study on network-assisted interference cancellation and suppression (NAIC) for LTE (release 12), 3GPP, Sophia Antipolis, France, Rep. TR 36.866, Mar 2014.

[110] A. R. Flores, R. C. De Lamare, and B. Clerckx, Linear precoding and stream combining for rate splitting in multiuser MIMO systems, *IEEE Commun. Lett.*, 24(4), 890–894, Apr 2020.

[111] C. Hao, B. Rassouli, and B. Clerckx, Achievable doF regions of MIMO networks with imperfect CSIT, *IEEE Trans. Inf. Theory*, 63(10), 6587–6606, Oct 2017.

[112] A. G. Davoodi and S. Jafar, Degrees of freedom region of the (M, N1, N2) MIMO broadcast channel with partial CSIT: An application of sum-set inequalities based on aligned image sets, *IEEE Trans. Inf. Theory*, 66(10), 6256–6279, Oct 2020.

[113] Z. Li, S. Yang, and S. Shamai, On linearly precoded rate splitting for Gaussian MIMO broadcast channels, *IEEE Trans. Inf. Theory*, 67(7), 4693–4709, Jul 2021.

[114] X. Su, Y. Yuan, and Q. Wang, Performance analysis of rate splitting in K-user interference channel under imperfect CSIT: Average sum rate, outage probability and SER, *IEEE Access*, 8, 136930–136946, 2020.

[115] J. Chen and V. K. N. Lau, Two-tier precoding for FDD multi-cell massive MIMO time-varying interference networks, *IEEE J. Sel. Areas Commun.*, 32(6), 1230–1238, Jun 2014.

[116] L. Li, K. Chai, J. Li, and X. Li, Resource allocation for multicarrier rate-splitting multiple access system, *IEEE Access*, 8, 174222–174232, 2020.

[117] H. Chen, D. Mi, B. Clerckx, Z. Chu, J. Shi, and P. Xiao, Joint power and subcarrier allocation optimization for multigroup multicast systems with rate splitting, *IEEE Trans. Veh. Technol.*, 69(2), 2306–2310, Feb 2020.

[118] O. Dizdar and B. Clerckx, Rate-splitting multiple access for communications and jamming in multi-antenna multi-carrier cognitive radio systems, *IEEE Trans. Inf. Forensics Security*, 17, 628–643, 2022.

[119] Study on downlink multiuser superposition transmission (MUST) for LTE (release 13), 3GPP, Sophia Antipolis, France, Rep. TR 36.859, Dec 2015

[120] E. Garro *et al.*, 5G mixed mode: NR multicast-broadcast services, *IEEE Trans. Broadcast.*, 66(2), 390–403, Jun 2020.

[121] J. Park, B. Lee, J. Choi, H. Lee, N. Lee, S-H. Park, K-J Lee, J. Choi, S. H. Chae, S-W. Jeon, K. S. Kwak, B. Clerckx, and W. Shin, Rate-splitting multiple access for 6G networks: Ten promising scenarios and applications, *IEEE Network*, 38(3), 128–136, May 2024.

Chapter 9

Intelligent Reflecting Surface-Enabled Rate-Splitting CR-NOMA Systems

Intelligent reflecting surface (IRS) technology has emerged as a promising approach to enhance wireless communication system performance. In this chapter, a rate-splitting (RS) scheme is presented for an IRS-assisted cognitive radio-inspired non-orthogonal multiple access (CR-NOMA) system. In this system, it is imperative to maintain the quality of service (QoS) requirements of the primary user (PU) equivalent to those in orthogonal multiple access (OMA) scenarios.

By leveraging IRS assistance, the threshold for the PU's acceptable interference power is enhanced, thereby enabling an increase in the achievable rate for the secondary user (SU). This RS scheme entails the joint design of optimal transmit power allocation, target rate allocation, and successive interference cancellation (SIC) decoding order. Additionally, the statistical characteristics of both the direct link and IRS-reflecting channels are considered.

Closed-form expressions for the outage probabilities of both the PU and SU are given, incorporating these channel statistics. Through extensive simulations, the superior outage performance achieved by this RS scheme compared to the existing CR-NOMA and IRS-assisted CR-NOMA schemes is demonstrated.

9.1 Introduction

IRS technology has garnered significant attention as a promising means to enhance system performance in wireless communications [1–4]. Essentially, an IRS comprises numerous low-cost reflecting elements capable of reconfiguring the wireless environment through beamforming reflection. RS finds application in various scenarios to achieve superior spectral efficiency (SE) and user fairness in NOMA systems [5–9]. For instance, proposals have been made to maximize the sum rate in uplink NOMA systems while adhering to individual power constraints [5]. Moreover, energy-efficient algorithms have been devised to strike a balance between maximizing sum rates and minimizing total power consumption [6]. The authors in Ref. [7] investigated the hardware architecture of IRS and highlighted its competitive advantages over existing technologies. Additionally, IRS was explored in a covert communication setting, where a legitimate NOMA transmitter communicates with a public user aided by an IRS [8]. Furthermore, research efforts have focused on developing resource allocation methods to enhance energy efficiency in IRS-assisted systems [9].

In NOMA systems, multiple users may encounter significant inter-user interference (IUI) when they concurrently occupy the same resource block. Typically, SIC is employed at a NOMA receiver to retrieve the desired signal, and the order in which SIC is applied plays a crucial role in enhancing decoding performance. In Ref. [10], an innovative SIC switching scheme was introduced to dynamically determine the SIC decoding order based on the ratio of the instantaneous received power to the target data rate of each user. Similarly, in Ref. [11], a transmit power allocation scheme was devised for the legitimate user to alleviate error propagation resulting from imperfect SIC.

For CR-NOMA systems, a QoS-SIC approach was proposed to enhance the SU's outage performance while ensuring that the PU's outage performance remains consistent with OMA systems [12, 13]. Furthermore, a novel hybrid SIC (NH-SIC) technique was introduced for CR-NOMA systems, incorporating both SIC decoding order and power control to further enhance the SU's outage performance [14].

Due to its potential to enhance SE, improve reliability, and reduce latency in multi-user transmissions, RS has garnered significant attention [15, 16]. In downlink multi-user transmissions, RS divides user messages into common and private streams, thereby bridging the gap between NOMA and space-division multiple access (SDMA) while offering a trade-off between system performance and implementation complexity [17].

In contrast to RS-aided downlink transmissions, RS-aided uplink transmissions involve the full decoding of data streams from all users at the base station (BS) using SIC [18–21]. Adhering to individual rate constraints,

RS has been employed to maximize the sum rate in uplink rate-splitting multiple access (RSMA) systems [18]. In single-input multiple-output (SIMO) NOMA systems, RS ensures maximum–minimum fairness among uplink users [21]. To circumvent exhaustive search requirements for optimal SIC decoding order determination in uplink RSMA, fixed RS and cognitive RS schemes have been proposed for CR-NOMA systems, resulting in improved outage performance and user fairness [22].

Recently, a CR-inspired RS scheme integrating transmit power allocation and SIC decoding order was introduced for CR-NOMA systems to enhance the SU's outage performance [23]. Additionally, the CR-inspired RS scheme has been deployed to enhance computational performance in mobile edge computing systems [24, 27].

In this chapter, a novel RS is utilized to enhance the outage performance of an IRS-assisted CR-NOMA system. In the system, the transmissions from the PU to the BS are facilitated by an IRS, thereby enhancing the equivalent channel gain from PU to BS and consequently improving the tolerable interference power threshold for PU. While maintaining the PU's outage performance, RS is implemented at the SU to enhance its outage performance. Through the combined use of IRS and RS, this scheme demonstrates superior outage performance compared to benchmark schemes where IRS and RS are not utilized [27].

9.2 System Model

An uplink CR-NOMA system configuration comprises a PU (i.e., U_1), an SU (i.e., U_2), and a BS, as illustrated in Fig. 9.1. An IRS is strategically positioned between the BS and PU to aid the PU's transmission. This arrangement allows the PU to transmit its signal to the BS using both the direct link and the IRS-reflected channels. It is assumed that the SU is situated in an area where no IRS is present nearby. For the purpose of outage performance analysis, each node is assumed to be equipped with a single antenna for transmission/reception, while the IRS comprises N reflecting elements.

As shown in the system, the direct link channel from $U_i (i = 1, 2)$ to the BS is represented by $\overline{h}_i = \sqrt{L_i} h_i$, where L_i is the distance-related path loss and h_i is the small-scale fading coefficient. The reflecting channels from the PU to the IRS and from IRS to BS are denoted by $\overline{h}_{UI} = \sqrt{L_{UI}} h_{UI}$ and $\overline{h}_{IB} = \sqrt{L_{IB}} h_{IB}$, respectively, where h_{UI} and h_{IB} are the $N \times 1$ vectors containing the small-scale fading coefficients, and

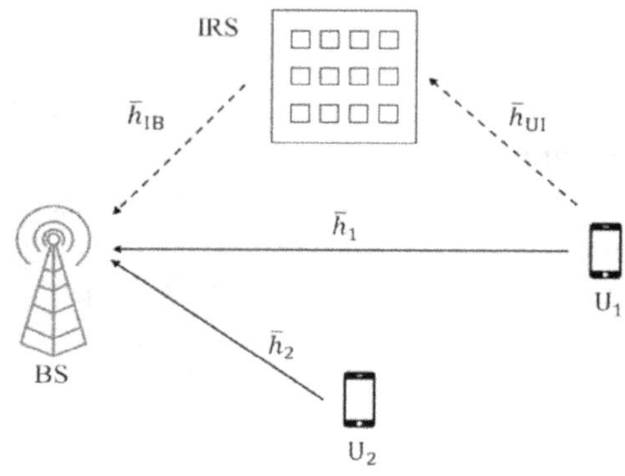

Fig. 9.1. System model.

L_{UI} and L_{IB} are the distance-related path loss associated with the reflecting channels. In this context, all the small-scale fading coefficients are assumed to follow independent and identically distributed Rayleigh fading, and the large-scale fading is modeled using the distance-related path loss, given by

$$L_i = \left(\frac{d_i}{d_0}\right)^{-\chi} \tag{9.1}$$

where $d_i \in \{d_1, d_2, d_{UI}, d_{IB}\}$ denotes the distance associated with the corresponding channel, d_0 denotes the reference distance, and χ is the path-loss exponential component.

For the IRS, a unit-reflecting amplitude is assumed for each element, and $\phi_n \in (-\pi, \pi]$ is used to denote the reflecting phase shift generated by the nth element ($n = 1, 2, \ldots, N$). To achieve the strongest channel gain for the PU's transmission, the ideal phase shift is adopted as $\phi_n = \arg(h_1) - \arg([h_{IB}]_n[h_{UI}]_n)$, where $[h_{IB}]_n$ and $[h_{UI}]_n$ are the nth elements of h_{IB} and h_{UI}, respectively. Then, the phase-shift matrix of the IRS can be expressed as $Q = \mathrm{diag}([e^{j\phi_1}, e^{j\phi_2}, \ldots, e^{j\phi_N}])$. For the PU's transmission to the BS, the equivalent channel gain can be written as

$$g_1 = |\sqrt{L_1}h_1 + \sqrt{L_R}h_{IB}^H Q h_{UI}|^2 \tag{9.2}$$

where $L_R = (d_{IB}d_{UI}/d_0^2)^{-\chi}$ is the distance-related path loss of the reflecting channels.

Before transmission, the BS sends an interference power threshold τ to the SU to ensure the QoS requirements of the PU. The interference power threshold is given by [13, 14]

$$\tau = \max\left\{0, \frac{P_1 g_1}{\epsilon_1} - \sigma^2\right\} \tag{9.3}$$

where P_i is the transmit power of U_i, σ^2 is the variance of the additive white Gaussian noise (AWGN) at the BS receiver, and $\epsilon_1 = 2^{\hat{R}_1} - 1$ with $\hat{R}_1$ denoting the target rate of U_1. The ideal phase shift achieves the maximum value of g_1 in equation (9.2), resulting in an increased interference power threshold by utilizing the IRS to assist the transmissions of PU.

To enhance the outage performance of the considered CR-NOMA system, an IRS is deployed between the BS and PU, and RS is implemented at the SU. Since $2K - 1$ split data streams are sufficient to approximate the full capacity region for the uplink RSMA with K users [25], the focus is solely on implementing RS at the SU, eliminating the need for RS at the PU.

In each transmission block, the message signal (x_2) of the SU can be split into two parts, x_{21} and x_{22}, to be transmitted. The signal received at the BS can be expressed as

$$\begin{aligned} y = &\left(\sqrt{L_1}h_1 + \sqrt{L_R}h_{IB}^H Q h_{UI}\right)\sqrt{P_1}x_1 + \sqrt{\alpha P_2 L_2}h_2 x_{21} \\ &+ \sqrt{(1-\alpha)P_2 L_2}h_2 x_{22} + n \end{aligned} \tag{9.4}$$

where x_1 is the message signal of the PU and P_i is the transmit power of U_i, n is the AWGN at the BS with zero mean and variance σ^2, and α is the transmit power allocation factor satisfying $0 \leq \alpha \leq 1$.

To achieve the maximum achievable rate for the SU, the SIC decoding order $x_{21} \rightarrow x_1 \rightarrow x_{22}$ is used at the BS receiver [23, 25]. Then, for the detection of x_{21}, x_1, and x_{22}, the received signal-to-noise ratios (SNRs) can be expressed, respectively, as

$$\gamma_{21} = \frac{\alpha \rho_2 g_2}{(1-\alpha)\rho_2 g_2 + \rho_1 g_1 + 1} \tag{9.5}$$

$$\gamma_1 = \frac{\rho_1 g_1}{(1-\alpha)\rho_2 g_2 + 1} \tag{9.6}$$

and

$$\gamma_{22} = (1-\alpha)\rho_2 g_2 \tag{9.7}$$

where $\rho_1 = P_1 g_1/\sigma^2$ and $\rho_2 = P_2 g_2/\sigma^2$ with $g_2 = L_2|h_2|^2$.

For the transmissions of x_{21}, x_1, and x_{22}, the achievable rates are then given by $R_{21} = \log_2(1+\gamma_{21})$, $R_1 = \log_2(1+\gamma_1)$, and $R_{22} = \log_2(1+\gamma_{22})$, respectively. In the considered system, the RS is designed to obtain the maximum achievable rate allowed for the SU, thereby improving the SU's outage performance. Corresponding to the SIC decoding order $x_{21} \to x_1 \to x_{22}$, the RS operations at the SU are determined by comparing the SU's channel gain to the interference power threshold, which are categorized into the following three cases:

(1) Case I: $\tau > 0$ and $\rho_2 g_2 \leq \tau$. In this scenario, the interference power resulting from the SU's transmission of x_{22} does not exceed the interference power threshold. Therefore, the SU does not perform RS and allocates all of its resources to transmit $x_2 (x_{22} = x_2)$, where x_2 represents the transmit signal generated by U_2 from its message signal without RS. Consequently, the transmit power allocation is set as $\alpha = 0$, and the SIC decoding order $x_{21} \to x_1 \to x_{22}$ degrades to $x_1 \to x_2$ (or equivalently $x_1 \to x_{22}$) at the BS receiver. In this case, the received SNRs for decoding x_1 and x_2 are given by $\gamma_1 = \rho_1 g_1/(\rho_2 g_2 + 1)$ and $\gamma_2 = \rho_2 g_2$, respectively, and the achievable rate for the SU is given by $R_2{}^{(I)} = \log_2(1 + \rho_2 g_2)$.

(2) Case II: $\tau > 0$ and $\rho_2 g_2 > \tau$. In this scenario, to achieve the maximum achievable rate for the SU, U_2 applies rate splitting by setting $\alpha = 1 - \tau/(\rho_2 g_2)$, resulting in the achievable rates $R_{21} = \log_2[1+(\rho_2 g_2 - \tau)/(\rho_1 g_1 + \tau + 1)]$ and $R_{22} = \log_2(1 + \tau)$. Since R_{22} achieves the maximum value allowed by the interference power threshold, the achievable rate $R_2{}^{(II)} = R_{21} + R_{22}$ of the SU is maximized in this case, which is given by $R_2{}^{(II)} = \log_2[1 + (\rho_2 g_2 - \tau)/(\rho_1 g_1 + \tau + 1)] + \log_2(1 + \tau)$. Corresponding to the transmissions of x_{21} and x_{22}, the target rates $\hat{R}_2$ of U_2 are split into $\hat{R}_{21}$ and $\hat{R}_{22}$, respectively. To ensure the successful detection of x_{22}, the split target rate is given by $\hat{R}_{22} = R_{22} = \log_2(1 + \tau)$. Then, the split target rate $\hat{R}_{21} = \hat{R}_2 - \hat{R}_{22}$ is set for the transmission of x_{21}. To avoid error propagation in SIC, it requires the successful detection of x_{21} in the first stage of the SIC decoding order $x_{21} \to x_1 \to x_{22}$. Thus, the transmission of the SU only occurs if $R_{21} \geq \hat{R}_{21}$. Otherwise, the SU remains silent, while the PU transmits alone in this case.

(3) Case III: $\tau = 0$. In this scenario, the PU's transmission encounters an outage due to a weak channel gain g_1. Thus, U_2 does not conduct RS and transmits the signal $x_{21} = x_2$ by setting $\alpha = 1$. Consequently, the SIC decoding order at the BS degrades to $x_2 \to x_1$ (or equivalently $x_{21} \to x_1$). The achievable rate of the SU is given by $R_2{}^{(III)} = \log_2[1 + \rho_2 g_2/(\rho_1 g_1 + 1)]$. Assisted by the IRS, the interference power threshold τ is improved. The achievable rate of U_2 attained by the proposed RS scheme can be expressed as follows:

$$R_2 = \begin{cases} R_2^{(\mathrm{I})}, & \tau > 0 \text{ and } \rho_2 g_2 \leq \tau \\ R_2^{(\mathrm{II})}, & \tau > 0 \text{ and } \rho_2 g_2 > \tau \\ R_2^{(\mathrm{III})}, & \tau = 0 \end{cases} \tag{9.8}$$

Corresponding to Cases I, II, and III, the achievable rates of U_1 can be characterized by $R_1 = \log_2[1 + \rho_1 g_1/(\rho_2 g_2 + 1)]$, $R_1 \geq \log_2\left[1 + \frac{\rho_1 g_1}{\tau+1}\right]$, and $R_1 \leq \log_2[1 + \rho_1 g_1]$, respectively.

The deployment of the IRS in the considered IRS-CR-NOMA system can increase the equivalent channel gain g_1 compared to the CR-NOMA system, benefiting the PU's outage probability improvement. However, this scheme also results in an increased τ, leading to a higher likelihood of the SU operating in Cases I and II. Consequently, the SU can achieve a higher achievable rate with the assistance of IRS and RS, thereby enhancing its outage performance.

9.3 Outage Performance

In this section, the outage performance of the PU and SU achieved by the RS scheme in the considered IRS-CR-NOMA system is analyzed. First, the distributions of the involved channel gains are derived. Then, closed-form expressions for the outage probabilities of the PU and SU are presented. Additionally, asymptotic approximations for the outage probabilities of the PU and SU in the high SNR region are discussed.

9.3.1 *Channel gain distribution analysis*

The equivalent channel gain g_1 for the PU's transmission, as given by equation (9.2), can be rewritten by substituting the ideal phase shift into the equation as follows:

$$g_1 = \left| \sqrt{L_1}|h_1| + \sqrt{L_R} \sum_{n=1}^{N} |[h_{IB}]_n||[h_{UI}]_n| \right| \tag{9.9}$$

The cumulative distribution function (CDF) of g_1 is denoted by $F_{g_1}(z) = P_r(g_1 \leq z)$, which has an analytical expression provided in the following theorem.

Theorem 1. *The CDF of the equivalent channel gain g_1 is given by*

$$F_{g_1}(z) = 1 - \Gamma(k, \sqrt{z}/\theta)/\Gamma(k) \tag{9.10}$$

where k and θ denote the shape and scale parameters of Gamma distribution, which are given, respectively, by

$$k = \frac{(\sqrt{L_1\pi} + 2N\xi\psi)^2}{4L_1 + 4N\xi\psi^2 - L_1\pi} \tag{9.11}$$

and

$$\theta = \frac{4L_1 + 4N\xi\psi^2 - L_1\pi}{2(\sqrt{L_1\pi} + 2N\xi\psi)} \tag{9.12}$$

with $\xi = \frac{\pi^2}{16-\pi^2}$ and $\psi = \left(4 - \frac{\pi^4}{4}\right)\sqrt{L_{IB}L_{UI}\pi}$.

Proof. See Appendix A.9.

Theorem 1 provides a closed-form expression for the CDF of g_1 in terms of the large-scale fading coefficients, facilitating the evaluation of the outage performance. The CDF of g_1 in equation 9.10 can be further rewritten as

$$F_{g_1}(z) = \frac{1}{\Gamma(k)}\gamma(k, \sqrt{z}/\theta) = \frac{1}{\theta^k\Gamma(k)}\sum_{m=0}^{\infty}\left(-\frac{1}{\theta}\right)^m \frac{z^{\frac{m+k}{2}}}{m!(m+k)} \tag{9.13}$$

In the high SNR region, the CDF of g_1 can be approximated as

$$F_{g_1}(z) = \frac{z^{\frac{k}{2}}}{k\theta^k\Gamma(k)} \tag{9.14}$$

With the above approximated CDF, the PDF of g_1 can be evaluated as

$$f_{g_1}(z) = \frac{z^{\frac{k}{2}-1}}{2\theta^k\Gamma(k)} \tag{9.15}$$

For the SU's transmission, since the direct link channel gain g_2 is an exponential RV, the CDF of g_2 can be written as

$$F_{g_2}(z) = 1 - e^{-\frac{z}{L_2}} \tag{9.16}$$

□

9.3.2 Outage probability analysis

In this section, the outage probabilities are derived for both the PU and SU achieved by the proposed RS scheme in the IRS-CR-NOMA system.

For the PU, the outage probability can be further evaluated as

$$P_{\text{out},1} = P_r(R_1 \leq \hat{R}_1)$$
$$P_{\text{out},1} = \mathrm{P}_r(\tau = 0) = \mathrm{P}_r(g_1 \leq \eta_1) = 1 - \Gamma(k, \sqrt{\eta_1}/\theta)/\Gamma(k) \tag{9.17}$$

where $\eta_1 = \epsilon_1/\rho_1$.

For the SU, the outage probability can be expressed as

$$P_{\text{out},2} = P^{(\mathrm{I})}_{\text{out},2} + P^{(\mathrm{II})}_{\text{out},2} + P^{(\mathrm{III})}_{\text{out},2} \tag{9.18}$$

where $P^{(\mathrm{I})}_{\text{out},2} = \mathrm{P}_r\{\tau > 0, \rho_2 g_2 \leq \tau, R^{(\mathrm{I})} < \hat{R}_2\}$, $P^{(\mathrm{II})}_{\text{out},2} = \mathrm{P}_r\{\tau > 0, \rho_2 g_2 > \tau, R^{(\mathrm{II})} < \hat{R}_2\}$, and $P^{(\mathrm{III})}_{\text{out},2} = \mathrm{P}_r\{\tau = 0, R^{(\mathrm{III})} < \hat{R}_2\}$ indicate the probabilities of outage for U_2 corresponding to Cases I, II, and III, respectively. The closed-form expressions for ${P_{\text{out},2}}^{(\mathrm{I})}$, ${P_{\text{out},2}}^{(\mathrm{II})}$, and ${P_{\text{out},2}}^{(\mathrm{III})}$ are provided in the following propositions, respectively.

Proposition 1. *The outage probability corresponding to the RS scheme's operation in Case II is given by*

$$P^{(\mathrm{II})}_{\text{out},2} = I_1 - I_2 \tag{9.19}$$

where I_1 and I_2 are given by

$$I_1 = \frac{e^{\frac{1}{\rho_2 L_2}}}{2\theta^k \Gamma(k)} \left(\frac{\rho_2 \epsilon_1 L_2}{\rho_1}\right)^{\frac{k}{2}} \left(\gamma\left(\frac{k}{2}, \frac{1+\epsilon_2}{\rho_2 L_2}\right) - \gamma\left(\frac{k}{2}, \frac{1}{\rho_2 L_2}\right)\right) \tag{9.20}$$

$$I_2 = \frac{e^{-\frac{\epsilon_1+\epsilon_2+\epsilon_1\epsilon_2}{\rho_2 L_2}}}{2\theta^k \Gamma(k)} \left(-\frac{\rho_2 L_2}{\rho_1}\right)^{\frac{k}{2}} \left(\gamma\left(\frac{k}{2}, -\frac{\rho_1 \eta_1}{\rho_2 L_2}\right) - \gamma\left(\frac{k}{2}, -\frac{\rho_1 \eta_1 (1+\epsilon_2)}{\rho_2 L_2}\right)\right) \tag{9.21}$$

Proof. See Appendix B.9. □

Proposition 2. *The outage probabilities corresponding to the RS scheme's operation in Case I and Case III are given by*

$$P_{\text{out},2}^{(\text{I})} = F_{g_1}(\eta_1(1+\epsilon_2))(1-F_{g_2}(\eta_2)) + F_{g_2}(\eta_2) - F_{g_1}(\eta_1) - I_3 \tag{9.22}$$

$$P_{\text{out},2}^{(\text{III})} = F_{g_1}(\eta_1) - I_4 \tag{9.23}$$

where I_3 and I_4 are given, respectively, by

$$I_3 = \frac{(\rho_2 L_2 \eta_1)^{\frac{k}{2}} e^{\frac{1}{\rho_2 L_2}}}{2\theta^k \Gamma(k)} \left(\gamma\left(\frac{k}{2}, \frac{1+\epsilon_2}{\rho_2 L_2}\right) - \gamma\left(\frac{k}{2}, \frac{1}{\rho_2 L_2}\right) \right) \tag{9.24}$$

and

$$I_4 = \frac{e^{-\frac{\epsilon_2}{\rho_2 L_2}}}{2\theta^k \Gamma(k)} \left(\frac{\rho_2 L_2}{\rho_1 \epsilon_2}\right)^{\frac{k}{2}} \left(\gamma\left(\frac{k}{2}, -\frac{\epsilon_1 \epsilon_2}{\rho_2 L_2}\right) \right) \tag{9.25}$$

Proof. See Appendix C.9. □

By combining (9.19), (9.22), and (9.23) together, the outage probability experienced by U_2 in the proposed RS scheme is given by

$$\begin{aligned} P_{\text{out},2} &= F_{g_1}(\eta_1(1+\epsilon_2))(1-F_{g_2}(\eta_2)) + F_{g_2}(\eta_2) + I_1 - I_2 - I_3 - I_4 \\ &= -\frac{e^{-\frac{\eta_2}{L_2}} \Gamma\left(k, \frac{\sqrt{\eta_1(1+\epsilon_2)}}{\theta}\right)}{\Gamma(k)} + I_1 - I_2 - I_3 - I_4 \end{aligned} \tag{9.26}$$

Remark 1. The following examples demonstrate that the proposed RS scheme can achieve a higher achievable rate than the IRS-CR-NOMA scheme, which does not deploy the RS.

Example 1. $\log_2(1+\tau) > \log_2(1+\frac{\rho_2 g_2}{\rho_1 g+1})$. In this example, we assume $g_1 = g_2 = 15$, $\rho_1 = 15$, $\rho_2 = 25$, and $\epsilon_1 = 4$. The interference threshold is given by $\tau = 225/4$. For the IRS-CR-NOMA scheme without RS, the achievable rate of U_2 is ${R_2}^{(\text{II})} = \log_2(1+625/226) \approx 0.58\,\text{BPCU}$. In contrast, the proposed scheme achieves with RS ${R_2}^{(\text{II})} = \log_2(1+318.75/282.25) + \log_2(1+225/4) \approx 2.09\,\text{BPCU}$.

Example 2. $\log_2(1+\tau) < \log_2(1+\frac{\rho_2 g_2}{\rho_1 g+1})$. In this example, we assume $g_1 = g_2 = 10$, $\rho_1 = 1$, $\rho_2 = 20$, and $\epsilon_1 = 3$. The interference threshold is given by $\tau = 7/3$. For the IRS-CR-NOMA scheme without RS, the achievable rate of U_2 is $R_2{}^{(\mathrm{II})} = \log_2\left(1+\frac{200}{11}\right) \approx 4.26\,\mathrm{BPCU}$. In contrast, the proposed scheme achieves with RS $R_2{}^{(\mathrm{II})} = \log_2(1+593/40) + \log_2\left(1+\frac{7}{3}\right) \approx 5.72\,\mathrm{BPCU}$.

9.4 Performance Results

In this section, numerical results are presented to clarify the outage performance achieved by the proposed RS scheme and to verify the accuracy of the developed analytical expressions. For comparison purposes, simulation results of the CR-NOMA and IRS-CR-NOMA schemes are also included. Unless otherwise specified, the following parameters are used throughout the simulation: $\sigma^2 = 10^{-8}$ W, $N = 32$, $\chi = 3$, $d_0 = 1\,\mathrm{m}$, $d_1 = 17\,\mathrm{m}$, $d_2 = 15\,\mathrm{m}$, $d_{UI} = 10\,\mathrm{m}$, $d_{IB} = 8\,\mathrm{m}$.

In Fig. 9.2, the outage probability of U_1 achieved by the IRS-CR-NOMA scheme is compared with that of the CR-NOMA scheme for various transmit power settings, where $P_1 = P_2$ and $N = 8$. Two target rate settings $\{\hat{R}_1 = 5\,\mathrm{BPCU}, \hat{R}_2 = 4\,\mathrm{BPCU}\}$ and $\{\hat{R}_1 = 4\,\mathrm{BPCU}, \hat{R}_2 = 3\,\mathrm{BPCU}\}$ are

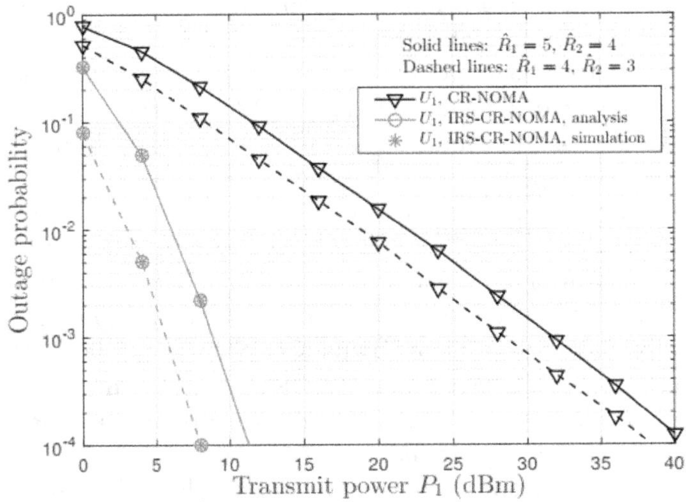

Fig. 9.2. PU's outage probability comparison.

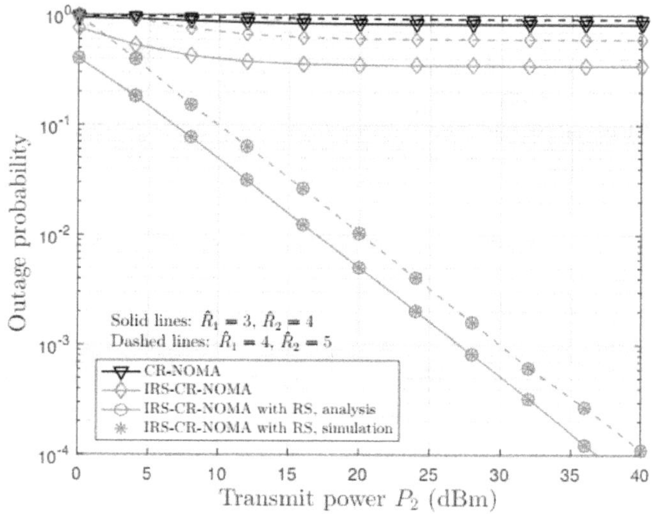

Fig. 9.3. SU's outage probability comparison.

considered. As shown in Fig. 9.2, the outage probabilities of U_1 achieved by the IRS-CR-NOMA and CR-NOMA schemes monotonically decrease with increasing transmit power. The results clearly illustrate that the IRS-CR-NOMA scheme outperforms the CR-NOMA scheme for U_1 across the entire transmit power region, indicating that the usage of IRS helps improve the outage performance of the PU.

In Fig. 9.3, the outage probability of U_2 achieved by the proposed RS scheme is compared with that of the CR-NOMA and IRS-CR-NOMA schemes for various transmit power settings, where $P_1 = P_2$. Two settings $\{\hat{R}_1 = 3\,\text{BPCU}, \hat{R}_2 = 4\,\text{BPCU}\}$ and $\{\hat{R}_1 = 4\,\text{BPCU}, \hat{R}_2 = 5\,\text{BPCU}\}$ are considered. From Fig. 9.3, it is evident that the proposed RS scheme exhibits the best outage performance among the three schemes.

Additionally, all three schemes approach an outage probability of 1 in the low transmit power region. However, as the transmit power increases, the outage probabilities achieved by the CR-NOMA and IRS-CR-NOMA schemes encounter an outage probability floor in the middle and high SNR regions. In contrast, the outage probability of the proposed RS scheme monotonically decreases with increasing transmit power.

In Fig. 9.4, the average throughput of U_1 achieved by the IRS-CR-NOMA scheme is compared with that of the CR-NOMA scheme for various

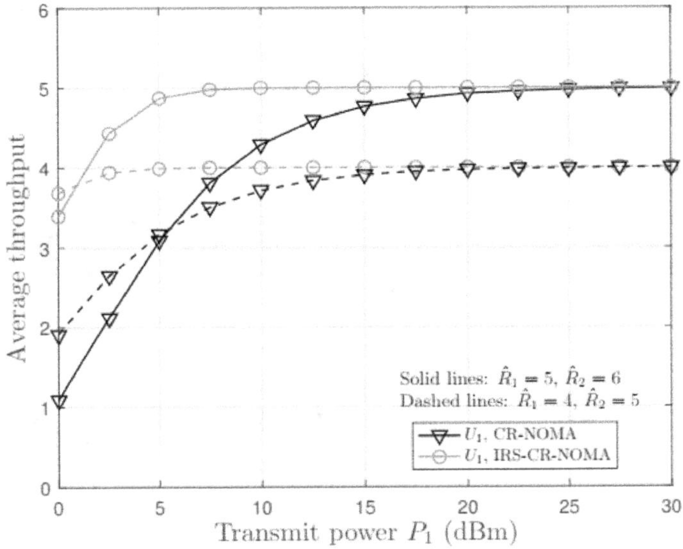

Fig. 9.4. Average throughput comparison for U_1.

transmit power settings, where $P_1 = P_2$ and $N = 8$. Two target rate pairs $\{\hat{R}_1 = 5\,\text{BPCU}, \hat{R}_2 = 6\,\text{BPCU}\}$ and $\{\hat{R}_1 = 4\,\text{BPCU}, \hat{R}_2 = 5\,\text{BPCU}\}$ are considered.

It is observed from Fig. 9.4 that the average throughputs of both schemes increase monotonically with increasing transmit power in the low transmit power region. As the transmit power increases, the average throughputs of both schemes approach the target rates in the middle and high transmit power regions. However, in the low and middle transmit power regions, the IRS-CR-NOMA scheme consistently achieves a higher average throughput than the CR-NOMA scheme, indicating the effectiveness of using the IRS in improving the PU's average throughput.

In Fig. 9.5, the average throughput of U_2 achieved by the proposed RS scheme is compared with the CR-NOMA and IRS-CR-NOMA schemes for various transmit power settings, where $P_1 = P_2$. Two target rate pairs $\{\hat{R}_1 = 3.5\,\text{BPCU}, \hat{R}_2 = 5.5\,\text{BPCU}\}$ and $\{\hat{R}_2 = 4\,\text{BPCU}, \hat{R}_2 = 5\,\text{BPCU}\}$ are considered.

It is evident from Fig. 9.5 that the proposed RS scheme consistently achieves the highest average throughputs among the three schemes for both target rate settings. Additionally, it is observed that the average

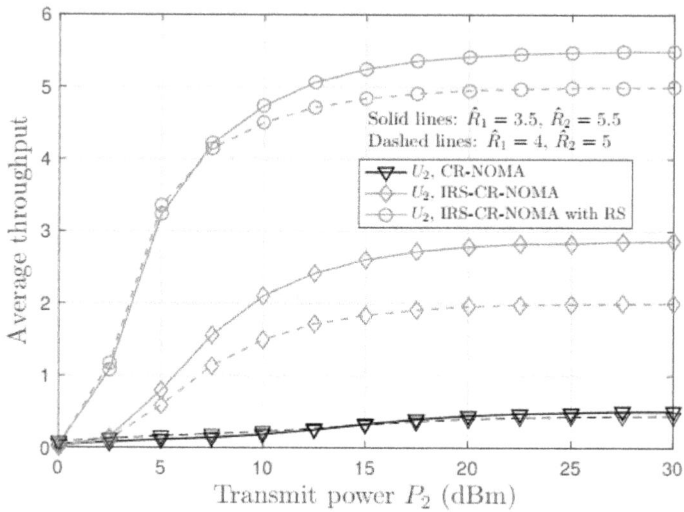

Fig. 9.5. Average throughput comparison for U_2.

throughput of all schemes approaches 0 in the low transmit power region and increases with increasing transmit power, stabilizing in the high transmit power region. However, only the average throughput achieved by the RS scheme approaches the target rate in the high transmit power region, while the average throughputs achieved by the CR-NOMA and IRS-CR-NOMA schemes remain lower than the target rate across all transmit power levels.

9.5 Summary

In this chapter, RS was considered to enhance the outage performance of the IRS-CR-NOMA system, facilitating simultaneous communication between the PU and SU with a BS. With IRS assistance, the PU benefited from an enhanced interference power threshold, leading to improved outage performance. Applying CR principles, RS was implemented at the SU to optimize its transmission. The design involved joint optimization of power allocation, target allocation, and SIC decoding order, ensuring the maximum achievable rate for the SU while maintaining PU's outage performance akin to OMA. The closed-form expressions for the outage probabilities of both PU and SU were derived. The numerical results demonstrated superior outage performance of the proposed RS scheme compared to CR-NOMA and IRS-CR-NOMA schemes without RS deployment.

Appendix A.9. A Proof of Theorem 1

Corresponding to g_1 in (9.9), we define $A \triangleq \sqrt{L_1}|h_1|$ and $B_n \triangleq \sqrt{L_R}|[\boldsymbol{h}_{IB}]_n||[\boldsymbol{h}_{UI}]_n|, \forall_n$. The RV A follows Rayleigh distribution with the mean $\sqrt{L_1\pi}/2$ and variance $(4-\pi)L_1/4$. Also, $|[\boldsymbol{h}_{IB}]_n|$ and $|[\boldsymbol{h}_{UI}]_n|$ follow Rayleigh distribution and the corresponding CDF and PDF are given, respectively, by

$$F_{[h_x]_n}(z) = 1 - e^{(-\frac{z^2}{L_x})} \tag{A.1}$$

and

$$f_{[h_x]_n}(z) = \frac{2z}{L_x}e^{(-\frac{z^2}{L_x})} \tag{A.2}$$

where $x \in \{IB, UI\}$. Then, the CDF of B_n can be expressed as

$$\begin{aligned} F_{B_n}(z) &= P_r(B_n < z) \\ &= P_r\left(|[\boldsymbol{h}_{IB}]_n| < \frac{z}{|[\boldsymbol{h}_{UI}]_n|} = \frac{z}{t}\right) \\ &= \int_0^\infty F_{[\boldsymbol{h}_{IB}]_n}\left(\frac{z}{t}\right) f_{[\boldsymbol{h}_{UI}]_n}(t)dt \\ &= 1 - \frac{2z}{\sqrt{L_{IB}L_{UI}}}K_1\left(\frac{2z}{\sqrt{L_{IB}L_{UI}}}\right) \end{aligned} \tag{A.3}$$

where $K_n(\cdot)$ is the modified Bessel function of the second kind with nth order. Referring to the table of integrals in Ref. [26], the PDF of B_n can be derived to follow a Gamma distribution with mean and variance, respectively, as

$$\mathbb{E}\{B_n\} = \frac{\pi\sqrt{L_{IB}L_{UI}}}{4} \tag{A.4}$$

and

$$\text{Var}\{B_n\} = \left(1 - \frac{\pi^2}{16}\right)L_{IB}L_{UI} \tag{A.5}$$

Following the approach of Ref. [28], we match B_n to Gamma distribution by using the moment-matching approach with $\xi\psi = \mathbb{E}\{B_n\}$ and $\xi\psi^2 = \text{Var}\{B_n\}$, where ξ and ψ are the shape and scale parameters of the Gamma distribution, respectively. For the RV $C = A + \sum_{n=1}^{N} B_n$, the mean and second moment can be expressed, respectively, as

$$\mathbb{E}\{C\} = \frac{\sqrt{\pi L_1}}{2} + N\xi\psi \tag{A.6}$$

and

$$\mathbb{E}\{C^2\} = L_1 + N\xi\psi(\psi + N\xi\psi + \sqrt{\pi L_1}) \tag{A.7}$$

Then, the variance of C is derived as

$$\text{Var}\{C\} = \mathbb{E}\{C^2\} - (\mathbb{E}\{C\})^2 = L_1 + N\xi\psi^2 - \frac{\pi L_1}{4} \tag{A.8}$$

Again, we match C to the Gamma distribution with the corresponding shape and scale parameters given by (9.11) and (9.12), respectively. Then, the CDF of $g_1 = C^2$ can be obtained as $F_{g_1}(z) = F_C(\sqrt{z})$ as shown in (9.10), which completes the proof.

Appendix B.9. A Proof of Proposition 1

Corresponding to Case II of the RS scheme, the probability $P_{\text{out}}^{(\text{II})}$ can be rewritten as

$$\begin{aligned} P_{\text{out}}^{(\text{II})} &= P_r\left(g_1 > \eta_1, g_2 > \frac{\rho_1\epsilon_1^{-1}g_1 - 1}{\rho^2}, g_2 < \frac{(1+\epsilon_1)(1+\epsilon_2) - (1+\rho_1 g_1)}{\rho_2}\right) \\ &= \underset{\eta_1 < g_1 < \eta_1(1+\epsilon_2) + \frac{\epsilon_2}{\rho_1}}{\mathbb{E}\{S\}} \end{aligned} \tag{B.1}$$

where S defined as

$$S \triangleq P_r\left(g_2 > \frac{\rho_1\epsilon_1^{-1}g_1 - 1}{\rho_2}, g_2 < \frac{(1+\epsilon_1)(1+\epsilon_2) - (1+\rho_1 g_1)}{\rho_2}\right) \tag{B.2}$$

Since g_2 is positive in practice, the inequality $\frac{(1+\epsilon_1)(1+\epsilon_2)-(1+\rho_1 g_1)}{\rho_2} > 0$ holds true for (B.1), which results in the constraint $g_1 < \eta_1(1 + \epsilon_2) + \frac{\epsilon_2}{\rho_1}$

for the expectation in (B.1). For the expression of S in (B.2), the upper bound on g_2 results in the following implicit constraint:

$$g_1 < \eta_1(1+\epsilon_2) \tag{B.3}$$

Therefore, the expectation for S in (B.1) should be taken over $\frac{\rho_1\epsilon_1^{-1}g_1-1}{\rho_2} < g_2 < \frac{(1+\epsilon_1)(1+\epsilon_2)-(1+\rho_1 g_1)}{\rho_2}$, and $P_{\text{out}}^{\text{II})}$ can be given by

$$\begin{aligned}
P_{\text{out}}^{(\text{II})} &= \int_{\eta_1}^{\eta_1(1+\epsilon_2)} (F_{g_2}(z_2) - F_{g_2}(z_1)) f_{g_1}(y)dy \\
&= \int_{\eta_1}^{\eta_1(1+\epsilon_2)} \left(1 - e^{-\frac{z_2}{L_2}}\right) f_{g_1}(y)dy - \int_{\eta_1}^{\eta_1(1+\epsilon_2)} \left(1 - e^{-\frac{z_1}{L_2}}\right) \\
&\quad f_{g_1}(y)dy \\
&= \int_{\eta_1}^{\eta_1(1+\epsilon_2)} e^{-\frac{z_1}{L_2}} \frac{y^{\frac{k}{2}-1}}{2\theta^k\Gamma(k)} dy - \int_{\eta_1}^{\eta_1(1+\epsilon_2)} e^{-\frac{z_2}{L_2}} \frac{y^{\frac{k}{2}-1}}{2\theta^k\Gamma(k)} dy \\
&= \underbrace{\frac{(L_2\rho_2\eta_1)^{\frac{k}{2}}}{2\theta^k\Gamma(k)} \int_0^{\frac{\epsilon_2}{L_2\rho_2}} \left(x + \frac{1}{L_2\rho_2}\right)^{\frac{k}{2}-1} e^{-x}dx}_{I_1} \\
&\quad - \underbrace{\frac{e^{-\frac{\epsilon_1+\epsilon_2+\epsilon_1\epsilon_2-\rho_1\eta_1(1+\epsilon_2)}{\rho_2 L_2}}}{2\theta^k\Gamma(k)} \left(-\frac{\rho_2 L_2}{\rho_1}\right)^{\frac{k}{2}} \times \int_0^{\frac{\rho_1\eta_1\epsilon_2}{\rho_2 L_2}} \left(x - \frac{\rho_1\eta_1(1+\epsilon_2)}{\rho_2 L_2}\right)^{\frac{k}{2}-1} e^{-x}dx}_{I_2}
\end{aligned} \tag{B.4}$$

where $z_1 = \frac{\rho_1\epsilon_1^{-1}y-1}{\rho_2}$ and $z_2 = \frac{(1+\epsilon_1)(1+\epsilon_2)-(1+\rho_1 y)}{\rho_2}$. Referring to the table of integrals in Ref. [26], I_1 and I_2 can be derived as (9.20) and (9.21), respectively.

Appendix C.9. A Proof of Proposition 2

Corresponding to Case I of the RS scheme, the probability $P_{\text{out}}^{(\text{I})}$ can be evaluated as

$$
\begin{aligned}
\mathrm{P}_{\text{out}}^{(\text{I})} &= P_r(\tau > 0, \rho_2 g_2 \le \tau, R^{(I)} < \hat{R}_2) \\
&= P_r\left(\frac{\rho_1 g_1}{\epsilon_1} - 1 > 0, \rho_2 g_2 < \min\left\{\frac{\rho_1 g_1}{\epsilon_1} - 1, \epsilon_2\right\}\right) \\
&= P_r\left(0 < \frac{\rho_1 g_1}{\epsilon_1} - 1 < \epsilon_2, \rho_2 g_2 < \frac{\rho_1 g_1}{\epsilon_1} - 1\right) \\
&\quad + P_r\left(\frac{\rho_1 g_1}{\epsilon_1} - 1 > \epsilon_2, \rho_2 g_2 < \epsilon_2\right) \\
&= \int_{\eta_1}^{\eta_1(1+\epsilon_2)} F_{g_2}\left(\frac{\rho_1 y}{\rho_2 \epsilon_1} - \frac{1}{\rho_2}\right) f_{g_1}(y) dy \\
&\quad + \int_{\eta_1(1+\epsilon_2)}^{\infty} F_{g_2}\left(\frac{\epsilon_2}{\rho_2}\right) f_{g_1}(y) dy \\
&= F_{g_1}(\eta_1(1+\epsilon_2))(1 - F_{g_2}(\eta_2)) + F_{g_2}(\eta_2) - F_{g_1}(\eta_1) \\
&\quad - \frac{e^{\frac{1}{\rho_2 L_2}}}{2\theta^k \Gamma(k)} \int_{\eta_1}^{\eta_1(1+\epsilon_2)} e^{-\frac{\rho_1 y}{\rho_2 \epsilon_1 L_2}} \mathrm{y}^{\frac{k}{2}-1} dy \\
&= F_{g_1}(\eta_1(1+\epsilon_2))(1 - F_{g_2}(\eta_2)) + F_{g_2}(\eta_2) - F_{g_1}(\eta_1) \\
&\quad \underbrace{- \frac{(\rho_2 L_2 \eta_1)^{\frac{k}{2}}}{2\theta^k \Gamma(k)} \int_0^{\frac{\epsilon_2}{\rho_2 L_2}} \left(z + \frac{1}{\rho_2 L_2}\right)^{\frac{k}{2}-1} e^{-x} dx}_{I_3}
\end{aligned} \tag{C.1}
$$

Referring to the table of integrals in Ref. [26], I_3 can be derived as (9.24).

Corresponding to Case III of the RS scheme, the probability $P_{\text{out}}^{(\text{III})}$ can be evaluated as

$$
\begin{aligned}
P_{\text{out}}^{(\text{III})} &= P_r(\tau = 0, R^{(\text{III})} < \hat{R}_2) \\
&= P_r\left(0 < g_1 < \eta_1, 0 < g_2 < \frac{(\rho_1 g_1 + 1)\epsilon_2}{\rho_2}\right)
\end{aligned}
$$

$$
\begin{aligned}
&= \int_0^{\eta_1} \left(F_{g_2} \left(\frac{(\rho_1 y + 1)\epsilon_2}{\rho_2} \right) \right) f_{g_1}(y) dy \\
&= F_{g_1}(\eta_1) - \underbrace{\frac{e^{-\frac{\epsilon_2}{\rho_2 L_2}}}{2\theta^k \Gamma(k)} \int_0^{\eta_1} e^{-\frac{\rho_1 \epsilon_2 y}{\rho_2 L_2}} y^{\frac{k}{2}-1} dy}_{I_4}
\end{aligned} \tag{C.2}
$$

where I_4 is characterized by the lower incomplete gamma function as shown in (9.25).

References

[1] B. Zheng and R. Zhang, Intelligent reflecting surface-enhanced OFDM: Channel estimation and reflection optimization, *IEEE Wireless Commun. Lett.*, 9(4), 518–522, 2020.

[2] C. You, B. Zheng, and R. Zhang, Wireless communication via double IRS: Channel estimation and passive beamforming designs, *IEEE Wireless Commun. Lett.*, 10(2), 431–435, Oct 2021.

[3] S. Zhang and R. Zhang, Intelligent reflecting surface aided multi-user communication: Capacity region and deployment strategy, *IEEE Trans. Commun.*, 69(9), 5790–5806, May 2021.

[4] B. Zheng, C. You, and R. Zhang, Efficient channel estimation for double-IRS aided multi-user MIMO system, *IEEE Trans. Commun.*, 69(6), 3818–3832, Mar 2021.

[5] M. Zeng, X. Li, G. Li, W. Hao, and O. A. Dobre, Sum rate maximization for IRS-assisted uplink NOMA, *IEEE Commun. Lett.*, 25(1), 234–238, Jan 2021.

[6] F. Fang, Y. Xu, Q.-V. Pham, and Z. Ding, Energy-efficient design of IRS-NOMA networks, *IEEE Trans. Veh. Technol.*, 69(11), 14088–14092, Nov 2020.

[7] Q. Wu and R. Zhang, Towards smart and reconfigurable environment: Intelligent reflecting surface aided wireless network, *IEEE Commun. Mag.*, 58(1), 106–112, Jan 2020.

[8] L. Lv, *et al.*, Achieving covert communication by IRS-NOMA, in *Proc. IEEE/CIC ICCC*, 421–426, Jul 2021.

[9] C. Huang, A. Zappone, G. C. Alexandropoulos, M. Debbah, and C. Yuen, Reconfigurable intelligent surfaces for energy efficiency in wireless communication, *IEEE Trans. Wireless Commun.*, 18(8), 4157–4170, Aug 2019.

[10] Z. Chen and B. Liu, Switching signal decoding order in uplink NOMA system with imperfect SIC, *IEEE Commun. Lett.*, 25(11), 3528–3532, Nov 2021.

[11] Z. Xiang, X. Tong, and Y. Cai, Secure transmission for NOMA systems with imperfect SIC, *China Commun.*, 17(11), 67–78, 2020.

[12] Z. Ding, R. Schober, and H. V. Poor, Unveiling the importance of SIC in NOMA systems — Part I: State of the art and recent findings, *IEEE Commun. Lett.*, 24(11), 2378–2382, Nov 2020.

[13] Z. Ding, R. Schober, and H. V. Poor, A new QoS-guarantee strategy for NOMA assisted semi-grant-free transmission, *IEEE Trans. Commun.*, 69(11), 7489–7503, 2021.

[14] Y. Sun, Z. Ding, and X. Dai, A new design of hybrid SIC for improving transmission robustness in uplink NOMA, *IEEE Trans. Veh. Tech.*, 70(5), 5083–5087, May 2021.

[15] Y. Mao, *et al.*, Rate-splitting multiple access: Fundamentals, Survey, and Future research trends, *IEEE Commun. Surv. Tutor.* 24(4), 2073–2126, Jul 2022.

[16] A. Mishra, Y. Mao, O. Dizdar, and B. Clerckx, Rate-splitting multiple access for 6G – Part I: Principles, applications and future works, *IEEE Commun. Lett.*, 26(10), 2232–2236, Oct 2022.

[17] B. Clerckx, H. Joudeh, C. Hao, M. Dai, and B. Rassouli, Rate splitting for MIMO wireless networks: A promising PHY-layer strategy for LTE evolution, *IEEE Commun. Mag.*, 54(5), 98–105, May 2016.

[18] Z. Yang, M. Chen, W. Saad, W. Xu, and M. Shikh-Bahaei, Sum-rate maximization of uplink rate splitting multiple access (RSMA) communication, *IEEE Trans. Mobile Comput.*, 21(7), 2596–2609, 2022.

[19] Y. Zhu, Z. Zhang, X. Wang, and X. Liang, A low-complexity non-orthogonal multiple access system based on rate splitting, in *Proc. WCSP*, 1–6, 2017.

[20] Y. Zhu, X. Wang, Z. Zhang, X. Chen, and Y. Chen, A rate-splitting non-orthogonal multiple access scheme for uplink transmission, in *Proc. WCSP*, 1–6, 2017.

[21] J. Zeng, *et al.*, Ensuring max-min fairness of UL SIMO-NOMA: A rate splitting approach, *IEEE Trans. Veh. Technol.*, 68(11), 11080–11093, Nov 2019.

[22] H. Liu, T. A. Tsiftsis, K. J. Kim, K. S. Kwak, and H. V. Poor, Rate splitting for uplink NOMA with enhanced fairness and outage performance, *IEEE Trans. Wireless Commun.*, 19(7), 4657–4670, Jul 2020.

[23] H. Liu, Z. Bai, H. Lei, G. Pan, K.J. Kim, T.A. Tsiftsis, A new rate splitting strategy for uplink CR-NOMA systems, *IEEE Trans. Veh. Tech.*, 71(7), 7947–7951, 2022.

[24] H. Liu, Y. Ye, Z. Bai, K. J. Kim, and T. Tsiftsis, Rate splitting multiple access aided mobile edge computing in cognitive radio networks, in *Proc. IEEE ICC Workshops*, 1–5, 2022.

[25] B. Rimoldi and R. Urbanke, A rate-splitting approach to the Gaussian multiple-access channel, *IEEE Trans. Inf. Theory*, 42(2), 364–375, Mar 1996.

[26] I. S. Gradshteyn and I. M. Ryzhik, Table of integrals, series, and products, *Math. Comput.*, 20(96), 1157–1160, 2007.

[27] H. You; H. Liu; Z. Bai, T. Tsiftsis, K. S. Kwak, Rate-splitting for intelligent reflecting surface-assisted CR-NOMA systems, *J. of Commun. and Netw.*, 25(1), 15–24 Feb, 2023.

[28] T. V. Chien, L. T. Tu, S. Chatzinotas, and B. Ottersten, Coverage probability and ergodic capacity of intelligent reflecting surface-enhanced communication systems, *IEEE Commun. Lett.*, 25(1), 69–73, Jan 2021.

Index

D

I

J

K

L

M

N

Z

www.ingramcontent.com/pod-product-compliance
Lightning Source LLC
LaVergne TN
LVHW011700200926
848484LV00004B/1013

* 9 7 8 9 8 1 9 8 0 1 0 6 0 *